SURGICAL ANATOMY

(ILLUSTRATED WITH SIXTY-EIGHT COLOURED PLATES)

FOREWORD

I INSCRIBE THIS WORK TO THE GENTLEMEN
WITH WHOM AS A FELLOW-STUDENT I WAS ASSOCIATED AT THE
London University College:

AND IN AN ESPECIAL MANNER, IN THEIR NAME AS WELL AS MY OWN,
I AVAIL MYSELF OF THE OPPORTUNITY TO RECORD, ON THIS PAGE,
ALBEIT IN CHARACTERS LESS IMPRESSIVE THAN THOSE WHICH ARE
WRITTEN ON THE LIVING TABLET OF MEMORY,
THE DEBT OF GRATITUDE WHICH WE OWE TO THE LATE

SAMUEL COOPER, F.R.S., AND ROBERT LISTON, F.R.S.,

TWO AMONG THE MANY DISTINGUISHED PROFESSORS OF THAT
INSTITUTION, WHOSE PUPILS WE HAVE BEEN,
AND FROM WHOM WE INHERIT THAT BETTER POSSESSION THAN LIFE
ITSELF, AN ASPIRATION FOR THE LIGHT OF SCIENCE.

JOSEPH MACLISE.

PREFACE.

The object of this work is to present to the student of medicine and the practitioner removed from the schools, a series of dissections demonstrative of the relative anatomy of the principal regions of the human body. Whatever title may most fittingly apply to a work with this intent, whether it had better be styled surgical or medical, regional, relative, descriptive, or topographical anatomy, will matter little, provided its more salient or prominent character be manifested in its own form and feature. The work, as I have designed it, will itself show that my intent has been to base the practical upon the anatomical, and to unite these wherever a mutual dependence was apparent.

That department of anatomical research to which the name topographical strictly applies, as confining itself to the mere account of the form and relative location of the several organs comprising the animal body, is almost wholly isolated from the main questions of physiological and transcendental interest, and cannot, therefore, be supposed to speak in those comprehensive views which anatomy, taken in its widest signification as a science, necessarily includes. While the anatomist contents himself with describing the form and position of organs as they appear exposed, layer after layer, by his dissecting instruments, he does not pretend to soar any higher in the region of science than the humble level of other mechanical arts, which merely appreciate the fitting arrangement of things relative to one another, and combinative to the whole design of the form or machine of whatever species this may be, whether organic or inorganic. The descriptive anatomist of the human body aims at no higher walk in science than this, and hence his nomenclature is, as it is, a barbarous jargon of words, barren of all truthful signification, inconsonant with nature, and blindly irrespective of the *cognitio certa ex principiis certis exorta.*

Still, however, this anatomy of form, although so much requiring purification of its nomenclature, in order to clothe it in the high reaching dignity of a science, does not disturb the medical or surgical practitioner, so far as *their* wants are concerned. Although it may, and actually does, trammel the votary who aspires to the higher generalizations and the development of a law of formation, yet, as this is not the object of the surgical anatomist, the nomenclature, such as it is, will answer conveniently enough the present purpose.

The anatomy of the human form, contemplated in reference to that of all other species of animals to which it bears comparison, constitutes the study of the comparative anatomist, and, as such, establishes the science in its full intent. But the anatomy of the human figure, considered as a species, *per se,* is confessedly the humblest walk of the understanding in a subject which, as anatomy, is relationary, and branches far and wide through all the domain of an animal kingdom. While restricted to the study of the isolated human species, the cramped judgment wastes in such narrow confine; whereas, in the expansive gaze over all allying and allied species, the intellect bodies forth to its vision the full appointed form of natural majesty; and after having experienced the manifold analogies and differentials of the many, is thereby enabled, when it returns to the study of the one, to view this *one* of human type under manifold points of interest, to the appreciation of which the understanding never wakens otherwise. If it did not happen that the study of the human form (confined to itself) had some practical bearing, such study could not deserve the name of anatomical, while anatomical means comparative, and whilst comparison implies inductive reasoning.

However, practical anatomy, such as it is, is concerned with an exact knowledge of the relationship of organs as they stand in reference to each other, and to the whole design of which these organs are theintegral parts. The figure, the capacity, and the contents of the thoracic and abdominal cavities, become a study of not more urgent concernment to the physician, than are the regions named cervical, axillary, inguinal, &c., to the surgeon. He who would combine both modes of a relationary practice, such as that of medicine and surgery, should be well acquainted with the form and structures characteristic of all regions of the human body; and it may be doubted whether he who pursues either mode of practice, wholly exclusive of the other, can do so with honest purpose and large range of understanding, if he be not equally well acquainted with the subject matter of both. It is, in fact, more triflingly fashionable than soundly reasonable, to seek to define the line of demarcation between the special callings of medicine and surgery, for it will ever be as vain an endeavour to separate the one from the other without extinguishing the vitality of both, as it would be to sunder the trunk from the head, and give to each a separate living existence. The necessary division of labour is the only reason that can be advanced in excuse of specialisms; but it will be readily agreed to, that that practitioner who has first laid within himself the foundation of a general knowledge of matters relationary to his subject, will always be found to pursue the speciality according to the light of reason and science.

Anatomy--the Γνῶθι σεαυτόν, the knowledge based on principle--is the foundation of the curativc art, cultivated as a science in all its branchings; and comparison is the nurse of reason, which we are fain to make our guide in bringing the practical to bear productively. The human body, in a state of health, is the standard whereunto we compare the same body in a state of disease. The knowledge of the latter can only exist by the knowledge of the former, and by the comparison of both.

Comparison may be fairly termed the pioneer to all certain knowledge. It is a potent instrument--the only one, in the hands of the pathologist, as well as in those of the philosophic generalizer of anatomical facts, gathered through the extended survey of an animal kingdom. We best recognise the condition of a dislocated joint after we have become well acquainted with the contour of its normal state; all abnormal conditions are best understood by a knowledge of what we know to be normal character. Every anatomist is a comparer, in a greater or lesser degree; and he is the greatest anatomist who compares the most generally.

Impressed with this belief, I have laid particular emphasis on imitating the character of the normal form of the human figure, taken as a whole; that of its several regions as parts of this whole, and that of the various organs (contained within those regions) as its integrals or elements. And in order to present this subject of relative anatomy in more vivid reality to the understanding of the student, I have chosen the medium of illustrating by figure rather than by that of written language, which latter, taken alone, is almost impotent in a study of this nature.

It is wholly impossible for anyone to describe form in words without the aid of figures. Even the mathematical strength of Euclid would avail nothing, if shorn of his diagrams. The professorial robe is impotent without its diagrams. Anatomy being a science existing by demonstration, (for as much as form in its actuality is the language of nature,) must be discoursed of by the instrumentality of figure.

An anatomical illustration enters the understanding straight-forward in a direct passage, and is almost independent of the aid of written language. A picture of form is a proposition which solves itself. It is an axiom encompassed in a frame-work of self-evident truth. The best substitute for Nature herself, upon which to teach the knowledge of her, is an exact representation of her form.

Every surgical anatomist will (if he examine himself) perceive that, previously to undertaking the performance of an operation upon the living body, he stands reassured and self-reliant in that degree in which he is capable of conjuring up before his mental vision a distinct picture of his subject. Mr. Liston could draw the same anatomical picture mentally which Sir Charles Bell's handicraft could draw in reality of form and figure. Scarpa was his own draughtsman.

If there may be any novelty now-a-days possible to be recognised upon the out-trodden track of human relative anatomy, it can only be in truthful and well-planned illustration. Under this view alone may the anatomist plead an excuse for reiterating a theme which the beautiful works of Cowper, Haller, Hunter, Scarpa, Soemmering, and others, have dealt out so respectably. Except the human anatomist turns now to what he terms the practical ends of his study, and marshals his little knowledge to bear upon those ends, one may proclaim anthropotomy to have worn itself out. Dissection can do no more, except to repeat Cruveilhier. And that which Cruveilhier has done for human anatomy, Muller has completed for the physiological interpretation of human anatomy; Burdach has philosophised, and Magendie has experimented to the full upon this theme, so far as it would permit. All have pushed the subject to its furthest limits, in one aspect of view. The narrow circle is footworn. All the needful facts are long since gathered, sown, and known. We have been seekers after those facts from the days of Aristotle. Are we to put off the day of attempting interpretation for three thousand years more, to allow the human physiologist time to slice the brain into more delicate atoms than he has done hitherto, in order to coin more names, and swell the dictionary? No! The work must now be retrospective, if we would render true knowledge progressive. It is not a list of new and disjointed facts that Science at present thirsts for; but she is impressed with the conviction that her wants can alone be supplied by the creation of a new and truthful theory,--a generalization which the facts already known are sufficient to supply, if they were well ordered according to their natural relationship and mutual dependence. "Le temps viendra peut-etre," says Fontenelle, "que 1'on joindra en un corps regulier ces membres epars; et, s'ils sont tels qu'on le souhaite, ils s'assembleront en quelque sorte d'eux-memes. Plusieurs verites separees, des qu'elles sont en assez grand nombre, offrent si vivement a 1'esprit leurs rapports et leur mutuelle dependance, qu'il semble qu'apres les avoir detachees par une espece de violence les unes des autres, elles cherchent naturellement a se reunir."--(Preface sur l'utilite des Sciences, &c.)

The comparison of facts already known must henceforward be the scalpel which we are to take in hand. We must return by the same road on which we set out, and reexamine the things and phenomena which, as novices, we passed by too lightly. The travelled experience may now sit down and contemplate. That which I have said and proved elsewhere in respect to the skeleton system may, with equal truth, be remarked of the nervous system--namely, that the question is not in how far does the limit of diversity extend through the condition of an evidently common analogy, but by what rule or law the uniform ens is rendered the diverse entity? The womb of anatomical science is pregnant of the true interpretation of the law of *unity in variety*; but the question is of longer duration than was the life of the progenitor. Though Aristotle and Linnaeus, and Buffon and Cuvier, and Geoffroy St. Hilaire and Leibnitz, and Gothe, have lived and spoken, yet the present state of knowledge proclaims the Newton of physiology to be as yet unborn. The iron scalpel has already made acquaintance with not only the greater parts, but even with the infinitesimals of the human body; and reason, confined to this narrow range of a subject, perceives herself to be imprisoned, and quenches her guiding light in despair. Originality has outlived itself; and discovery is a long-forgotten enterprise, except as pursued in the microcosm on the field of the microscope, which, it must be confessed, has drawn forth demonstrations only commensurate in importance with the magnitude of the littleness there seen.

The subject of our study, whichever it happen to be, may appear exhausted of all interest, and the promise of valuable novelty, owing to two reasons:--It may be, like descriptive human anatomy, so cold, poor and sterile in its own nature, and so barren of product, that it will be impossible for even the genius of Promethean fire to warm it; or else, like existing physiology, the very point of view from which the mental eye surveys the theme, will blight the fair prospect of truth, distort induction, and clog up the paces of ratiocination. The physiologist of the present day is too little of a comparative anatomist, and far too closely enveloped in the absurd jargon of the anthropotomist, ever to hope to reveal any great truth for science, and dispel the mists which still hang over the phenomena of the nervous system. He is steeped too deeply in the base nomenclature of the antique school, and too indolent to question the import of Pons, Commissure, Island, Taenia, Nates, Testes, Cornu, Hippocamp, Thalamus, Vermes, Arbor Vitro, Respiratory Tract, Ganglia of Increase, and all such phrase of unmeaning sound, ever to be productive of lucid interpretation of the cerebro-spinal ens. Custom alone sanctions his use of such names; but;

"Custom calls him to it!
What custom wills; should custom always do it,
The dust on antique time would lie unswept,
And mountainous error be too highly heaped,
For truth to overpeer."

Of the illustrations of this work I may state, in guarantee of their anatomical accuracy, that they have been made by myself from my own dissections, first planned at the London University College, and afterwards realised at the Ecole Pratique, and School of Anatomy adjoining the Hospital La Pitie, Paris, a few years since. As far as the subject of relative anatomy could admit of novel treatment, rigidly confined to facts unalterable, I have endeavoured to give it.

The unbroken surface of the human figure is as a map to the surgeon, explanatory of the anatomy arranged beneath; and I have therefore left appended to the dissected regions as much of the undissected as was necessary. My object was to indicate the interior through the superficies, and thereby illustrate the whole living body which concerns surgery, through its dissected dead counterfeit. We dissect the dead animal body in order to furnish the memory with as clear an account of the structure contained in its living representative, which we are not allowed to analyse, as if this latter were perfectly translucent, and directly demonstrative of its component parts.

J. M

THE FORM OF THE THORAX, AND THE RELATIVE POSITION OF ITS CONTAINED PARTS--THE LUNGS, HEART, AND LARGER BLOOD VESSELS.

The structure, mechanism, and respiratory motions of the thoracic apparatus. Its varieties in form, according to age and sex. Its deformities. Applications to the study of physical diagnosis.

In the human body there does not exist any such space as *cavity*, properly so called. Every space is occupied by its contents. The thoracic space is completely filled by its viscera, which, in mass, take a perfect cast or model of its interior. The thoracic viscera lie so closely to one another, that they respectively influence the form and dimensions of each other. That space which the lungs do not occupy is filled by the heart, &c., and *vice versa*. The thoracic apparatus causes no vacuum by the acts of either contraction or dilatation. Neither do the lungs or the heart. When any organ, by its process of growth, or by its own functional act, forces a space for itself, it immediately inhabits that space entirely at the expense of neighbouring organs. When the heart dilates, the pulmonary space contracts; and when the thoracic space increases, general space diminishes in the same ratio.

The mechanism of the functions of respiration and circulation consists, during the life of the animal, in a constant oscillatory *nisus* to produce a vacuum which it never establishes. These vital forces of the respiratory and circulatory organs, so characteristic of the higher classes of animals, are opposed to the general forces of surrounding nature. The former vainly strive to make exception to the irrevocable law, that "*nature abhors a vacuum.*" This act of opposition between both forces constitutes the respiratory act, and thus the respiratory thoracic being (like a vibrating pendulum) manifests respiratory motion, not as an effort of volition originating solely with itself, but according to the measure of the force of either law; as entity is relationary, so is functionality likewise. The being is functional by relationship; and just as a pendulum is functional, by reason of the counteraction of two opposing forces,--viz., the force of motion and the force of gravity,--so is a thoracic cavity (considering it as a mechanical apparatus) functional by two opposing forces--the vital force and the surrounding physical force. The inspiration of thoracic space is the expiration of general space, and reciprocally.

The thoracic space is a symmetrical enclosure originally, which aftercoming necessities modify and distort in some degree. The spaces occupied by the opposite lungs in the adult body do not exactly correspond as to capacity, O O, Plate 1. Neither is the cardiac space, A E G D, Plate 1, which is traversed by the common median line, symmetrical. The asymmetry of the lungs is mainly owing to the form and position of the heart; for this organ inclines towards the left thoracic side. The left lung is less in capacity than the right, by so much space as the heart occupies in the left pulmonary side. The general form of the thorax is that of a cone, I I N N, Plate 1, bicleft through its perpendicular axis, H M. The line of bicleavage is exactly median, and passes through the centre of the sternum in front, and the centres of the dorsal vertebral behind. Between the dorsal vertebral and the sternum, the line of median cleavage is maintained and sketched out in membrane. This membranous middle is formed by the adjacent sides of the opposite pleural or enveloping bags in which the lungs are enclosed. The heart, A, Plate 1, is developed between these two pleural sacs, F F, and separates them from each other to a distance corresponding to its own size. The adjacent sides of the two pleural sacs are central to the thorax, and form that space which is called mediastinum; the heart is located in this mediastinum, U E, Plate 1. The extent of the thoracic region ranges *perpendicularly* from the root of the neck, Q, Plate 1, to the roof of the abdomen--viz., the diaphragm, P, *transversely* from the ribs of one side, I N, Plate 1, to those of the other, and *antero-posteriorly* from the sternum, H M, to the vertebral column. All this space is pulmonary, except the cardiac or median space, which, in addition to the heart, A, Plate 1, and great bloodvessels, G C B, contains the oesophagus, bronchi, &c. The ribs are the true enclosures of thoracic space, and, generally, in mammalian forms, they fail or degenerate at that region of the trunk which is not pulmonary or respiratory. In human anatomy, a teleological reason is given for this--namely, that of the ribs being mechanically subservient to the function of respiration alone. But the transcendental anatomists interpret this fact otherwise, and refer it to the operation of a higher law of formation.

The capacity of the thorax is influenced by the capacity of the abdomen and its contents. In order to admit of full inspiration and pulmonary expansion, the abdominal viscera recede in the same ratio as the lungs dilate. The diaphragm, P P, Plate 1, or transverse musculo-membranous partition which divides the pulmonary and alimentary cavities, is, by virtue of its situation, as mechanically subservient to the abdomen as to the thorax. And under general notice, it will appear that even the abdominal muscles are as directly related to the respiratory act as those of the thorax. The connexion between functions is as intimate and indissoluble as the connexion between organs in the same body. There can be no more striking proof of the divinity of design than by such revelations as anatomical science everywhere manifests in facts such as this--viz., that each organ serves in most cases a double, and in many a triple purpose, in the animal economy.

The apex of the lung projects into the root of the neck, even to a higher level, Q, Plate 1, than that occupied by the sternal end of the clavicle, K. If the point of a sword were pushed through the neck above the clavicle, at K, Plate 1, it would penetrate the apex of the right lung, where the subclavian artery, Q, Plate 1, arches over it. In connexion with this fact, I may mention it as very probable that the *bruit*, or continuous murmur which we hear through the stethoscope, in chlorotic females, is caused by the pulsation of the subclavian artery against the top of the lung. The stays or girdle which braces the loins of most women prevents the expansion of the thoracic apparatus, naturally attained by the descent of the diaphragm; and hence, no doubt, the lung will distend inordinately above towards the neck. It is an interesting fact for those anatomists who study the higher generalizations of their science, that at those very localities--viz., the neck and loins, where the lungs by their own natural effort are prone to extend themselves in forced inspiration--happen the "anomalous" creations of cervical and lumbar ribs. The subclavian artery is occasionally complicated by the presence of these costal appendages.

If the body be transfixed through any one of the intercostal spaces, the instrument will surely wound some part of the lung. If the thorax be pierced from any point whatever, provided the instrument be directed towards a common centre, A, Plate 1, the lung will suffer lesion; for the heart is, almost completely, in the healthy living body, enveloped in the lungs. So true is it that all the costal region (the asternal as well as the sternal) is a pulmonary enclosure, that any instrument which pierces intercostal space must wound the lung.

As the sternal ribs degenerate into the "false" asternal or incomplete ribs from before, obliquely backward down to the last dorsal vertebra, so the thoracic space takes form. The lungs range through a much larger space, therefore, posteriorly than they do anteriorly.

The form of the thorax, in relation to that of the abdomen, may be learned from the fact that a gunshot, which shall enter a little below N, Plate 1, and, after traversing the body transversely, shall pass out at a corresponding point at the opposite side, would open the thorax and the abdomen into a common cavity; for it would pierce the thorax at N, the arching diaphragm at the level of M, and thereat enter the belly; then it would enter the thorax again at P, and make exit below N, opposite. If a cutting instrument were passed horizontally from before backward, a little below M, it would first open the abdomen, then Pierce the arching diaphragm, and pass into the thorax, opposite the ninth or eighth dorsal vertebra.

The outward form or superficies masks in some degree the form of the interior. The width of the thorax above does not exceed the diameter between the points I I, of Plate 1, or the points W W, of Plate 2. If we make percussion directly from before backwards at any place external to I, Plate 1, we do not render the lung vibrative. The diameters between I I and N N, Plate 1, are not equal; and these measures will indicate the form of the thorax in the living body, between the shoulders above and the loins below.

The position of the heart in the thorax varies somewhat with several bodies. The size of the heart, even in a state of perfect health, varies also in subjects of corresponding ages, a condition which is often mistaken for pathological. For the most part, its form occupies a space ranging from two or three lines right of the right side of the sternum to the middle of the shafts of the fifth and sixth ribs of the left side. In general, the length of the osseous sternum gives the exact perpendicular range of the heart, together with its great vessels.

The aorta, C, Plates 1 and 2, is behind the upper half of the sternum, from which it is separated by the pericardium, D, Plate 1, the thin edge of the lung, and the mediastinal pleurae, U E, Plate 1, &c. If the heart be injected from the abdominal aorta, the aortal arch will flatten against the sternum. Pulmonary space would not be opened by a penetrating instrument passed into the root of the neck in the median line above the sternum, at L, Plate 1. But the apices of both lungs would be wounded if the same instrument entered deeply on either side of this median line at K K. An instrument which would pierce the sternum opposite the insertion of the second, third, or fourth costal cartilage, from H downwards, would transfix some part of the arch of the aorta, C, Plate 1. The same instrument, if pushed horizontally backward through the second, third, or fourth interspaces of the costal cartilages close to the sternum, would wound, on the right of the sternal line, the vena cava superior, G, Plate 1; on the left, the pulmonary artery, B, and the descending thoracic aorta. In the healthy living body, the thoracic sounds heard in percussion, or by means of the stethoscope, will vary according to the locality operated upon, in consequence of the variable thickness of those structures (muscular and osseous, &c.,) which invest the thoracic walls. Uniformity of sound must, owing to these facts, be as materially interrupted, as it certainly is, in consequence of the variable contents of the cavity. The variability of the healthy thoracic sounds will, therefore, be too often likely to be mistaken for that of disease, if we forget to admit these facts, as instanced in the former state. Considering the form of the thoracic space in reference to the general form of the trunk of the living body, I see reason to doubt whether the practitioner can by any boasted delicacy of manipulation, detect an abnormal state of the pulmonary organs by percussion, or the use of the stethoscope, applied at those regions which he terms coracoid, scapulary, subclavian, &c., if the line of his examination be directed from before backwards. The scapula, covered by thick carneous masses, does not lie in the living body directly upon the osseous-thorax, neither does the clavicle. As all antero-posterior examination in reference to the lungs external to the points, I I, between the shoulders cannot, in fact, concern the pulmonary organs, so it cannot be diagnostic of their state either in health or disease. The difficulties which oppose the practitioner's examination of the state of the thoracic contents are already numerous enough, independent of those which may arise from unanatomical investigation.

DESCRIPTION OF PLATES 1 & 2.

PLATE 1.

A. Right ventricle of the heart.

B. Origin of pulmonary artery.

C. Commencement of the systemic aorta, ascending part of aortic arch.

D. Pericardium investing the heart and the origins of the great bloodvessels.

E. Mediastinal pleura, forming a second investment for the heart, bloodvessels, &c.

F. Costal pleura, seen to be continuous above with that which forms the mediastinum.

G. Vena cava superior, entering pericardium to join V, the right auricle.

H. Upper third of sternum.

I I. First ribs.

K K. Sternal ends of the clavicles.

L. Upper end of sternum.

M. Lower end of sternum.

N N. Fifth ribs.

O O. Collapsed lungs.

P P. Arching diaphragm.

Q. Subclavian artery.

R. Common carotid artery, at its division into internal and external carotids.

S S. Great pectoral muscles.

T T. Lesser pectoral muscles.

U. Mediastinal pleura of right side.

V. Right auricle of the heart.

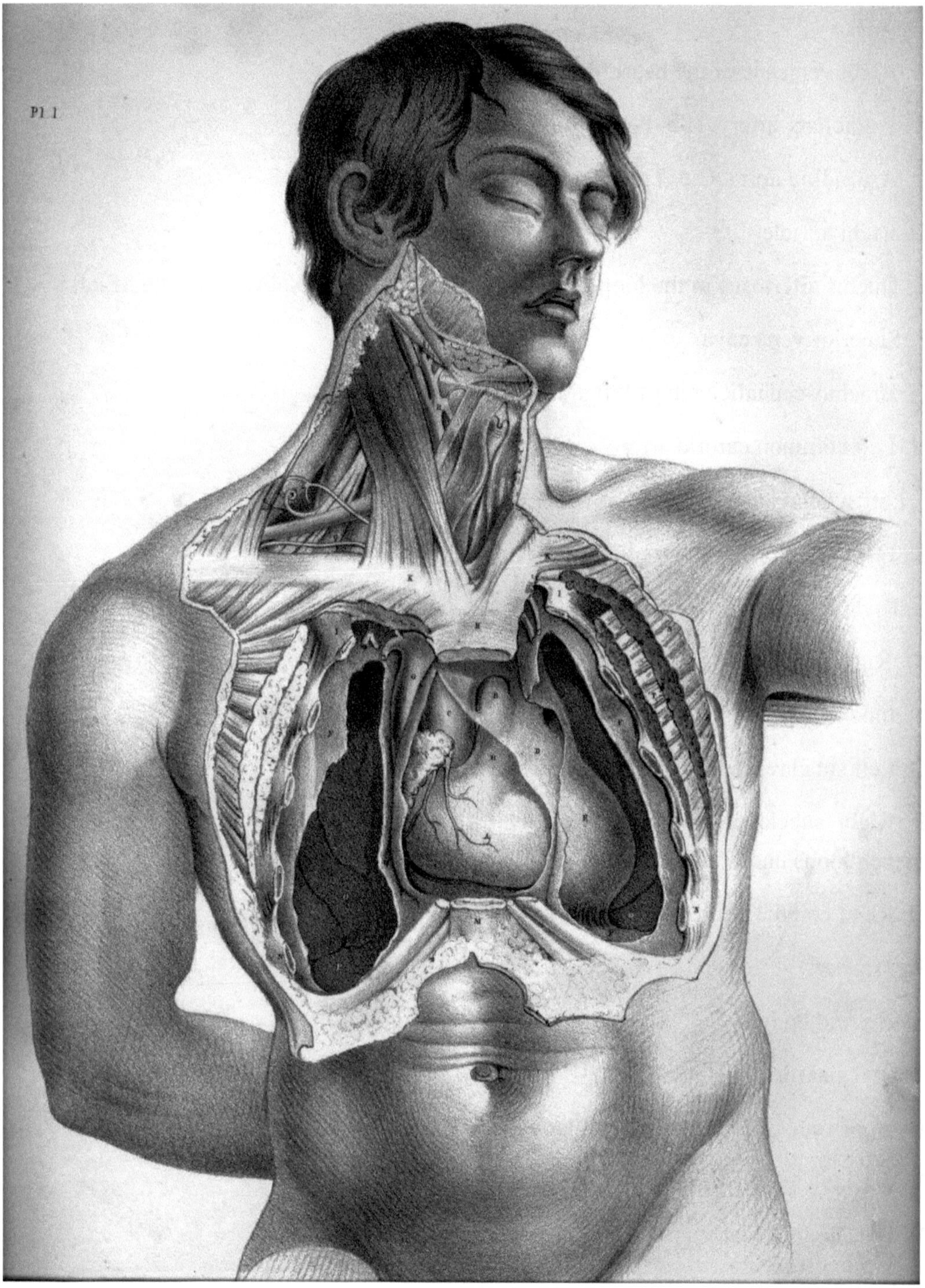

Plate 1

W. Submaxillary gland.

X. Digastric muscle.

Y. Lymphatic body.

Z. Hyoid bone.

1. Thyroid cartilage.

2. Superior thyroid artery.

3. Anterior jugular vein.

4. Hyoid half of omo-hyoid muscle.

5. Sterno-hyoid muscle.

6. Top of the sternum.

7. Clavicle.

8. Trapezius muscle.

9. Splenius capitis and colli muscle.

10. Occipital half of occipito-frontalis muscle.

11. Levator auris muscle.

12. Frontal half of occipito-frontalis muscle.

13. Orbicularis oculi muscle.

14. Zygomaticus major muscle.

15. Buccinator muscle.

16. Depressor anguli oris muscle.

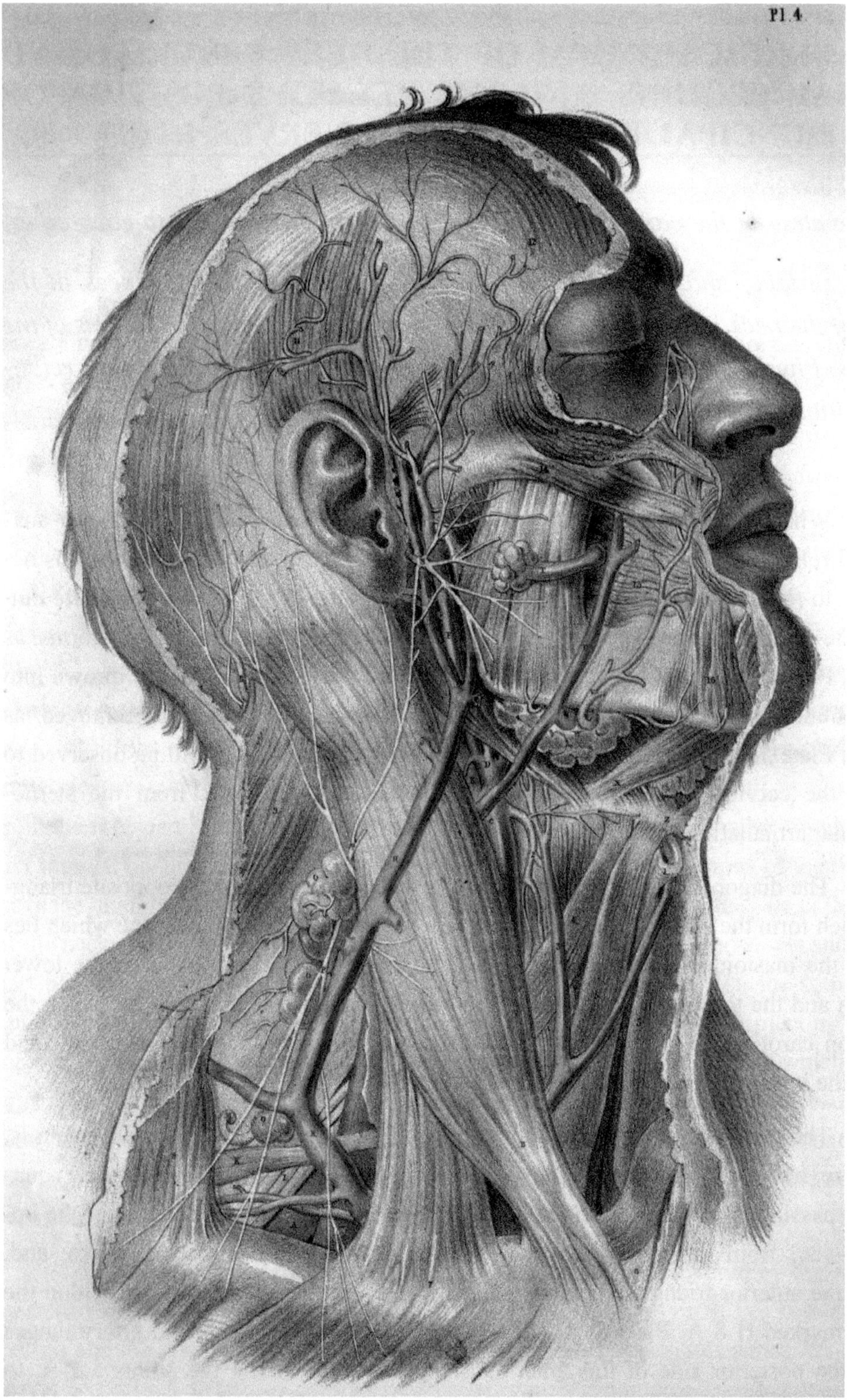

Plate 4

le is substantial and fleshy. The united spaces occupied by these structures give the depth of the subclavian artery in the middle part of its course.

The length of the subclavian artery between its point of branching from the innominate and that where it gives off its first branches varies in different bodies, but is seldom so extensive as to assure the operator of the ultimate success of the process of ligaturing the vessel. Above and below D, Plate 6, the thyroid axis, come off the vertebral and internal mammary arteries internal and anterior to the scalenus muscle. External and posterior to the scalenus, a large vessel, the post scapular, G, Plate 6, R, Plate 5, arises. If an aneurism attack any part of this subclavian arch, it must be in close connexion with some one of these branches. If a ligature is to be applied to any part of the arch, it will seldom happen that it can be placed farther than half an inch from some of these principal collateral branches.

When the shoulder is depressed, the clavicle follows it, and the subclavian artery will be more exposed and more easily reached than if the shoulder be elevated, as this latter movement raises the clavicle over the locality of the vessel. Dupuytren alludes practically to the different depths of the subclavian artery in subjects with short necks and high shoulders, and those with long necks and pendent shoulders. When the clavicle is depressed to the fullest extent, if then the sterno-cleido-mastoid and scalenus muscles be relaxed by inclining the head and neck towards the artery, I believe it may be possible to arrest the flow of blood through the artery by compressing it against the first rib, and this position will also facilitate the operation of ligaturing the vessel.

The subclavian vein, W, Plate 5, is removed to some distance from the artery, Q, Plate 5. The width of the scalenus muscle, X, separates the vein from the artery. An instance is recorded by Blandin in which the vein passed in company with the artery under the scalenus muscle.

DESCRIPTION OF PLATES 5 & 6.

PLATE 5.

A. Innominate artery at its point of bifurcation.

B. Subclavian artery crossed by the vagus nerve.

C. Common carotid artery with the vagus nerve at its outer side, and the descendens noni nerve lying on it.

D. External carotid artery.

E. Internal carotid artery with the descendens noni nerve lying on it.

F. Lingual artery passing under the fibres of the hyo-glossus muscle.

G. Tortuous facial artery.

H. Temporo-maxillary artery.

I. Occipital artery crossing the internal carotid artery and jugular vein.

K. Internal jugular vein crossed by some branches of the cervical plexus, which join the descendens noni nerve.

L. Spinal accessory nerve, which pierces the sterno-mastoid muscle, to be distributed to it and the trapezius.

M. Cervical plexus of nerves giving off the phrenic nerve to descend the neck on the outer side of the internal jugular vein and over the scalenus muscle.

N. Vagus nerve between the carotid artery and internal jugular vein.

O. Ninth or hypoglossal nerve distributed to the muscles of the tongue.

P P. Branches of the brachial plexus of nerves.

Q. Subclavian artery in connexion with the brachial plexus of nerves.

R R. Post scapular artery passing through the brachial plexus.

S. Transversalis humeri artery.

T. Transversalis colli artery.

U. Union of the post scapular and external jugular veins, which enter the subclavian vein by a common trunk.

V. Post-half of the omo-hyoid muscle.

W. Part of the subclavian vein seen above the clavicle.

X. Scalenus muscle separating the subclavian artery from vein.

Y. Clavicle.

Z. Trapezius muscle.

1. Sternal origin of sterno-mastoid muscle of left side.

2. Clavicular origin of sterno-mastoid muscle of right side turned down.

3. Scalenus posticus muscle.

4. Splenius muscle.

5. Mastoid insertion of sterno-mastoid muscle.

6. Internal maxillary artery passing behind the neck of lower jaw-bone.

7. Parotid duct.

8. Genio-hyoid muscle.

9. Mylo-hyoid muscle, cut and turned aside.

10. Superior thyroid artery.

11. Anterior half of omo-hyoid muscle.

12. Sterno-hyoid muscle, cut.

13. Sterno-thyroid muscle, cut.

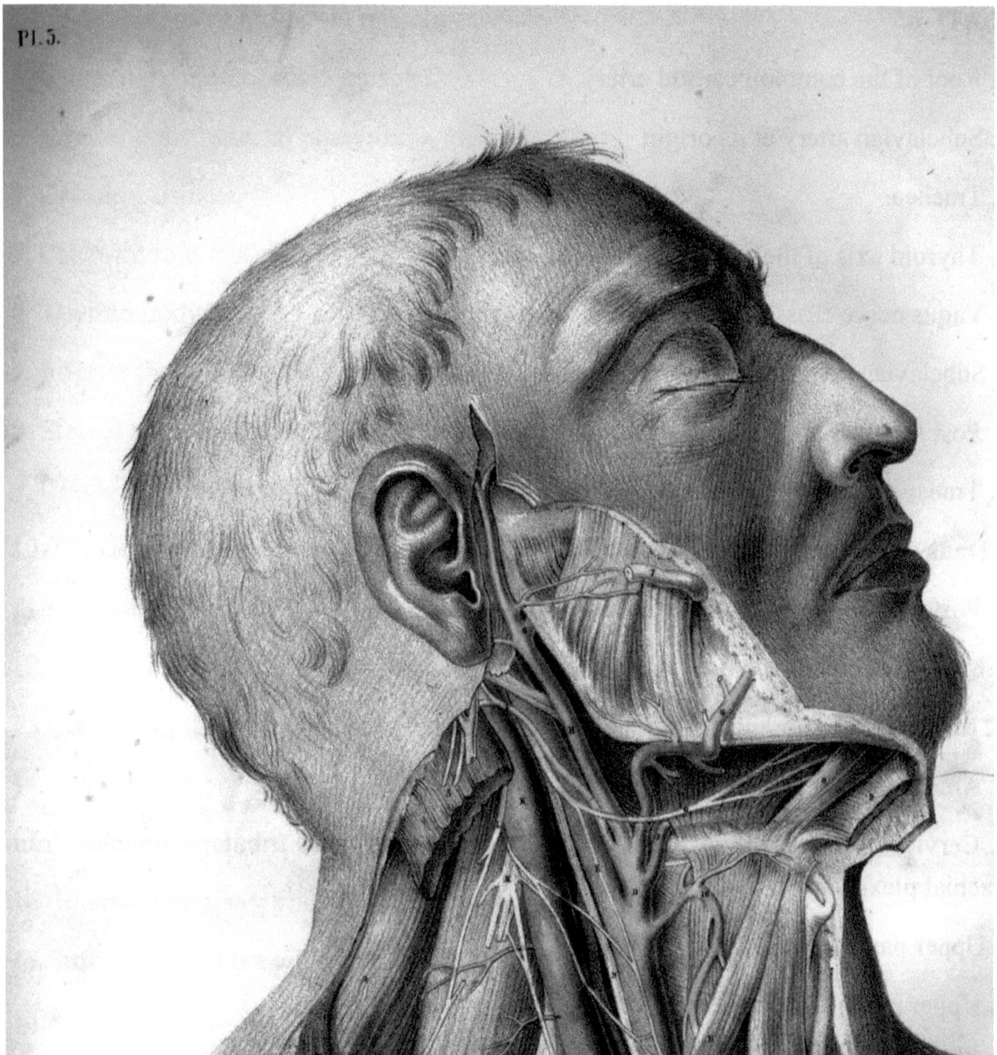

Plate 5

O. Omo-hyoid muscle.

P. Sterno-thyroid muscle.

Q. Fascia proper of the vessels.

R. Layer of the cervical fascia beneath the sterno-mastoid muscle.

S. Portion of the same fascia.

T. External jugular vein injected beneath the skin.

U. Clavicle at the mid-point, where the subclavian artery passes beneath it.

V. Locality of the subclavian artery in the third part of its course.

W. Prominence of the trapezius muscle.

X. Prominence of the clavicular portion of the sterno-cleido-mastoid muscle.

Y. Place indicating the interval between the clavicular and sternal insertions of sterno-cleido-mastoid muscle.

Z. Projection of the sternal portion of the sterno-cleido-mastoid muscle.

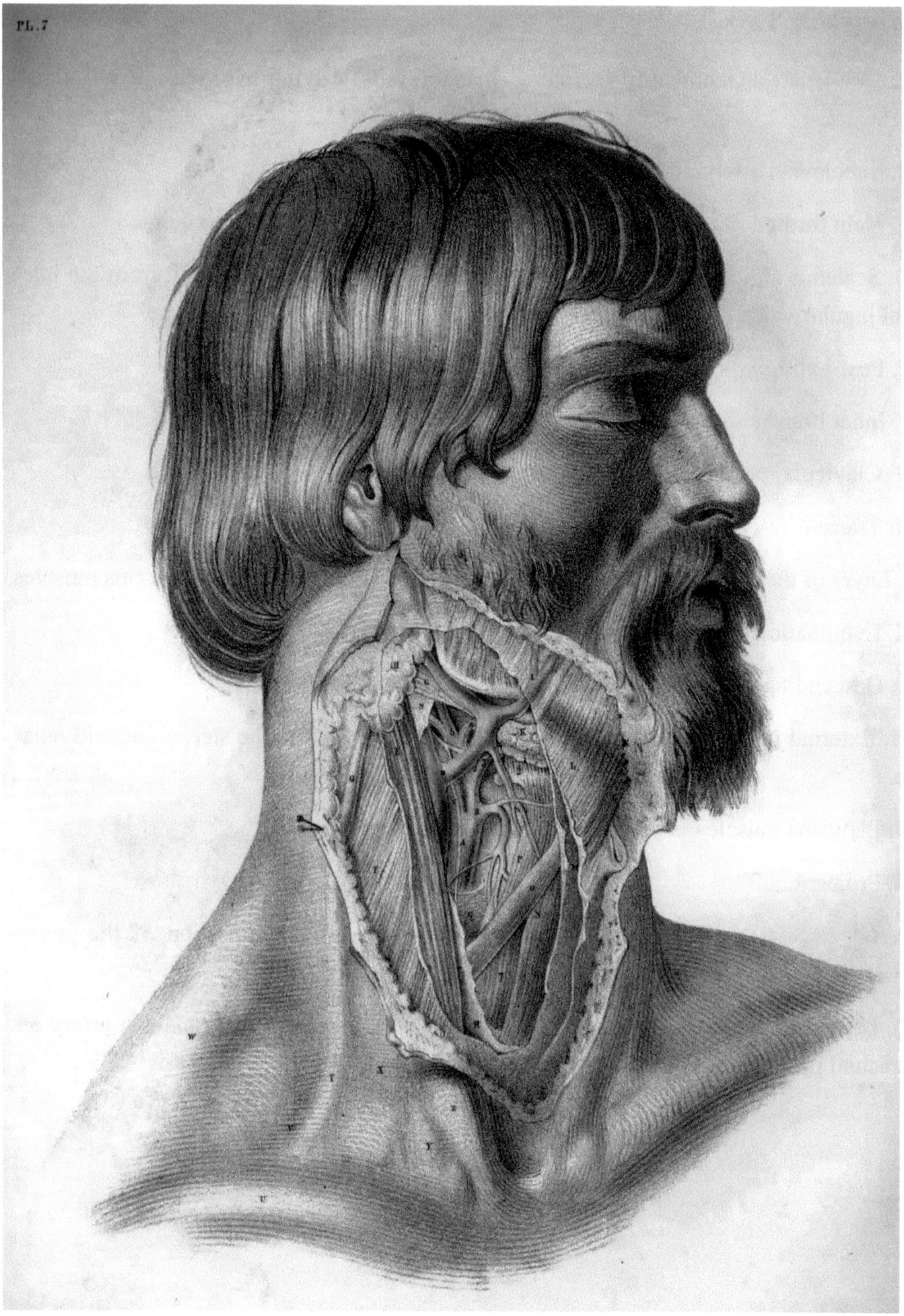

Plate 7

In the abstract or general survey of all those peculiarities of length to which the innominate artery, A, Plate 9, is subject, I here lay it down as a proposition, that they occur as graduated phases of the bicleavage of this innominate trunk from the level of A, to the aortic arch, in which latter phasis the aorta gives a separate origin to the carotid and subclavian vessels of the right side as well as the left. On the other hand, I observe that the peculiarities to the normal separate condition of A and B, the carotid and subclavian arteries of Plate 10, display, in the relationary aggregate, a phasial gradation of A and B joining into a common trunk union, in which state we then find the aorta giving origin to a right and left innominate artery. Between these two forms of development--viz., that where the four vessels spring separately from the aortic arch, and that where two innominate or brachio-cephalic arteries arise from the same--may be read all the sum of variation to which these vessels are liable. It is true that there are some states of these vessels which cannot be said to be naturally embraced in the above generalization; but though I doubt not that these might be encompassed in a higher generalization; still, for all practical ends, the lesser general rule is all-sufficient.

In many instances, the innominate artery, A, Plate 9, is of such extraordinary length, that it rises considerably (for an inch, or even more) above the level of C, the sternal end of the clavicle. In other cases, the innominate artery bifurcates soon after it leaves the first part of the aortic arch; and between these extremes as to length, the vessel varies infinitesimally.

The innominate artery lies closer behind the right sterno-clavicular junction than the left carotid or subclavian arteries lie in relation to the left sterno-clavicular articulation; and this difference of depth between the vessel of the right side and those of the left is mainly owing to the form and direction of the aortic arch from which they take origin. The aortic arch ranges, not alone transversely, but also from before backward, and to the left side of the dorsal spine; and consequently, as the innominate artery, A, Plate 9, springs from the first or fore part of the aorta, while the left carotid and subclavian arteries arise from the second and deeper part of its arch, the vessels of both sides rising into the neck perpendicularly from the root in the thorax, will still, in the cervical region, manifest a considerable difference as to antero-posterior depth. The depth of the left subclavian artery, B, Plate 10, from cervical surface, is even greater than that of the left common carotid, A, Plate 10, and this latter, at its root in the aortic arch, is deeper than the innominate artery. Both common carotids, A A, Plates 9 and 10, hold nearly the same antero-posterior depth on either side of the trachea, M, Plate 10, and D, Plate 9. Although the relative depth of the arterial vessels on both sides of the trachea is different, still they are covered by an equal number of identical structures, taking the same order of superposition.

On either side of the episternal cervical pit, which, even in the undissected body of male or female, infant or adult, is always a well-marked surgical feature, may be readily recognised the converging sternal attachments of the sterno-mastoid muscles, L G, Plate 10; and midway between these symmetrical muscular prominences in the neck, but holding a deeper level than them, is situated that part of the trachea which is generally the subject of the operation of tracheotomy. The relative anatomy of the trachea, M, Plate 10, D, Plate 9, at this situation requires therefore to be carefully considered. The trachea is said to incline rather to the right side of the median line; but perhaps this observation would be more true to nature if it were accompanied by the remark, that this seeming inclination to the right side is owing to the fact, that the innominate artery, A, Plate 9, lies obliquely over its fore part, near the sternum. However this may be, it certainly will be the safer step in the operation to regard the median position of the trachea as fixed, than to encroach upon the locality of the carotid vessels; and to make the incision longitudinally and exactly through the median line, while the neck is extended backwards, and the chin made to correspond with the line of incision. And when the operator takes into consideration the situation of the vessel A, Plate 9, and A, Plate 10, at this region of the neck, he will at once own to the necessity of opening the trachea, D, Plate 9, M, Plate 10, at a situation nearer the larynx than the point marked in the figures. The course taken by the common carotid arteries is, in respect to the trachea, divergent from below upwards; and as these vessels will consequently be found to stand wider apart at the level of K, I, Plate 10, than they do at the level of M, Plate 10, so the farther upwards from the sternum we choose the point at which to open the trachea, the less likely are we to endanger the great arterial vessels.

In addition to the fact, that the carotid arteries at an inch above the sternum lie nearer the median line than they do higher up in the neck, it should always be remembered, that the trachea itself is situated much deeper at the point M, Plate 10, D, Plate 9, than it is opposite the points F and K of the same figures. The laryngo-tracheal line is, in the lateral view of the neck, downwards and backwards, and therefore it will be found always at a considerable depth from cervical surface, as it passes behind the first bone of the sternum, midway between both sterno-mastoid muscles.

In the operation of tracheotomy, the cutting instrument divides the following named structures as they lie beneath the common integument: If the incision be made directly upon the median line, the muscles F, sterno-hyoid, and E, sterno-thyroid, Plate 9, are not necessarily divided, as these structures and their fellows hold a somewhat lateral position opposite to each other. Beneath these muscles and above them, thus encasing them, the cervical fascia, *f f,* Plate 10, is required to be divided, in order to expose the trachea. Beneath *f f* the cervical fascia, will next be felt the rounded bilobed mass of the thyroid body, lying on the forepart of the trachea; above the thyroid body, the cricoid and some tracheal cartilaginous rings will be felt; and since the thyroid body varies much as to bulk in several individuals of the same and different sexes, as also from a consideration that its substance is traversed by large arterial and venous vessels, it will be therefore preferable to open the trachea above it, than through it or below it.

On the forepart of the tracheal median line, either superficial to, or deeper than, the cervical fascia, the tracheotomist occasionally meets with a chain of lymphatic glands or a plexus of veins, which latter, when divided, will trammel the operation by the copious haemorrhage which all veins at this region of the neck are prone to supply, owing to their direct communication with the main venous trunks of the heart; and not unfrequently the inferior thyroid artery overlies the trachea at the point D, Plate 9, when this thyroid vessel arises directly from the arch of the aorta, between the roots of the innominate and left common carotid, or when it springs from the innominate itself. The inferior thyroid vein, sometimes single and sometimes double, overlies the trachea at the point D, Plate 9, when this vein opens into the left innominate venous trunk, as this latter crosses over the root of the main arteries springing from the aorta.

Laryngotomy is, anatomically considered, a far less dangerous operation than tracheotomy, for the above-named reasons; and the former should always be preferred when particular circumstances do not render the latter operation absolutely necessary. In addition to the fact, that the carotid arteries lie farther apart from each other and from the median place--viz., the crico-thyroid interval, which is the seat of laryngotomy--than they do lower down on either side of the trachea, it should also be noticed that the tracheal tube being more moveable than the larynx, is hence more liable to swerve from the cutting instrument, and implicate the vessels. Tracheotomy on the infant is a far more anxious proceeding than the same operation performed on the adult; because the trachea in the infant's body lies more closely within the embrace of the carotid arteries, is less in diameter, shorter, and more mobile than in the adult body.

The episternal or interclavicular region is a locality traversed by so many vitally important structures gathered together in a very limited space, that all operations which concern this region require more steady caution and anatomical knowledge than most surgeons are bold enough to test their possession of. The reader will (on comparing Plates 9 and 10) be enabled to take account of those structures which it is necessary to divide in the operation required for ligaturing the innominate artery, A, Plate 9, or either of those main arterial vessels (the right common carotid and subclavian) which spring from it; and he will also observe that, although the same number and kind of structures overlie the carotid and subclavian vessels, A B, of the left side, Plate 10, still, that these vessels themselves, in consequence of their separate condition, will materially influence the like operation in respect to them. An aneurism occurring in the first part of the course of the right subclavian artery, at the locality *a*, Plate 9, will lie so close to the origin of the right common carotid as to require a ligature to be passed around the innominate common trunk, thus cutting off the flow of blood from both vessels; whereas an aneurism implicating either the left common carotid at the point A, or the left subclavian artery at the point B, does not, of course, require that both vessels should be included in the same ligature. There seems to be, therefore, a greater probability of effectually treating an aneurism of the left brachio-cephalic vessels by ligature than attaches to those of the right side; for if space between collateral branches, and also a lesser caliber of arterial trunk, be advantages, allowing the ligature to hold more firmly, then the vessels of the left side of the root of the neck manifest these advantages more frequently than those of the right, which spring from a common trunk. Whenever, therefore, the "peculiarity" of a separate aortic origin of the right carotid and subclavian arteries occurs, it is to be regarded more as a happy advantage than otherwise.

DESCRIPTION OF PLATES 9 & 10.

PLATE 9.

A. Innominate artery, at its point of bifurcation.

B. Right internal jugular vein, joining the subclavian vein.

C. Sternal end of the right clavicle.

D. Trachea.

E. Right sterno-thyroid muscle, cut.

F. Right sterno-hyoid muscle, cut.

G. Right sterno-mastoid muscle, cut.

a. Right vagus nerve, crossing the subclavian artery.

b. Anterior jugular vein, piercing the cervical fascia to join the subclavian vein.

Plate 9

PLATE 10.

A. Common carotid artery of left side.

B. Left subclavian artery, having *b*, the vagus nerve, between it and A.

C. Lower end of left internal jugular vein, joining--

D. Left subclavian vein, which lies anterior to *d*, the scalenus anticus muscle.

E. Anterior jugular vein, coursing beneath sterno-mastoid muscle and over the fascia.

F. Deep cervical fascia, enclosing in its layers *f f f,* the several muscles.

G. Left sterno-mastoid muscle, cut across, and separated from *g g*, its sternal and clavicular attachments.

H. Left sterno-hyoid muscle, cut.

I. Left sterno-thyroid muscle, cut.

K. Right sterno-hyoid muscle.

L. Right sterno-mastoid muscle.

M. Trachea.

N. Projection of the thyroid cartilage.

O. Place of division of common carotid.

P. Place where the subclavian artery passes beneath the clavicle.

Q. Sternal end of the left clavicle.

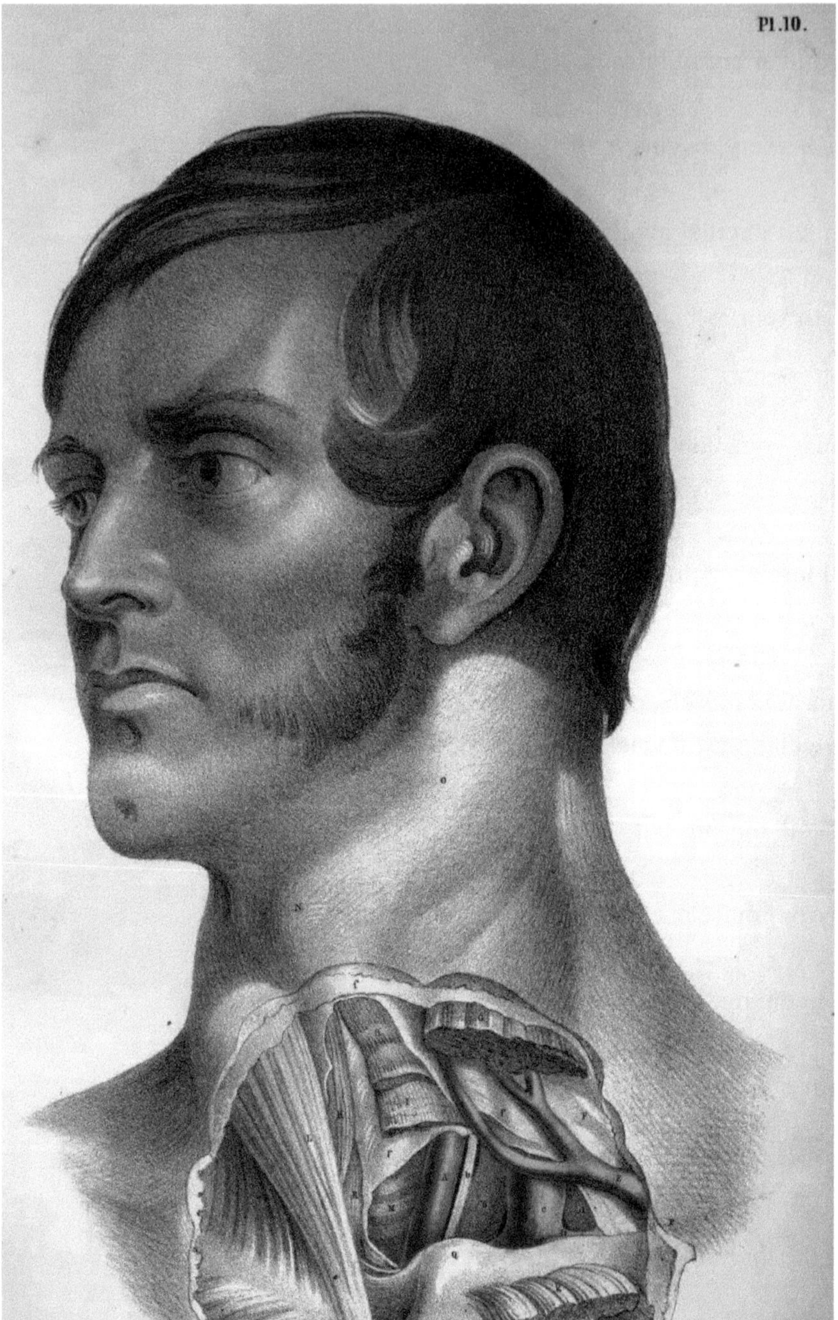

Plate 10

THE SURGICAL DISSECTION OF THE AXILLARY AND BRACHIAL REGIONS, DISPLAYING THE RELATIVE POSITION OF THEIR CONTAINED PARTS.

The operation for tying the axillary artery. Remarks on fractures of the clavicle and dislocation of the humerus in reference to the axillary vessels. The operation for tying the brachial artery near the axilla. Mode of compressing this vessel against the humerus.

All surgical regions have only artificial boundaries; and these, as might be expected, do not express the same meaning while viewed from more points than one. These very boundaries themselves, being moveable parts, must accordingly influence the relative position of the structures which they bound, and thus either include within or exclude from the particular region those structures wholly or in part which are said to be proper to it. Of this kind of conventional surgical boundary the moveable clavicle is an example; and the bloodvessels which it overarches manifest consequently neither termination nor origin except artificially from the fixed position which the bone, R, assumes, as in Plate 11, or *c**, Plate 12. In this position of the arm in relation to the trunk, the subclavian artery, B, terminates at the point where, properly speaking, it first takes its name; and from this point to the posterior fold of the axilla formed by the latissimus dorsi muscle, O, Plate 11, N, Plate 12, and the anterior fold formed by the great pectoral muscle, K, Plate 11, I, Plate 12, the continuation of the subclavian artery is named axillary. From the posterior fold of the axilla, O P, Plate 11, to the bend of the elbow, the same main vessels take the name of brachial.

When the axillary space is cut into from the forepart through the great pectoral muscle, H K, Plate 11, and beneath this through the lesser pectoral muscle, L I, together with the fascial processes which invest these muscles anteriorly and posteriorly, the main bloodvessels and nerves which traverse this space are displayed, holding in general that relative position which they exhibit in Plate 11. These vessels, with their accompanying nerves, will be seen continued from those of the neck; and thus may be attained in one view a comparative estimate of the cervical and axillary regions, together with their line of union beneath the clavicle, *c**, Plate 12, R, Plate 11, which serves to divide them surgically.

In the neck, the subclavian artery, B, Plate 11, is seen to be separated from the subclavian vein, A, by the breadth of the anterior scalenus muscle, D, as the vessels arch over the first rib, F. In this region of the course of the vessels, the brachial plexus of nerves, C, ranges along the outer border of the artery, B, and is separated by the artery from the vein, A, as all three structures pass beneath the clavicle, R, and the subclavius muscle, E. From this latter point the vessels and nerves take the name axillary, and in this axillary region the relative position of the nerves and vessels to each other and to the adjacent organs is somewhat changed. For now in the axillary region the vein, *a*, is in direct contact with the artery, *b,* on the forepart and somewhat to the inner side of which the vein lies; while the nerves, D, *d,* Plate 12, embrace the artery in a mesh or plexus of chords, from which it is often difficult to extricate it, for the purpose of ligaturing, in the dead subject, much less the living. The axillary plexus of nerves well merits the name, for I have not found it in any two bodies assuming a similar order or arrangement. Perhaps the order in which branches spring from the brachial plexus that is most constantly met with is the one represented at D, Plate 12, where we find, on the outer border of B, the axillary artery, a nervous chord, *d,* giving off a thoracic branch to pass behind H, the lesser pectoral muscle, while the main chord itself, *d*, soon divides into two branches, one the musculo-cutaneous, *e,* which pierces G, the coraco-brachialis muscle, and the other which forms one of the roots of the median nerve, *h*. Following that order of the nerves as they are shown in Plate 12, they may be enumerated from without inwards as follows:--the external or musculo-cutaneous, *e*; the two roots of the median, *h*; the ulnar, *f*; the musculo-spiral, *g*; the circumflex, *i*; close to which are seen the origins of the internal cutaneous, the nerve of Wrisberg, some thoracic branches, and posteriorly the subscapular nerve not seen in this view of the parts.

The branches which come off from the axillary artery are very variable both as to number and place of origin, but in general will be found certain branches which answer to the names thoracic, subscapular, and circumflex. These vessels, together with numerous smaller arteries, appear to be confined to no fixed point of origin, and on this account the place of election for passing a ligature around the main axillary artery sufficiently removed from collateral branches must be always doubtful. The subscapular artery, Q, Plate 12, is perhaps of all the other branches that one which manifests the most permanent character; its point of origin being in general opposite the interval between the latissimus and sub-scapular muscles, but I have seen it arise from all parts of the axillary main trunk. If it be required to give, in a history of the arteries, a full account of all the deviations from the so-called normal type to which these lesser branches here and elsewhere are subject, such account can scarcely be said to be called for in this place.

The form of the axillary space is conical, while the arm is abducted from the side, and while the osseous and muscular structures remain entire. The apex of the cone is formed at the root of the neck beneath the clavicle, R, Plate 11, and the subclavious muscle, E, and between the coracoid process, L*, of the scapula and the serratus magnus muscle, as this lies upon the thoracic side; at this apex the subclavian vessels, A B, enter the axillary space. The base of the cone is below, looking towards the arm, and is formed in front by the pectoralis major, K H, and behind by the latissimus dorsi, O, and teres muscles, P, together with a dense thick fascia; at this base the axillary vessels, *a b*, pass out to the arm, and become the brachial vessels, *a**b***. The anterior side of the cone is formed by the great pectoral muscle, H K, Plate 11, and the lesser pectoral, L I. The inner side is formed by the serratus magnus muscle, M, Plate 12, on the side of the thorax; the external side is formed by the scapular and humeral insertion of the subscapular muscle, the humerus and coraco-brachialis muscle; and the posterior side is formed by the latissimus dorsi, the teres and body of the subscapular muscle.

In this axillary region is contained a complicated mass of bloodvessels, nerves, and lymphatic glands, surrounded by a large quantity of loose cellular membrane and adipose tissue. All the arterial branches here found are given off from the axillary artery; and the numerous veins which accompany these branches enter the axillary vein. Nerves from other sources besides those of the axillary plexus traverse the axillary space; such nerves, for example, as those named intercosto-humeral, seen lying on the latissimus tendon, O, Plate 11. The vein named cephalic, S, enters the axillary space at that cellular interval occurring between the clavicular origin of the deltoid muscle, G, and the humeral attachment of the pectoralis major, H, which interval marks the place of incision for tying the axillary artery.

The general course of the main vessels through the axillary space would be indicated with sufficient accuracy by a line drawn from the middle of the clavicle, R R, Plate 11, to the inner border of the biceps muscle, N. In this direction of the axillary vessels, the coracoid process, L*, from which arises the tendon of the pectoralis minor muscle, L, is to be taken as a sure guide to the place of the artery, *b*, which passes, in general, close to the inner side of this bony process. Even in the undissected body the coracoid process may be felt as a fixed resisting point at that cellular interval between the clavicular attachments of the deltoid and great pectoral muscles. Whatever necessity shall require a ligature to be placed around the axillary in preference to the subclavian artery, must, of course, be determined by the particular case; but certain it is that the main artery, at the place B, a little above the clavicle, will always be found freer and more isolated from its accompanying nerves and vein, and also more easily reached, owing to its comparatively superficial situation, than when this vessel has become axillary. The incision required to be made, in order to reach the axillary artery, *b*, from the forepart, through the skin, both pectoral muscles, and different layers of fasciae, must be very deep, especially in muscular, well-conditioned bodies; and even when the level of the vessel is gained, it will be found much complicated by its own branches, some of which overlie it, as also by the plexus of nerves, D, Plate 12, which embraces it on all sides, while the large axillary vein, *a*, Plate 11, nearly conceals it in front. This vein in Plate 11 is drawn somewhat apart from the artery.

Sometimes the axillary artery is double, in consequence of its high division into brachial branches. But as this peculiarity of premature division never takes place so high up as where the vessel, B, Plate 11, overarches the first rib, F, this circumstance should also have some weight with the operator.

When we view the relative position of the subclavian vessels, A B, Plate 11, to the clavicle, R, we can readily understand why a fracture of the middle of this bone through that arch which it forms over the vessels, should interfere with the free circulation of the blood which these vessels supply to the arm. When the clavicle is severed at its middle, the natural arch which the bone forms over the vessels and nerves is lost, and the free moving broken ends of the bone will be acted on in opposing directions by the various muscles attached to its sternal and scapular extremities. The outer fragment follows more freely than the inner piece the action of the muscles; but, most of all, the weight of the unsupported shoulder and arm causes the displacement to which the outer fragment is liable. The subclavius muscle, E, like the pronator quadratus muscle of the forearm, serves rather to further the displacement of the broken ends of the bone than to hold them in situ.

If the head of the humerus be dislocated forwards beneath L, Plate 11, the coracoid attachment of the pectoralis minor muscle, it must press out of their proper place and put tensely upon the stretch the axillary vessels and plexus of nerves. So large and resistent a body as the head of the humerus displaced forwards, and taking the natural position of these vessels and nerves, will accordingly be attended with other symptoms--such as obstructed circulation and pain or partial paralysis, besides those physical signs by which we distinguish the presence of it as a new body in its abnormal situation.

When the main vessels and nerves pass from the axillary space to the inner side of the arm, they become comparatively superficial in this latter situation. The inner border of the biceps muscle is taken as a guide to the place of the brachial artery for the whole extent of its course in the arm. In plate 11, the artery, *b**, is seen in company with the median nerve, which lies on its fore part, and with the veins called *comites* winding round it and passing with it and the nerve beneath the fascia which encases in a fold of itself all three structures in a common sheath. Though the axillary vein is in close contact with the axillary artery and nerves, yet the basilic vein, *d**, the most considerable of those vessels which form the axillary vein, is separated from the brachial artery by the fascia. The basilic vein, however, overlies the brachial artery to its inner side, and is most commonly attended by the internal cutaneous nerve, seen lying upon it in Plate 11, as also by that other cutaneous branch of the brachial plexus, named the nerve of Wrisberg. If a longitudinal incision in the course of the brachial artery be made (avoiding the basilic vein) through the integument down to the fascia of the arm, and the latter structure be slit open on the director, the artery will be exposed, having the median nerve lying on its outer side in the upper third of the arm, and passing to its inner side towards the bend of the elbow, as at *b**, Plate 12. The superior and inferior profunda arteries, seen springing above and below the point *b*, Plate 12, are those vessels of most importance which are given off from the brachial artery, but the situation of their origin is very various. The ulnar nerve, *f*, lies close to the inner side of the main arterial trunk, as this latter leaves the axilla, but from this place to the inner condyle, Q, behind which the ulnar nerve passes into the forearm, the nerve and artery become gradually more and more separated from each other in their descent. The musculo-spiral nerve, *g*, winds under the brachial artery at the middle of the arm, but as this nerve passes deep between the short and long heads of the triceps muscle, P, and behind the humerus to gain the outer aspect of the limb, a little care will suffice for avoiding the inclusion of it in the ligature.

The brachial artery may be so effectually compressed by the fingers on the tourniquet, against the humerus in any part of its course through the arm, as to stop pulsation at the wrist.

The tourniquet is a less manageable and not more certain compressor of the arterial trunk than is the hand of an intelligent assistant. At every region of the course of an artery where the tourniquet is applicable, a sufficient compression by the hand is also attainable with greater ease to the patient; and the hand may compress the vessel at certain regions where the tourniquet would be of little or no use, or attended with inconvenience, as in the locality of the subclavian artery, passing over the first rib, or the femoral artery, passing over the pubic bone, or the carotid vessels in the neighbourhood of the trachea, as they lie on the fore part of the cervical spinal column.

DESCRIPTION OF PLATES 11 & 12.

PLATE 11.

A. Subclavian vein, crossed by a branch of the brachial plexus given to the subclavius muscle; *a*, the axillary vein; *a* *, the basilic vein, having the internal cutaneous nerve lying on it.

B. Subclavian artery, lying on F, the first rib; *b*, the axillary artery; *b* *, the brachial artery, accompanied by the median nerve and venae comites.

C. Brachial plexus of nerves; *c**, the median nerve.

D. Anterior scalenus muscle.

E. Subclavius muscle.

F F. First rib.

G. Clavicular attachment of the deltoid muscle.

H. Humeral attachment of the great pectoral muscle.

I. A layer of fascia, encasing the lesser pectoral muscle.

K. Thoracic half of the great pectoral muscle.

L. Coracoid attachment of the lesser pectoral muscle.

L*. Coracoid process of the scapula.

M. Coraco-brachialis muscle.

N. Biceps muscle.

O. Tendon of the latissimus dorsi muscle, crossed by the intercosto-humeral nerves.

P. Teres major muscle, on which and O is seen lying Wrisberg's nerve.

Q. Brachial fascia, investing the triceps muscle. .

R R. Scapular and sternal ends of the clavicle.

S. Cephalic vein, coursing between the deltoid and pectoral muscles, to enter at their cellular interval into the axillary vein beneath E, the subclavius muscle.

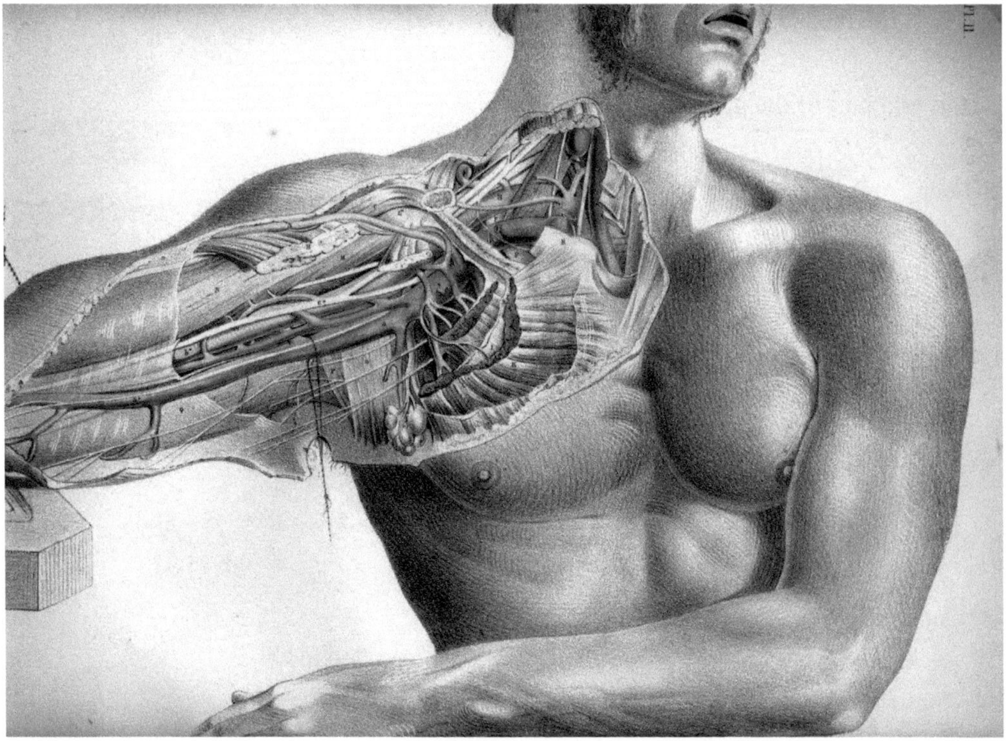

Plate 11

PLATE 12.

A. Axillary vein, cut and tied; *a*, the basilic vein, cut.

B. Axillary artery; *b*, brachial artery, in the upper part of its course, having *h*, the median nerve, lying rather to its outer side; *b**, the artery in the lower part of its course, with the median nerve to its inner side.

C. Subclavius muscle.

C*. Clavicle.

D. Axillary plexus of nerves, of which *d* is a branch on the coracoid border of the axillary artery; *e*, the musculo-cutaneous nerve, piercing the coraco-brachialis muscle; *f*, the ulnar nerve; *g*, musculo-spiral nerve; *h*, the median nerve; *i*, the circumflex nerve.

E. Humeral part of the great pectoral muscle.

F. Biceps muscle.

G. Coraco-brachialis muscle.

H. Thoracic half of the lesser pectoral muscle.

I. Thoracic half of the greater pectoral muscle.

K. Coracoid attachment of the lesser pectoral muscle.

K*. Coracoid process of the scapula.

L. Lymphatic glands.

M. Serratus magnus muscle.

N. Latissimus dorsi muscle.

O. Teres major muscle.

P. Long head of triceps muscle.

Q. Inner condyle of humerus.

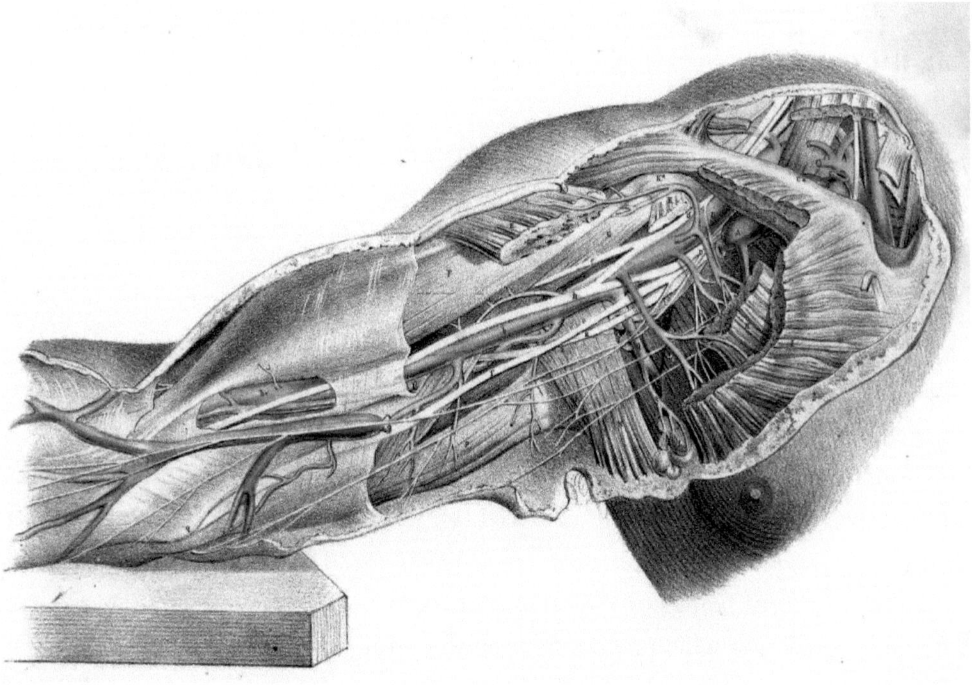

Plate 12

THE SURGICAL FORMS OF THE MALE AND FEMALE AXILLAE COMPARED.

The mammary and axillary glands in health and disease. Excision of these glands. Axillary abscess. General surgical observations on the axilla.

Certain characteristic features mark those differences which are to be found in all corresponding regions of both sexes. Though the male and female bodies, in all their regions, are anatomically homologous or similar at basis, yet the constituent and corresponding organs of each are gently diversified by the plus or minus condition, the more or the less, which the development of certain organs exhibits; and this diversity, viewed in the aggregate, constitutes the sexual difference. That diversity which defines the sexual character of beings of the same species, is but a link in that extended chain of differential gradation which marks its progress through the whole animal kingdom. The female breast is a plus glandular organ, situated, pendent, in that very position where, in a male body, the unevolved mamma is stil rudimentarily manifested.

The male and female axillae contain the same number and species of organs; and the difference by which the external configuration of both are marked mainly arises from the presence of the enlarged mammary gland, which, in the female, Plate 14, masks the natural outline of the pectoral muscle, E, whose axillary border is overhung by the gland; and thus this region derives its peculiarity of form, contrasted with that of the male subject.

When the dissected axilla is viewed from below, the arm being raised, and extended from the side, its contained parts, laid deeply in their conical recess, are sufficiently exposed, at the same time that the proper boundaries of the axillary cavity are maintained. In this point of view from which the axillary vessels are now seen, their relative position, in respect to the thorax and the arm, are best displayed. The thickness of that fleshy anterior boundary formed by both pectoral muscles, E F, Plate 13, will be marked as considerable; and the depth at which these muscles conceal the vessels, A B, in the front aspect of the thoracico-humeral interval, will prepare the surgeon for the difficulties he is to encounter when proceeding to ligature the axillary artery at the incision made through the anterior or pectoral wall of this axillary space.

The bloodvessels of the axilla follow the motions of the arm; and according to the position assumed by the arm, these vessels describe various curves, and lie more or less removed from the side of the thorax. While the arm hangs close to the side, the axillary space does not (properly speaking) exist; and in this position, the axillary vessels and nerves make a general curve from the clavicle at the point K, Plate 14, to the inner side of the arm, the concavity of the curve being turned towards the thoracic side. But when the arm is abducted from the side, and elevated, the vessels which are destined to supply the limb follow it, and in this position they take, in reality, a serpentine course; the first curve of which is described, in reference to the thorax, from the point K to the head of the humerus; and the next is that bend which the head of the humerus, projecting into the axilla in the elevated position of the member, forces them to make around itself in their passage to the inner side of the arm. The vessels may be readily compressed against the upper third of the humerus by the finger, passed into the axilla, and still more effectually if the arm be raised, as this motion will rotate the tuberous head of the humerus downwards against them.

The vessels and nerves of the axilla are bound together by a fibrous sheath derived from the membrane called costo-coracoid; and the base or humeral outlet of this axillary space, described by the muscles C, K, E, G, Plate 13, is closed by a part of the fascial membrane, *g,* extended across from the pectoral muscle, E, to the latissimus dorsi tendon, K. In the natural position of the vessels at that region of their course represented in the Plates, the vein A overlies the artery B, and also conceals most of the principal nerves. In order to show some of these nerves, in contact with the artery itself, the axillary vein is drawn a little apart from them.

The axillary space gives lodgment to numerous lymphatic glands, which are either directly suspended from the main artery, or from its principal branches, by smaller branches, destined to supply them. These glands are more numerous in the female axilla, Plate 14, than in the male, Plate 13, and while they seem to be, as it were, indiscriminately scattered here and there through this region, we observe the greater number of them to be gathered together along the axillary side of the great pectoral muscle; at which situation, *h,* in the diseased condition of the female breast, they will be felt to form hard, nodulated masses, which frequently extend as far up through the axillary space as the root of the neck, involving the glands of this latter region also in the disease.

The contractile motions of the pectoral muscle, E, of the male body, Plate 13, are during life readily distinguishable; and that boundary which it furnishes to the axillary region is well defined; but in the female form, Plate 14, the general contour of the muscle E, while in motion, is concealed by the hemispherical mammary gland, F, which, surrounded by its proper capsule, lies loosely pendent from the fore part of the muscle, to which, in the healthy state of the organ, it is connected only by free-moving bonds of lax cellular membrane. The motions of the shoulder upon the trunk do not influence the position of the female mammary gland, for the pectoral muscle acts freely beneath it; but when a scirrhus or other malignant growth involves the mammary organ, and this latter contracts, by the morbid mass, a close adhesion to the muscle, then these motions are performed with pain and difficulty.

When it is required to excise the diseased female breast, (supposing the disease to be confined to the structure of the gland itself,) the operation may be performed confidently and without difficulty, in so far as the seat of operation does not involve the immediate presence of any important nerves or bloodvessels. But when the disease has extended to the axillary glands, the extirpation of these (as they lie in such close proximity to the great axillary vessels and their principal branches) requires cautious dissection. It has more than once happened to eminent surgeons, that in searching for and dissecting out these diseased axillary glands, H, *h*, Plate 14, the main artery has been wounded.

As the coracoid process points to the situation of the artery in the axilla, so the coraco-brachialis muscle, C, marks the exact locality of the vessel as it emerges from this region; the artery ranges along the inner margin of both the process and the muscle, which latter, in fleshy bodies, sometimes overhangs and conceals it. When the vessel has passed the insertion of the coraco-brachialis, it becomes situated at the inner side of the biceps, which also partly overlaps it, as it now lies on the forepart of the brachialis anticus. As the general course of the artery, from where it leaves the axilla to the bend of the elbow, is one of winding from the inner side to the forepart of the limb, so should compression of the vessel, when necessary, be directed in reference to the bone accordingly--viz., in the upper or axillary region of the arm, from within outwards, and in the lower part of the arm, from before backwards.

All incised, lacerated, or contused wounds of the arm and shoulder, happening by pike, bayonet, sabre, bullet, mace, or arrow, on the outer aspect of the limb, are (provided the weapon has not broken the bones) less likely to implicate the great arteries, veins, and nerves. These instruments encountering the inner or axillary aspect of the member, will of course be more likely to involve the vessels and nerves in the wound. In severe compound fractures of the humerus occurring from force applied at the external side of the limb, the brachial vessels and nerves have been occasionally lacerated by the sharp jagged ends of the broken bone,--a circumstance which calls for immediate amputation of the member.

The axilla becomes very frequently the seat of morbid growths, which, when they happen to be situated beneath the dense axillary fascia, and have attained to a large size, will press upon the vessels and nerves of this region, and cause very great inconvenience. Adipose and other kind of tumours occurring in the axilla beneath the fascia, and in close contact with the main vessels, have been known to obstruct these vessels to such a degree, as to require the collateral or anastomatic circulation to be set up for the support; of the limb. When abscesses take place in the axilla, beneath the fascia, it is this structure which will prevent the matter from pointing; and it is required, therefore, to lay this fascia freely open by a timely incision. The accompanying Plates will indicate the proper direction in which such incision should be made, so as to avoid the vessels A, B. When the limb is abducted from the side, the main vessels and nerves take their position parallel with the axis of the arm. The axillary vessels and nerves being thus liable to pressure from the presence of large tumours happening in their neighbourhood, will suggest to the practitioner the necessity for fashioning of a proper form and size all apparatus, which in fracture or dislocation of the shoulder-bones shall be required to bear forcibly against the axillary region. While we know that the locality of the main vessels and nerves is that very situation upon which a pad or fulcrum presses, when placed in the axilla for securing the reduction of fractures of the clavicle, the neck of the humerus, or scapula, so should this member of the fracture apparatus be adapted, as well to obviate this pressure upon these structures, as to give the needful support to the limb in reference to the clavicle, &c. The habitual use, for weeks or more, of a hard, resisting fulcrum in the axilla, must act in some degree like the pad of a tourniquet, arresting the flow of a vigorous circulation, which is so essential to the speedy union of all lesions of bones. And it should never be lost sight of, that all grievously coercive apparatus, which incommode the suffering patient, under treatment, are those very instruments which impede the curative process of Nature herself.

The anatomical mechanism of the human body, considered as a whole, or divisible into regions, forms a study so closely bearing upon practice, that the surgeon, if he be not also a mechanician, and fully capable of making his anatomical knowledge suit with the common principles of mechanics, while devising methods for furthering the efforts, of Nature curatively, may be said to have studied anatomy to little or no purpose. The shoulder apparatus, when studied through the principle of mechanics, derives an interest of practical import which all the laboured description of the schools could never supply to it, except when illustrating this principle.

The disposal of the muscular around the osseous elements of the shoulder apparatus, forms a study for the surgeon as well in the abnormal condition of these parts, as in their normal arrangement; for in practice he discovers that that very mechanical principle upon which both orders of structures (the osseous and muscular) are grouped together for normal articular action, becomes, when the parts are deranged by fracture or, other accident, the chief cause whereby rearrangement is prevented, and the process of reunion obstructed. When a fracture happens in the shaft of the humerus, above or below the insertions of the pectoral and latissimus dorsi muscles, these are the very agents which when the bone possessed its integrity rendered it functionally fitting, and which, now that the bone is severed, produce the displacement of the lower fragment from the upper one. To counteract this source of derangement, the surgeon becomes the mechanician, and now, for the first time, he recognises the necessity of the study of topographical anatomy.

When a bone is fractured, or dislocated to a false position and retained there by the muscular force, the surgeon counteracts this force upon mechanical principle; but while he puts this principle in operation, he also acknowledges to the paramount necessity of ministering to the ease of Nature as much as shall be consistent with the effectual use of the remedial agent; and in the present state of knowledge, it is owned, that that apparatus is most efficient which simply serves both objects, the one no less than the other. And, assuming this to be the principle which should always guide us in our treatment of fractures and dislocations, I shall not hesitate to say, that the pad acting as a fulcrum in the axilla, or the perineal band bearing as a counterextending force upon the groin (the suffering body of the patient being, in both instances, subjected for weeks together to the grievous pressure and irritation of these members of the apparatus), do not serve both objects, and only one incompletely; I say incompletely, for out of every six fractures of either clavicle or thigh-bone, I believe that, as the result of our treatment by the present forms of mechanical contrivances, there would not be found three cases of coaptation of the broken ends of the bone so complete as to do credit to the surgeon. The most pliant and portable of all forms of apparatus which constitute the hospital armamentaria, is the judgment; and this cannot give its approval to any plan of instrument which takes effect only at the expense of the patient.

DESCRIPTION OF PLATES 13 & 14.

PLATE 13.

A. Axillary vein, drawn apart from the artery, to show the nerves lying between both vessels. On the bicipital border of the vein is seen the internal cutaneous nerve; on the tricipital border is the nerve of Wrisberg, communicating with some of the intercosto-humeral nerves; *a*, the common trunk of the venae comites, entering the axillary vein.

B. Axillary artery, crossed by one root of the median nerve; *b*, basilic vein, forming, with *a*, the axillary vein, A.

C. Coraco-brachialis muscle.

D. Coracoid head of the biceps muscle.

E. Pectoralis major muscle.

F. Pectoralis minor muscle.

G. Serratus magnus muscle, covered by *g*, the axillary fascia, and perforated, at regular intervals, by the nervous branches called intercosto-humeral.

H. Conglobate gland, crossed by the nerve called "external respiratory" of Bell, distributed to the serratus magnus muscle. This nerve descends from the cervical plexus.

I. Subscapular artery.

K. Tendon of latissimus dorsi muscle.

L. Teres major muscle.

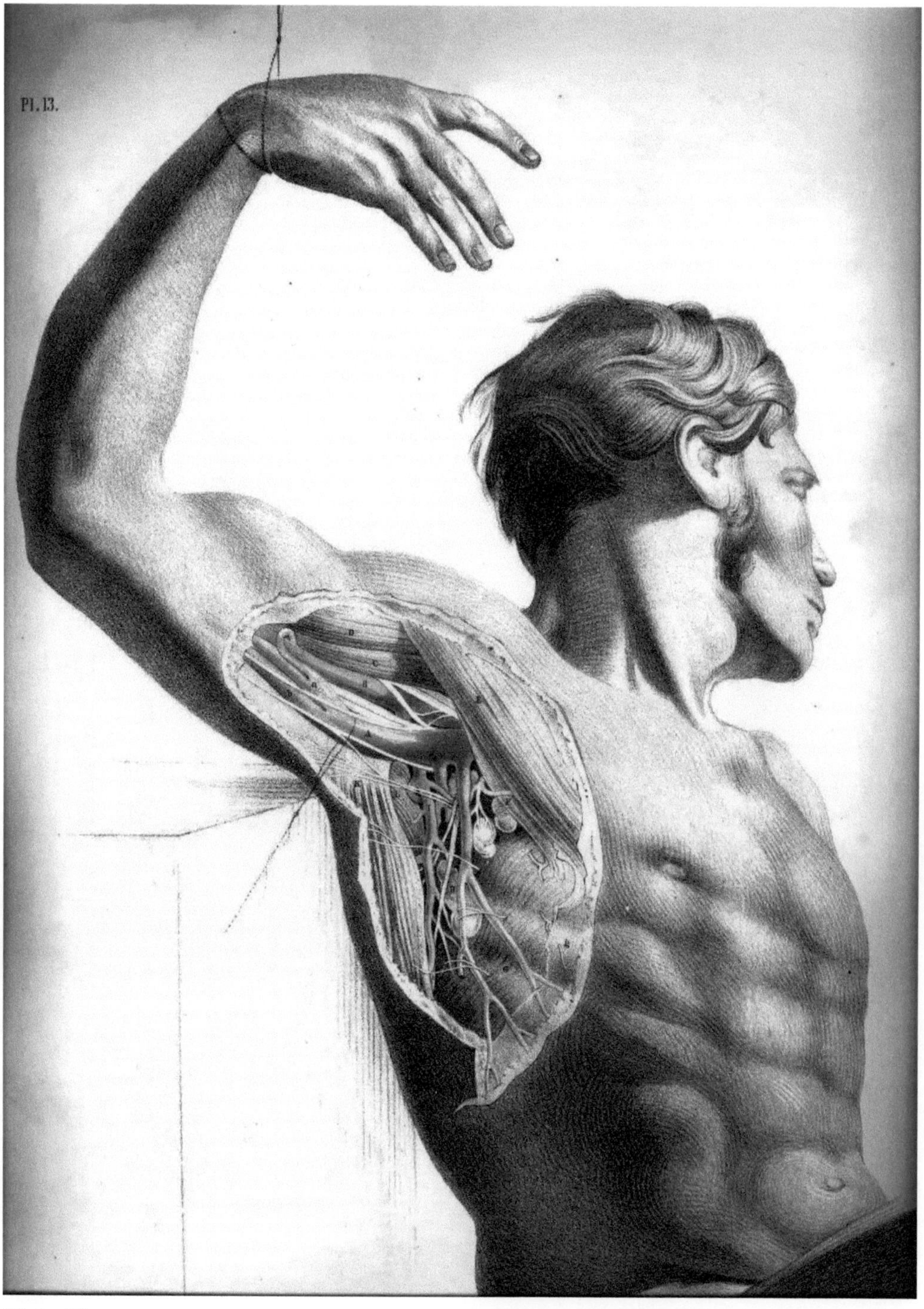

Plate 13

PLATE 14.

A. Axillary vein.

B. Axillary artery.

C. Coraco-brachialis muscle.

D. Short head of the biceps muscle.

E. Pectoralis major muscle.

F. Mammary gland, seen in section.

G. Serratus magnus muscle.

H. Lymphatic gland; *h h*, other glands of the lymphatic class.

I. Subscapular artery, crossed by the intercosto-humeral nerves and descending parallel to the external respiratory nerve. Beneath the artery is seen a subscapular branch of the brachial plexus, given to the latissimus dorsi muscle.

K. Locality of the subclavian artery.

L. Locality of the brachial artery at the bend of the elbow.

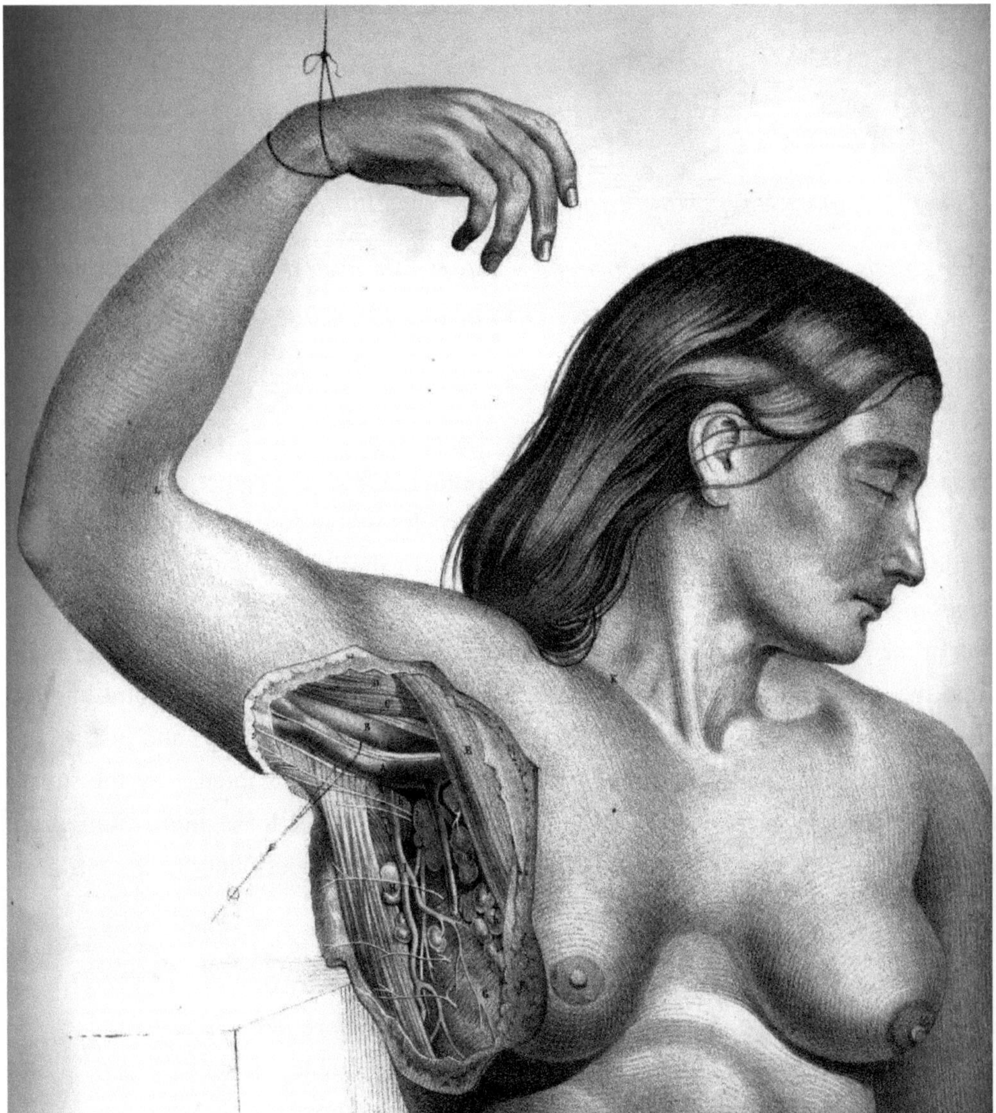

Plate 14

THE SURGICAL DISSECTION OF THE BEND OF THE ELBOW AND THE FOREARM, SHOWING THE RELATIVE POSITION OF THE VESSELS AND NERVES.

General remarks. Operation for tying the brachial artery at its middle and lower thirds. Varieties of the brachial artery. Venesection at the bend of the elbow. The radial and ulnar pulse. Operations for tying the radial and ulnar arteries in several parts.

The farther the surgical region happens to be removed from the centre of the body, the less likely is it that all accidents or operations which involve such regions will concern the life immediately. The limbs undergo all kinds of mutilation, both by accident and intention, and yet the patient survives; but when the like happens at any region of the trunk of the body, the life will be directly and seriously threatened. It seems, therefore, that in the same degree as the living principle diverges from the body's centre into the outstanding members, in that degree is the life weakened in intensity; and just as, according to physical laws, the ray of light becomes less and less intense by the square of the distance from the central source, so the vital ray, or vis, loses momentum in the same ratio as it diverges from the common central line to the periphery.

The relative anatomy of every surgical region becomes a study of more or less interest to the surgeon, according to the degree of importance attaching to the organs contained, or according to the frequency of such accidents as are liable to occur in each. The bend of the elbow is a region of anatomical importance, owing to the fact of its giving passage to C, Plate 15, the main artery of the limb, and also because in it are located the veins D, B, E, F, which are frequently the subject of operation. The anatomy of this region becomes, therefore, important; forasmuch as the operation which is intended to concern the veins alone, may also, by accident, include the main arterial vessel which they overlie. The nerves, which are seen to accompany the veins superficially, as well as that which accompanies the more deeply-situated artery, are, for the same reason, required to be known.

The course of the brachial artery along the inner border of the biceps muscle is comparatively superficial, from the point where it leaves the axilla to the bend of the elbow. In the whole of this course it is covered by the fascia of the arm, which serves to isolate it from the superficial basilic vein, B, and the internal cutaneous nerve, both of which nevertheless overlie the artery. The median nerve, *d*, Plate 15, accompanies the artery in its proper sheath, which is a duplication of the common fascia; and in this sheath are also situated the venae comites, making frequent loops around the artery. The median nerve itself, D, Plate 16, takes a direct course down the arm; and the different relative positions which this nerve holds in reference to the artery, C, at the upper end, the middle, and the lower end of the arm, occur mainly in consequence of the undulating character of the vessel itself.

When it is required to ligature the artery in the middle of the arm, the median nerve will be found, in general, at its outer side, between it and the biceps; but as the course of the artery is from the inner side of the biceps to the middle of the bend of the elbow, so we find it passing under the nerve to gain this locality, C, Plate 16, where the median nerve, D, then becomes situated at the inner side of the vessel. The median nerve, thus found to be differently situated in reference to the brachial artery, at the upper, the middle, and the lower part of the arm, is (with these facts always held in memory) taken as the guide to that vessel. An incision made of sufficient length (an inch and a half, more or less) over the course of the artery, and to the outer side of the basilic vein, B, Plate 16, will divide the skin, subcutaneous adipose membrane, which varies much in thickness in several individuals, and will next expose the common fascial envelope of the arm. When this fascia is opened, by dividing it on the director, the artery becomes exposed; the median nerve is then to be separated from the side of the vessel by the probe or director, and, with the precaution of not including the venal comites, the ligature may now be passed around the vessel. In the lower third of the arm it is not likely that the operator will encounter the ulnar nerve, and mistake it for the median, since the former, *d*, Plate 16, is considerably removed from the vessel. If the incision be made precisely in the usual course of the brachial artery, the ulnar nerve will not show itself. It will be well, however, to bear in mind the possible occurrence of some of those anomalies to that normal relative position of the artery, the median, and the ulnar nerve, which the accompanying Plates represent.

The median nerve, D, Plate 16, is sometimes found to lie beneath the artery in the middle and lower third of the arm. At other times it is found far removed to the inner side of the usual position of the vessel, and lying in close contact with the ulnar nerve, *d*. Or the brachial artery may take this latter position, while the median nerve stands alone at the position of D, Plate 16. Or both the main artery and the median nerve may course much to the inner side of the biceps muscle, A, Plate 16, while in the usual situation of the nerve and vessel there is only to be found a small arterial branch (the radial), which springs from the brachial, high up in the arm. Or the nerve and vessel may be lying concealed beneath a slip of the brachialis anticus muscle, E, Plate 16, in which case no appearance of them will be at all manifested through the usual place of incision made for the ligature of the brachial vessel. Or, lastly, there may be found more arteries than the single main brachial appearing at this place in the arm, and such condition of a plurality of vessels occurs in consequence of a high division of the brachial artery. Each of these variations from the normal type is more or less frequent; and though it certainly is of practical import to bear them in mind, still, as we never can foretell their occurrence by a superficial examination of the limb, or pronounce them to be present till we actually encounter them in operation, it is only when we find them that we commence to reason upon the facts; but even at this crisis the knowledge of their anatomy may prevent a confusion of ideas.

That generalization of the facts of such anomalies as are liable to occur to the normal character of the brachial artery, represented in Plates 15 and 16, which appears to me as being most inclusive of all their various conditions, is this--viz., that the point of division into radial, ulnar, and interosseous, which F, Plate 16, usually marks, may take place at any part of the member between the bend of the elbow and the coracoid process in the axillary space.

At the bend of the elbow, the brachial artery usually occupies the middle point between *e,* the inner condyle of the humerus and the external margin of the supinator radii longus muscle, G. The structures which overlie the arterial vessel, C, Plate 16, at this locality, numbering them from its own depth to the cutaneous surface, are these--viz., some adipose cellular membrane envelopes the vessel, as it lies on E, the brachialis anticus muscle, and between the two accompanying veins; at the inner side of the artery, but separated from it by a small interval occupied by one of the veins, is situated the median nerve *d,* Plate 15. Above all three structures is stretched that dense fibrous band of the fascia, H, Plate 16, which becomes incorporated with the common fascial covering of the forearm. Over this fascial process lies the median basilic vein, F B, Plate 15, accompanying which are seen some branches of the internal cutaneous nerve. The subcutaneous adipose tissue and common integument cover these latter. If it be required to ligature the artery at this locality, an incision two inches and a half in length, made along the course of the vessel, and avoiding the superficial veins, will expose the fascia; and this being next divided on the director, the artery will be exposed resting on the brachialis anticus, and between the biceps tendon and pronator teres muscle. As this latter muscle differs in width in several individuals, sometimes lying in close contact with the artery, and at other times leaving a considerable interval between the vessel and itself, its outer margin is not, therefore, to be taken as a sure guide to the artery. The inner border of the biceps indicates much more generally the situation of the vessel.

The bend of the elbow being that locality where the operation of phlebotomy is generally performed, it is therefore required to take exact account of the structures which occupy this region, and more especially the relation which the superficial veins hold to the deeper seated artery. In Plate 15, the artery, C, is shown in its situation beneath the fascial aponeurosis, which comes off from the tendon of the biceps, a portion of which has been cut away; and the venous vessel, F B, which usually occupies the track of the artery, is pushed a little to the inner side. While opening any part of the vessel, F B, which overlies the artery, it is necessary to proceed with caution, as well because of the fact that between the artery, C, and the vein, F B, the fascia alone intervenes, as also because the ulnar artery is given off rather frequently from the main vessel at this situation, and passes superficial to the fascia and flexors of the forearm, to gain its usual position at K, Plate 15. I have met with a well marked example of this occurrence in the living subject.

The cephalic vein, D, is accompanied by the external cutaneous nerve, which branches over the fascia on the outer border of the forearm. The basilic vein, B, is accompanied by the internal cutaneous nerve, which branches in a similar way over the fascia of the inner and fore part of the forearm. The numerous branches of both these nerves interlace with the superficial veins, and are liable to be cut when these veins are being punctured. Though the median basilic, F, and the basilic vein, B, are those generally chosen in the performance of the operation of bleeding, it will be seen, in Plate 15, that their contiguity to the artery necessarily demands more care and precision in that operation executed upon them, than if D, the cephalic vein, far removed as it is from the course of the artery, were the seat of phlebotomy.

As it is required, in order to distend the superficial veins, D, B, F, that a band should be passed around the limb at some locality between them and the heart, so that they may yield a free flow of blood on puncture, a moderate pressure will be all that is needful for that end. It is a fact worthy of notice, that the excessive pressure of the ligaturing band around the limb at A B, Plate 15, will produce the same effect upon the veins near F, as if the pressure were defective, for in the former case the ligature will obstruct the flow of blood through the artery; and the vein, F, will hence be undistended by the recurrent blood, just as when, in the latter case, the ligature, making too feeble a pressure on the vein, B, will not obstruct its current in that degree necessary to distend the vessel, F.

Whichever be the vein chosen for phlebotomy at the bend of the elbow, it will be seen, from an examination of Plates 15 and 16, that the opening may be made with most advantage according to the longitudinal axis of the vessel; for the vessel while being cut open in this direction, is less likely to swerve from the point of the lancet than if it were to be incised across, which latter mode is also far more liable to implicate the artery. Besides, as the nerves course along the veins from above downwards--making, with each other, and with the vessels, but very acute angles--all incisions made longitudinally in these vessels, will not be so likely to divide any of these nerves as when the instrument is directed to cut crossways.

The brachial artery usually divides, at the bend of the elbow, into the radial, the ulnar, and the interosseous branches. The point F, Plate 16, is the common place of division, and this will be seen in the Plate to be somewhat below the level of the inner condyle, *e.* From that place, where the radial and ulnar arteries spring, these vessels traverse the forearm, in general under cover of the muscles and fascia, but occasionally superficial to both these structures. The radial artery, F N, Plate 16, takes a comparatively superficial course along the radial border of the forearm, and is accompanied, for the upper two-thirds of its length, by the radial branch of the musculo-spiral nerve, seen in Plate 16, at the outer side of the vessel. The supinator radii longus muscle in general overlaps, with its inner border, both the radial artery and nerve. At the situation of the radial pulse, I, Plate 15, the artery is not accompanied by the nerve, for this latter will be seen, in plate 16, to pass outward, under the tendon of the supinator muscle, to the integuments.

The ulnar artery, whose origin is seen near F, Plate 16, passes deeply beneath the superficial flexor muscles, L M K, and the pronator teres, I, and first emerges from under cover of these at the point O, from which point to S, Plate 16, the artery may be felt, in the living body, obscurely beating as the ulnar pulse. On the inner border of the ulnar artery, and in close connexion with it, the ulnar nerve may be seen looped round by small branches of the vessel.

The radial and ulnar arteries may be exposed and ligatured in any part of their course; but of the two, the radial vessel can be reached with greater facility, owing to its comparatively superficial situation. The inner border of the supinator muscle, G, Plate 16, is the guide to the radial artery; and the outer margin of the flexor carpi ulnaris muscle, K, Plate 16, indicates the locality of the ulnar artery. Both arteries, I, K, Plate 15, at the wrist, lie beneath the fascia. If either of these vessels require a ligature in this region of the arm, the operation may be performed with little trouble, as a simple incision over the track of the vessels, through the skin and the fascia, will readily expose each.

Whenever circumstances may call for placing a ligature on the ulnar artery, as it lies between the superficial and deep flexor muscles, in the region of I L M, Plate 16, the course of the vessel may be indicated by a line drawn from a central point of the forearm, an inch or so below the level of the inner condyle--viz., the point F, and carried to the pisiform bone, T. The line of incision will divide obliquely the superficial flexors; and, on a full exposure of the vessel in this situation, the median nerve will be seen to cross the artery at an acute angle, in order to gain the mid-place in the wrist at Q. The ulnar nerve, *d,* Plate 16, passing behind the inner condyle, *e,* does not come into connexion with the ulnar artery until both arrive at the place O. It will, however, be considered an awkward proceeding to subject to transverse section so large a mass of muscles as the superficial flexors of the forearm, when the vessel may be more readily reached elsewhere, and perhaps with equal advantage as to the locality of the ligature.

When either the radial or ulnar arteries happen to be completely divided in a wound, both ends of the vessel will bleed alike, in consequence of the free anastomosis of both arteries in the hand.

DESCRIPTION OF PLATES 15 & 16.

PLATE 15.

A. Fascia covering the biceps muscle.

B. Basilic vein, with the internal cutaneous nerve.

C. Brachial artery, with the venae comites.

D. Cephalic vein, with the external cutaneous nerve; *d*, the median nerve.

E. A communicating vein, joining the venae comites.

F. Median basilic vein.

G. Lymphatic gland.

H. Radial artery at its middle.

I. Radial artery of the pulse.

K. Ulnar artery, with ulnar nerve.

L. Palmaris brevis muscle.

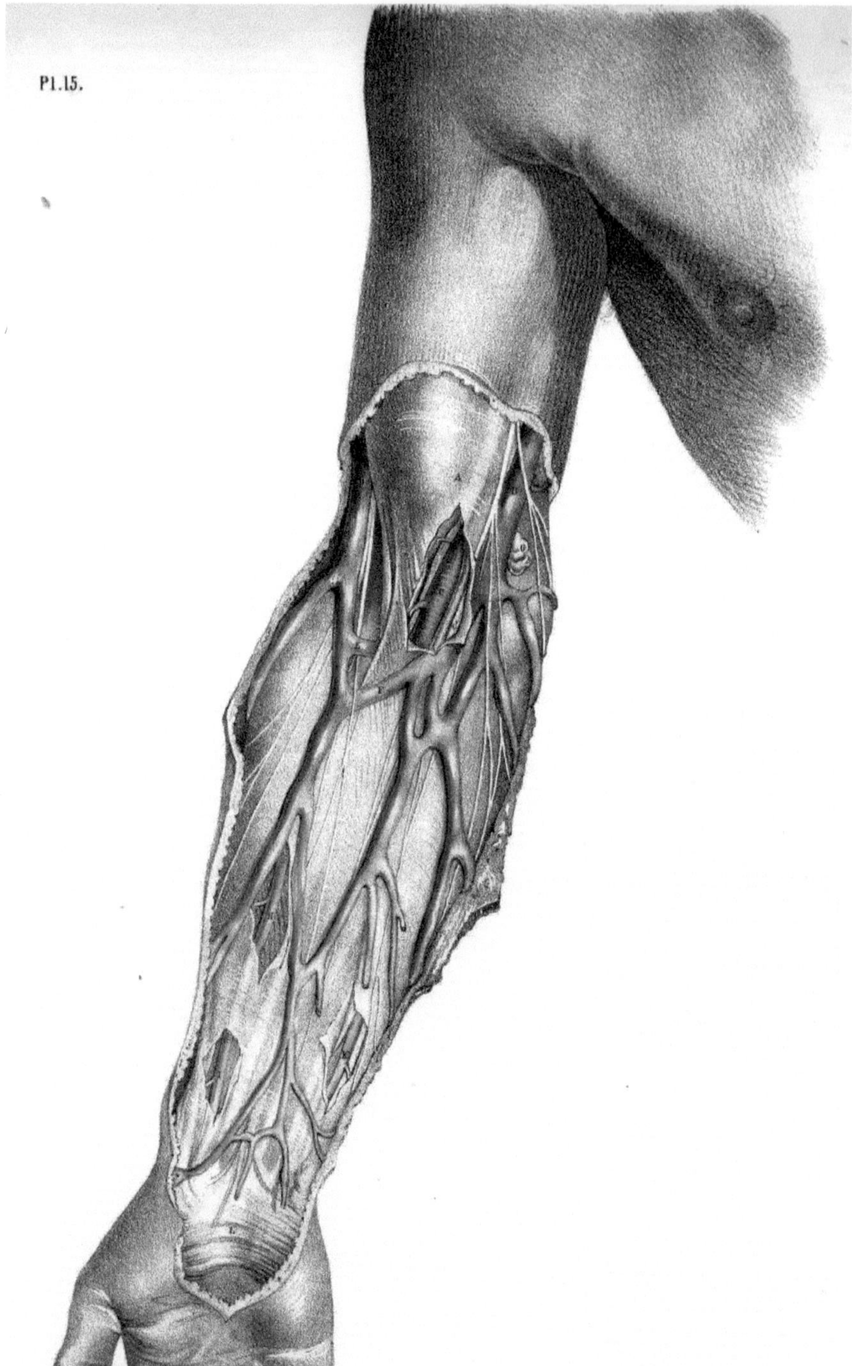

Plate 15

PLATE 16.

A. Biceps muscle.

B. Basilic vein, cut.

C. Brachial artery.

D. Median nerve; *d,* the ulnar nerve.

E. Brachialis anticus muscle; *e*, the internal condyle.

F. Origin of radial artery.

G. Supinator radii longus muscle.

H. Aponeurosis of the tendon of the biceps muscle.

I. Pronator teres muscle.

K. Flexor carpi ulnaris muscle.

L. Flexor carpi radialis muscle.

M. Palmaris longus muscle.

N. Radial artery, at its middle, with the radial nerve on its outer side.

O. Flexor digitorum sublimis.

P. Flexor pollicis longus.

Q. Median nerve.

R. Lower end of radial artery.

S. Lower end of ulnar artery, in company with the ulnar nerve.

T. Pisiform bone.

U. Extensor metacarpi pollicis.

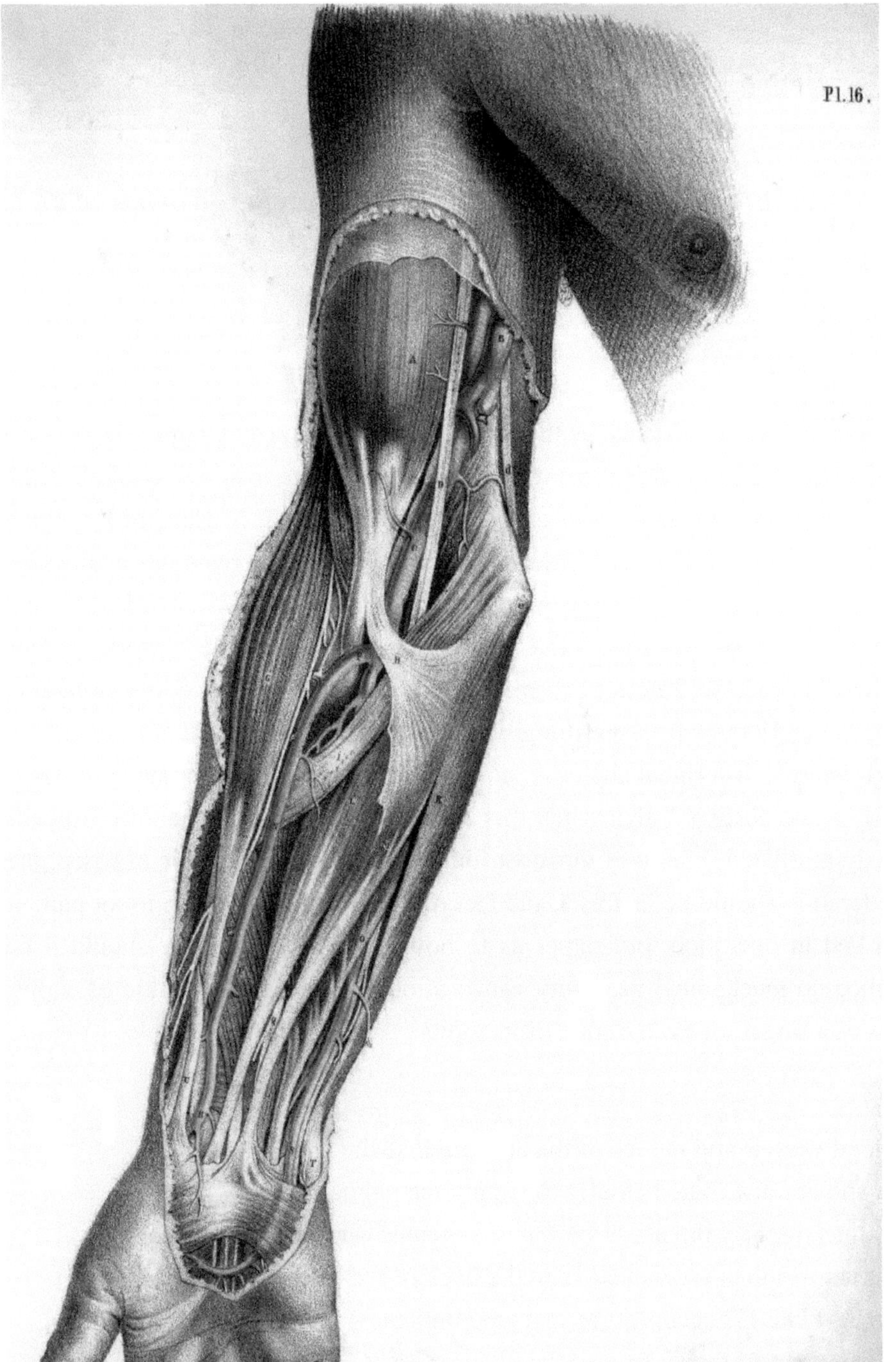

Plate 16

THE SURGICAL DISSECTION OF THE WRIST AND HAND.

General observations. Superficial and deep palmar arches. Wounds of these vessels requiring a ligature to be applied to both ends. General surgical remarks on the arteries of the upper limb. Palmar abscess, &c.

A member of such vast importance as the human hand necessarily claims a high place in regard to surgery. The hand is typical of the mind. It is the material symbol of the immaterial spirit, It is the prime agent of the will; and it is that instrument by which the human intellect manifests its presence in creation. The human hand has a language of its own. While the tongue demonstrates the thought through the word, the hand realizes and renders visible the thought through the work. This organ, therefore, by whose fitness of form the mind declares its own entity in nature, by the invention and creation of the thing, which is, as it were, the mind's autograph, claims a high interest in surgical anatomy; and accordingly the surgeon lays it down as a rule, strictly to be observed, that when this beautiful and valuable member happens to be seriously mutilated, in any of those various accidents to which it is exposed, the prime consideration should be, not as to the fact of how much of its quantity or parts it can be deprived in operation, but rather as to how little of its quantity should it be deprived, since no mechanical ingenuity can fashion an apparatus, capable of supplying the loss of a finger, or even of one of its joints.

The main blood vessels and nerves of the arm traverse the front aspect of the wrist, and are distributed chiefly to supply the palmar surface of the hand, since in the palm are to be found a greater variety and number of structures than are met with on the back of the hand. The radial artery, A, Plate 17, occupies (as its name indicates) the radial border of the forepart of the wrist, and the ulnar artery, C, Plate 17, occupies the ulnar border; both vessels in this region of their course lie parallel to each other; both are comparatively superficial, but of the two, the radial artery is the more superficial and isolated, and thereby occasions the radial pulse. The anatomical situation of the radial artery accounts for the fact, why the pulsation of this vessel is more easily felt than that of the ulnar artery.

The radial vessel, A, Plate 17, at the wrist, is not accompanied by the radial nerve; for this nerve, C, Plate 19, passes from the side of the artery, at a position, C, Plate 19, varying from one to two or more inches above the wrist, to gain the dorsal aspect of the hand. The ulnar artery, C, Plate 17, is attended by the ulnar nerve, D, in the wrist, and both these pass in company to the palm. The ulnar nerve, D E, lies on the ulnar border of the artery, and both are in general to be found ranging along the radial side of the tendon of the flexor carpi ulnaris muscle, T, and the pisiform bone, G. The situation of the radial artery is midway between the flexor carpi radialis tendon, I, and the outer border of the radius. The deep veins, called *comites*, lie in close connexion with the radial and ulnar arteries. When it is required to lay bare the radial or ulnar artery, at the wrist, it will be sufficient for that object to make a simple longitudinal incision (an inch or two in length) over the course of the vessel A or C, Plate 17, through the integument, and this incision will expose the fascia, which forms a common investment for all the structures at this region. When this fascia has been cautiously slit open on the director, the vessels will come into view. The ulnar artery, however, lies somewhat concealed between the adjacent muscles, and in order to bring this vessel fully into view, it will be necessary to draw aside the tendon of the flexor ulnaris muscle, T.

The radial artery, A, Plate 18, passes external to the radial border of the wrist, beneath the extensor tendons, B, of the thumb; and after winding round the head of the metacarpal bone of the thumb, as seen at E, Plate 19, forms the deep palmar arch E, Plate 18. This deep palmar arch lies close upon the forepart of the carpo-metacarpal joints; it sends off branches to supply the deeply situated muscles, and other structures of the palm; and from it are also derived other branches, which pierce the interosseal spaces, and appear on the back of the hand, Plate 19. The deep palmar arch, E, Plate 18, inosculates with a branch of the ulnar artery, I, Plate 18, whilst its dorsal interosseal branches, Plate 19, communicate freely with the dorsal carpal arch, which is formed by a branch of the radial artery E, Plate 19, and the terminal branch of the posterior interosseous vessel.

The ulnar artery, C, Plate 17, holds a direct and superficial course, from the ulnar border of the forearm through the wrist; and still remains superficial in the palm, where it forms the superficial palmar arch, F. From this arch arise three or four branches of considerable size, which are destined to supply the fingers. A little above the interdigital clefts, each of these digital arteries divides into two branches, which pass along the adjacent sides of two fingers--a mode of distribution which also characterises the digital branches of the median, *b b*, and ulnar nerves, *e e*. The superficial palmar arch of the ulnar vessel anastomoses with the deep arch of the radial vessel. The principal points of communication are, first, by the branch, (ramus profundus,) I, Plate 18, which passes between the muscles of the little finger to join the deep arch beneath the long flexor tendons. 2nd, by the branch (superficialis volae) which springs from the radial artery, A, Plate 17, and crosses the muscles of the ball of the thumb, to join the terminal branch of the superficial arch, F, Plate 17. 3rd, by another terminal branch of the superficial arch, which joins the arteries of the thumb, derived from the radial vessel, as seen at *e*, Plate 18.

The frequent anastomosis thus seen to take place between the branches of the radial, the ulnar, and the interosseous arteries in the hand, should be carefully borne in mind by the surgeon. The continuity of the three vessels by anastomosis, renders it very difficult to arrest a haemorrhage occasioned by a wound of either of them. It will be at once seen, that when a haemorrhage takes place from any of these larger vessels of the hand, the bleeding will not be commanded by the application of a ligature to either the radial, the ulnar, or the interosseous arteries in the forearm; and for this plain reason, viz., that though in the arm these arteries are separate, in the hand their communication renders them as one.

If a haemorrhage therefore take place from either of the palmar vessels, it will not be sufficient to place a ligature around the radial or the ulnar artery singly, for if F, Plate 17, bleeds, and in order to arrest that bleeding we tie the vessel C, Plate 17, still the vessel F will continue to bleed, in consequence of its communication with the vessel E, Plate 18, by the branch 1, Plate 18, and other branches above mentioned. If E, Plate 18, bleeds, a ligature applied to the vessel A, Plate 18, will not stop the flow of blood, because of the fact that E anastomoses with G, by the branch I and other branches, as seen in Plates 17 and 19.

Any considerable haemorrhage, therefore, which may be caused by a wound of the superficial or deep palmar arches, or their branches, and which we are unable to arrest by compression, applied directly to the patent orifices of the vessel, will in general require that a ligature be applied to both the radial and ulnar arteries at the wrist; and it occasionally happens that even this proceeding will not stop the flow of blood, for the interosseous arteries, which also communicate with the vessels of the hand, may still maintain the current of circulation through them. These interosseous arteries being branches of the ulnar artery, and being given off from the vessel at the bend of the elbow, if the bleeding be still kept up from the vessel wounded in the hand, after the ligature of the ulnar and radial arteries is accomplished, are in all probability the channels of communication, and in this case the brachial artery must be tied. A consideration of the above mentioned facts, proper to the normal distribution of the vessels of the upper extremity, will explain to the practitioner the cause of the difficulty which occasionally presents itself, as to the arrest of haemorrhage from the vessels of the hand. In addition to these facts he will do well to remember some other arrangements of these vessels, which are liable to occur; and upon these I shall offer a few observations.

While I view the normal disposition of the arteries of the arm as a whole, (and this view of the whole great fact is no doubt necessary, if we would take within the span and compass of the reason, all the lesser facts of which the whole is inclusive,) I find that as one main vessel (the brachial) divides into three lesser branches, (the ulnar, radial and interosseous,) so, therefore, when either of these three supplies the haemorrhage, and any difficulty arises preventing our having access at once to the open orifices of the wounded vessel, we can command the flow of blood by applying a ligature to the main trunk--the brachial. If this measure fail to command the bleeding, then we may conclude that the wounded vessel (whichever it happen to be, whether the radial, the ulnar, or the interosseous) arises from the brachial artery, higher up in the arm than that place whereat we applied the ligature. To this variety as to the place of origin, the ulnar, radial, and interosseous arteries are individually liable.

Again, as the single brachial artery divides into the three arteries of the forearm, and as these latter again unite into what may (practically speaking) be termed a single vessel in the hand, in consequence of their anastomosis, so it is obvious that in order to command a bleeding from any of the palmar arteries, we should apply a ligature upon each of the vessels of the forearm, or upon the single main vessel in the arm. When the former proceeding fails, we have recourse to the latter, and when this latter fails (for fail it will, sometimes,) we then reasonably arrive at the conclusion that some one of the three vessels of the forearm, springs higher up than the place of the ligature on the main brachial vessel.

But however varied as to the normal locality of their origin, at the bend of the elbow, these vessels of the forearm may at times manifest themselves, still one point is quite fixed and certain, viz., that they communicate with each other in the hand. Hence, therefore, it becomes evident, that in order to command, at once and effectually, a bleeding, either from the palmar arteries, or those of the forearm, we attain to a more sure and successful result, the nearer we approach the fountain-head and place a ligature on it--the brachial artery. It is true that to stop the circulation through the main vessel of the limb, is always attended with danger, and that such a proceeding is never to be adopted but as the lesser one of two great hazards. It is also true that to tie the main brachial artery for a haemorrhage of anyone of its terminal branches, may be doing too much, while a milder course may serve; or else that even our tying the brachial may not suffice, owing to a high distribution of the vessels of the arm, in the axilla, above the place of the ligature. Thus doubt as to the safest measure, viz., that which is sufficient and no more, enveils the proper place whereat to apply a ligature on the principal vessel; but whatever be the doubt as to this particular, there can be none attending the following rule of conduct, viz., that in all cases of haemorrhage, caused by wounds of the vessels of the upper limb, we should, if at all practicable, endeavour to stop the flow of blood from the divided vessels *in the wound itself,* by ligature or otherwise; and both ends of the divided vessel require to be tied. Whenever this may be done, we need not trouble ourselves concerning the anomaly in vascular distribution.

The superficial palmar arch, F, Plate 17, lies beneath the dense palmar fascia; and whenever matter happens to be pent up by this fascia, and it is necessary that an opening be made for its exit, the incision should be conducted at a distance from the locality of the vessel. When matter forms beneath the palmar fascia, it is liable, owing to the unyielding nature of this fibrous structure, to burrow upwards into the forearm, beneath the annular ligament D, Plates 17 and 18. All deep incisions made in the median line of the forepart of the wrist are liable to wound the median nerve B, Plate 17. When the thumb, together with its metacarpal bone, is being amputated, the radial artery E, Plate 19, which winds round near the head of that bone, may be wounded. It is possible, by careful dissection, to perform this operation without dividing the radial vessel.

DESCRIPTION OF PLATES 17, 18, & 19.

PLATE 17.

A. Radial artery.

B. Median nerve; *b b b b*, its branches to the thumb and fingers.

C. Ulnar artery, forming F, the superficial palmar arch.

D. Ulnar nerve; E *e e*, its continuation branching to the little and ring fingers, &c.

G. Pisiform bone.

H. Abductor muscle of the little finger.

I. Tendon of flexor carpi radialis muscle.

K. Opponens pollicis muscle.

L. Flexor brevis muscle of the little finger.

M. Flexor brevis pollicis muscle.

N. Abductor pollicis muscle.

OOOO. Lumbricales muscles.

P P P P. Tendons of the flexor digitorum sublimis muscle.

Q. Tendon of the flexor longus pollicis muscle.

R. Tendon of extensor metacarpi pollicis.

S. Tendons of extensor digitorum sublimis; P P P, their digital prolongations.

T. Tendon of flexor carpi ulnaris.

U. Union of the digital arteries at the tip of the finger.

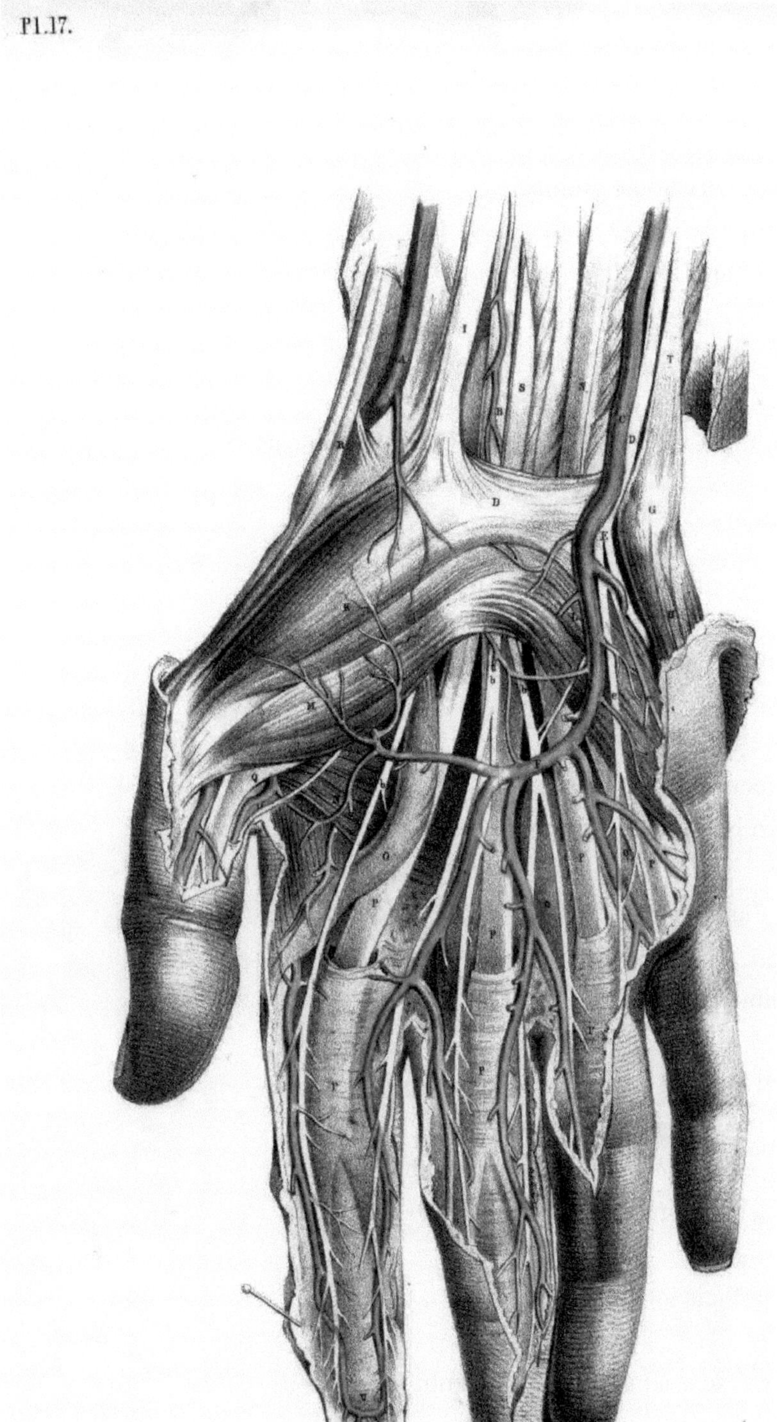

Plate 17

PLATE 18.

A. Radial artery.

B. Tendons of the extensors of the thumb.

C. Tendon of extensor carpi radialis.

D. Annular ligament.

E. Deep palmar arch, formed by radial artery giving off *e*, the artery of the thumb.

F. Pisiform bone.

G. Ulnar artery, giving off the branch I to join the deep palmar arch E of the radial artery.

H. Ulnar nerve; *h*, superficial branches given to the fingers. Its deep palmar branch is seen lying on the interosseous muscles, M M.

K. Abductor minimi digiti.

L. Flexor brevis minimi digiti.

M. Palmar interosseal muscles.

N. Tendons of flexor digitorum sublimis and profundus, and the lumbricales muscles cut and turned down.

O. Tendon of flexor pollicis longus.

P. Carpal end of the metacarpal bone of the thumb.

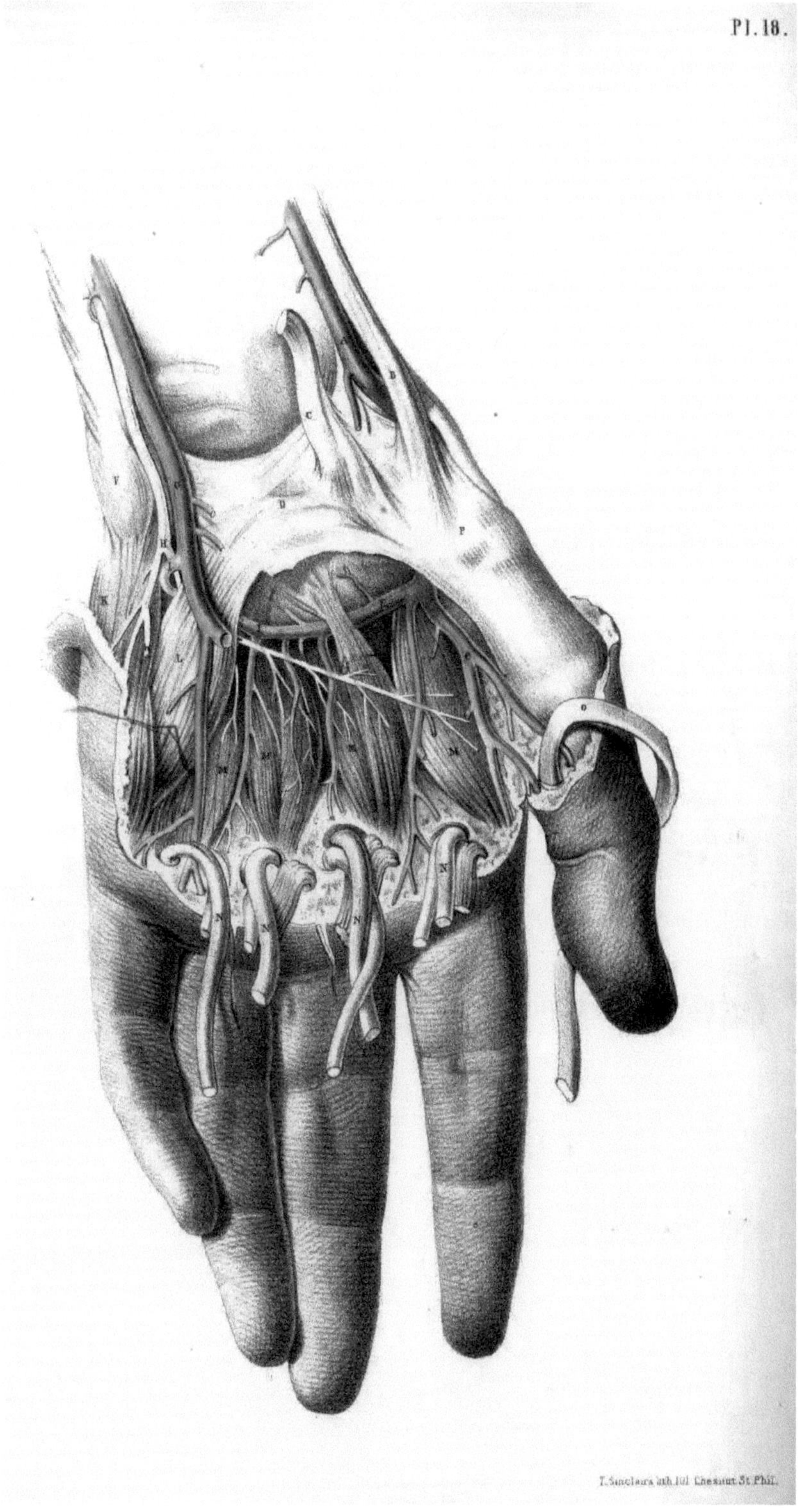

Plate 18

PLATE 19.

AAA. Tendons of extensor digitorum communis; A*, tendon overlying that of the indicator muscle.

B. Dorsal part of the annular ligament.

C. End of the radial nerve distributed over the back of the hand, to two of the fingers and the thumb.

D. Dorsal branch of the ulnar nerve supplying the back of the hand and the three outer fingers.

E. Radial artery turning round the carpal end of the metacarpal bone of the thumb.

F. Tendon of extensor carpi radialis brevis.

G. Tendon of extensor carpi radialis longus.

H. Tendon of third extensor of the thumb.

I. Tendon of second extensor of the thumb.

K. Tendon of extensor minimi digiti joining a tendon of extensor communis.

Plate 19

THE RELATIVE POSITION OF THE CRANIAL, NASAL, ORAL, AND PHARYNGEAL CAVITIES, ETC.

Fractures of the cranium, and the operation of trephining anatomically considered. Instrumental measures in reference to the fauces, tonsils, oesophagus, and lungs.

On making a section (vertically through the median line) of the cranio-facial and cervico-hyoid apparatus, the relation which these structures bear to each other in the osseous skeleton reminds me strongly of the great fact enunciated by the philosophical anatomists, that the facial apparatus manifests in reference to the cranial structures the same general relations which the hyoid apparatus bears to the cervical vertebrae, and that these relations are similar to those which the thoracic apparatus bears to the dorsal vertebrae. To this anatomical fact I shall not make any further allusions, except in so far as the acknowledgment of it shall serve to illustrate some points of surgical import.

The cranial chamber, A A H, Plate 20, is continuous with the spinal canal C. The osseous envelope of the brain, called calvarium, Z B, holds serial order with the cervical spinous processes, E I, and these with the dorsal spinous processes. The dura-matral lining membrane, A A A*, of the cranial chamber is continuous with the lining membrane, C, of the spinal canal. The brain is continuous with the spinal cord. The intervertebral foramina of the cervical spine are manifesting serial order with the cranial foramina. The nerves which pass through the spinal region of this series of foramina above and below C are continuous with the nerves which pass through the cranial region. The anterior boundary, D I, of the cervical spine is continuous with the anterior boundary, Y F, of the cranial cavity. And this common serial order of osseous parts--viz., the bodies of vertebrae, serves to isolate the cranio-spinal compartment from the facial and cervical passages. Thus the anterior boundary, Y F D I, of the cranio-spinal canal is also the posterior boundary of the facial and cervical cavities.

Now as the cranio-spinal chamber is lined by the common dura-matral membrane, and contains the common mass of nervous structure, thus inviting us to fix attention upon this structure as a whole, so we find that the frontal cavity, Z, the nasal cavity, X W, the oral cavity, 4, 5, S, the pharyngeal and oesophageal passages 8 Q, are lined by the common mucous membrane, and communicate so freely with each other that they may be in fact considered as forming a common cavity divided only by partially formed septa, such as the one, U V, which separates to some extent the nasal fossa from the oral fossa.

As owing to this continuity of structure, visible between the head and spine, we may infer the liability which the affections of the one region have to pass into and implicate the other, so likewise by that continuity apparent between all compartments of the face, fauces, oesophagus, and larynx, we may estimate how the pathological condition of the one region will concern the others.

The cranium, owing to its comparatively superficial and undefended condition, is liable to fracture. When the cranium is fractured, in consequence of force applied to its anterior or posterior surfaces, A or B, Plate 20, the fracture will, for the most part, be confined to the place whereat the force has been applied, provided the point opposite has not been driven against some resisting body at the same time. Thus when the point B is struck by a force sufficient to fracture the bone, while the point A is not opposed to any resisting body, then B alone will yield to the force applied; and fracture thus occurring at the point B, will have happened at the place where the applied force is met by the force, or weight, or inertia of the head itself. But when B is struck by any ponderous body, while A is at the same moment forced against a resisting body, then A is also liable to suffer fracture. If fracture in one place be attended with counter-fracture in another place, as at the opposite points A and B, then the *fracture* occurs from the force impelling, while the *counter-fracture* happens by the force resisting.

Now in the various motions which the cranium A A B performs upon the top of the cervical spine C, motions backwards, forwards, and to either side, it will follow that, taking C as a fixed point, almost all parts of the cranial periphery will be brought vertical to C in succession, and therefore whichever point happens at the moment to stand opposite to C, and has impelling force applied to it, then C becomes the point of resistance, and thus counter-fractures at the cranial base occur in the neighbourhood of C. When force is applied to the cranial vertex, whilst the body is in the erect posture, the top of the cervical spine, E D C, becomes the point of resistance. Or if the body fall from a height upon its cranial vertex, then the propelling force will take effect at the junction of the spine with the cranial base, whilst the resisting force will be the ground upon which the vertex strikes. In either case the cranial base, as well as the vertex, will be liable to fracture.

The anatomical form of the cranium is such as to obviate a frequent liability to fracture. Its rounded shape diffuses, as is the case with all rotund forms, the force which happens to strike upon it. The mode in which the cranium is set upon the cervical spine serves also to diffuse the pressure at the points where the two opposing forces meet--viz., at the first cervical vertebra E and the cranial basilar process F. This fact might be proved upon mechanical principle.

The tegumentary envelope of the head, as well as the dura-matral lining, serves to damp cranial vibration consequent upon concussion; while the sutural isolation of the several component bones of the cranium also prevents, in some degree, the extension of fractures and the vibrations of concussion. The contents of the head, like the contents of all hollow forms, receive the vibratory influence of force externally applied. The brain receives the concussion of the force applied to its osseous envelope; and when this latter happens to be fractured, the danger to life is not in proportion to the extent of the fracture here, any more than elsewhere in the skeleton fabric, but is solely in proportion to the amount of shock or injury sustained by the nervous centre.

When it is required to trephine any part of the cranial envelope, the points which should be avoided, as being in the neighbourhood of important bloodvessels, are the following--the occipital protuberance, B, within which the "torcular Herophili" is situated, and from this point passing through the median line of the vertex forwards to Z the frontal sinus, the trephine should not be applied, as this line marks the locality of the superior longitudinal sinus. The great lateral sinus is marked by the superior occipital ridge passing from the point B outwards to the mastoid process. The central point B of the side of the head, Plate 21, marks the locality of the root of the meningeal artery within the cranium, and from this point the vessel branches forwards and backwards over the interior of the cranium.

The nasal fossae are situated on either side of the median partition formed by the vomer and cartilaginous nasal septum. Both nasal fossae are open anteriorly and posteriorly; but laterally they do not, in the normal state of these parts, communicate. The two posterior nares answering to the two nasal fossae open into the upper part of the bag of the pharynx at 8, Plate 20, which marks the opening of the Eustachian tube.

The structures observable in both the nasal fossae absolutely correspond, and the foramina which open into each correspond likewise. All structures situated on either side of the median line are similar. And the structure which occupies the median line is itself double, or duality fused into symmetrical unity. The osseous nasal septum is composed of two laminae laid side by side. The spongy bones, X W, are attached to the outer wall of the nasal fossa, and are situated one above the other. These bones are three in number, the uppermost is the smallest. The outer wall of each naris is grooved by three fossae, called meatuses, and these are situated between the spongy bones. Each meatus receives one or more openings of various canals and cavities of the facial apparatus. The sphenoidal sinus near F opens into the upper meatus. The frontal, Z, and maxillary sinuses open into the middle meatus, and the nasal duct opens into the inferior sinus beneath the anterior inferior angle of the lower spongy bone, W.

In the living body the very vascular fleshy and glandular Schneiderian membrane which lines all parts of the nasal fossa almost completely fills this cavity. When polypi or other growths occupy the nasal fossae, they must gain room at the expense of neighbouring parts. The nasal duct may have a bent probe introduced into it by passing the instrument along the outer side of the floor of the nasal fossa as far back as the anterior inferior angle of the lower spongy bone, W, at which locality the duct opens. An instrument of sufficient length, when introduced into the nostrils in the same direction, will, if passed backwards through the posterior nares, reach the opening of the Eustachian tube, 8.

While the jaws are closed, the tongue, R, Plate 20, occupies the oral cavity almost completely. When the jaws are opened they form a cavity between them equal in capacity to the degree at which they are sundered from each other. The back of the pharynx can be seen when the jaws are widely opened if the tongue be depressed, as R, Plate 20. The hard palate, U, which forms the roof of the mouth, is extended further backwards by the soft palate, V, which hangs as the loose velum of the throat between the nasal fossae above and the fauces below. Between the velum palati, V, and the root of the tongue, we may readily discern, when the jaws are open, two ridges of arching form, 5, 6, on either side of the fauces. These prominent arches and their fellows are named the pillars of the fauces. The anterior pillar, 5, is formed by the submucous palato-glossus muscle; the posterior pillar, 6, is formed by the palato-pharyngeus muscle. Between these pillars, 5 and 6, is situated the tonsil, S, beneath the mucous membrane. When the tonsils of opposite sides become inflamed and suppurate, an incision may be made into either gland without much chance of wounding the internal carotid artery; for, in fact, this vessel lies somewhat removed from it behind. In Plate 21, that point of the superior constrictor of the pharynx, marked D, indicates the situation of the tonsil gland; and a considerable interval will be seen to exist between D and the internal carotid vessel F.

If the head be thrown backwards the nasal and oral cavities will look almost vertically towards the pharyngeal pouch. When the juggler is about to "swallow the sword," he throws the head back so as to bring the mouth and fauces in a straight line with the pharynx and oesophagus. And when the surgeon passes the probang or other instruments into the oesophagus, he finds it necessary to give the head of the person on whom he operates the same inclination backwards. When instruments are being passed into the oesophagus through the nasal fossa, they are not so likely to encounter the rima glottidis below the epiglottis, 9, as when they are being passed into the oesophagus by the mouth. The glottis may be always avoided by keeping the point of the instrument pressing against the back of the pharynx during its passage downwards.

When in suspended animation we endeavour to inflate the lungs through the nose or mouth, we should press the larynx, 10, 11,12, backwards against the vertebral column, so as to close the oesophageal tube.

DESCRIPTION OF PLATES 20 & 21.

PLATE 20.

A A. The dura-matral falx; A*, its attachment to the tentorium.

B. Torcular Herophili.

C. Dura-mater lining the spinal canal.

D D*. Axis vertebra.

E E*. Atlas vertebra.

F F*. Basilar processes of the sphenoid and occipital bones.

G. Petrous part of the temporal bone.

H. Cerebellar fossa.

I I*. Seventh cervical vertebra.

K K*. First rib surrounding the upper part of the pleural sac.

L L*. Subclavian artery of the right side overlying the pleural sac.

M M*. Right subclavian vein.

N. Right common carotid artery cut at its origin.

O. Trachea.

P. Thyroid body.

Q. Oesophagus.

R. Genio-hyo-glossus muscle.

S. Left tonsil beneath the mucous membrane.

T. Section of the lower maxilla.

U. Section of the upper maxilla.

V. Velum palati in section.

W. Inferior spongy bone.

X. Middle spongy bone.

Y. Crista galli of oethmoid bone.

Z. Frontal sinus.

2. Anterior cartilaginous part of nasal septum.

3. Nasal bone.

4. Last molar tooth of the left side of lower jaw.

5. Anterior pillar of the fauces.

6. Posterior pillar of the fauces.

7. Genio-hyoid muscle.

8. Opening of Eustachian tube.

9. Epiglottis.

10. Hyoid bone.

11. Thyroid bone.

12. Cricoid bone.

13. Thyroid axis.

14. Part of anterior scalenus muscle.

15. Humeral end of the clavicle.

16. Part of posterior scalenus muscle.

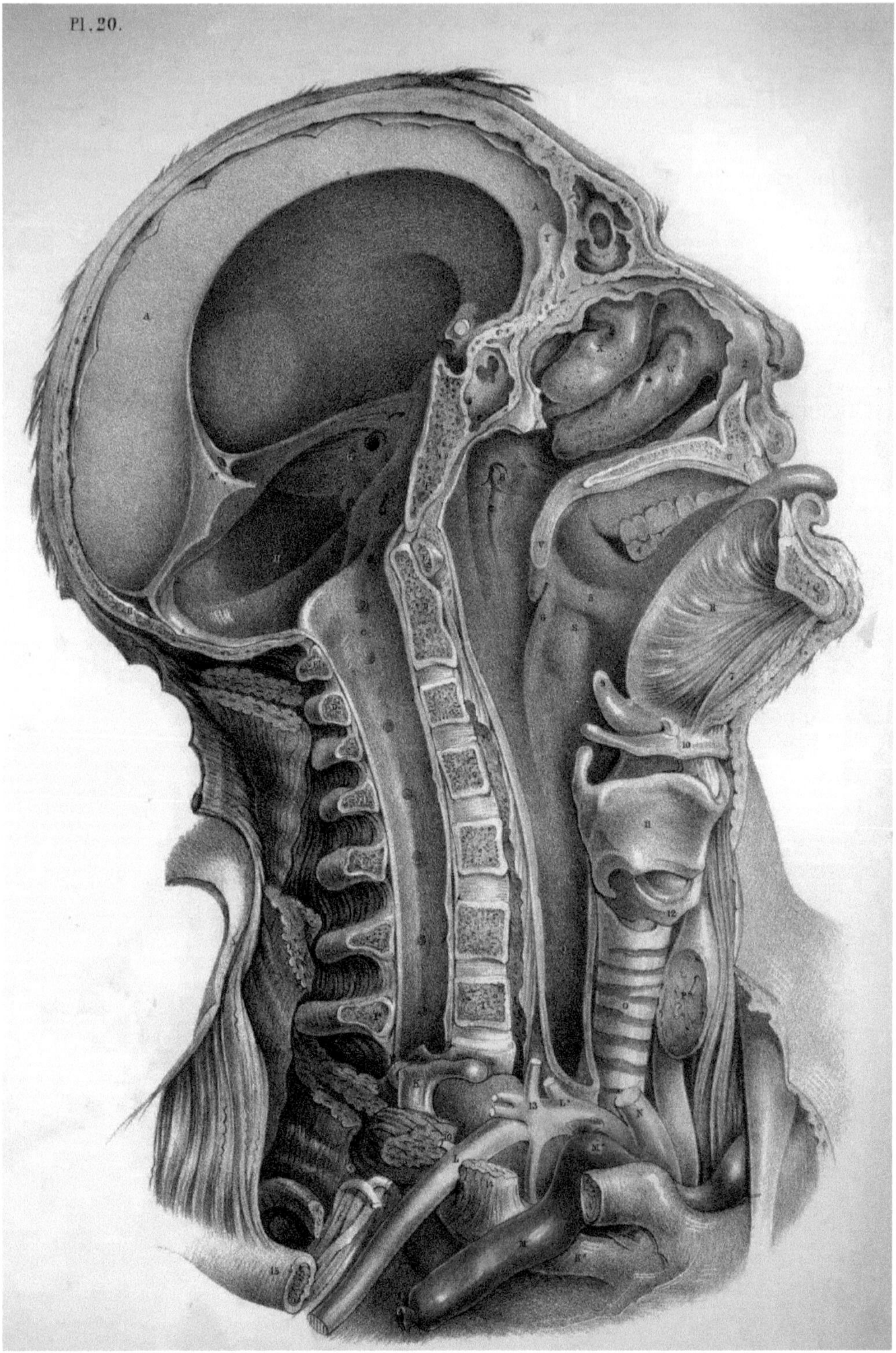

Plate 20

PLATE 21.

A. Zygoma.

B. Articular glenoid fossa of temporal bone.

C. External pterygoid process lying on the levator and tensor palati muscles.

D. Superior constrictor of pharynx.

E. Transverse process of the Atlas.

F. Internal carotid artery. Above the point F, is seen the glosso-pharyngeal nerve; below F, is seen the hypoglossal nerve.

G. Middle constrictor of pharynx.

H. Internal jugular vein.

I. Common carotid cut across.

K. Rectus capitis major muscle.

L. Inferior constrictor of pharynx.

M. Levator anguli scapulae muscle.

N. Posterior scalenus muscle.

O. Anterior scalenus muscle.

P. Brachial plexus of nerves.

Q. Trachea.

R R*. Subclavian artery.

S. End of internal jugular vein.

T. Bracheo-cephalic artery.

U U*. Roots of common carotid arteries.

V. Thyroid body.

W. Thyroid cartilage.

X. Hyoid bone.

Y. Hyo-glossus muscle.

Z. Upper maxillary bone.

2. Inferior maxillary branch of fifth cerebral nerve.

3. Digastric muscle cut.

4. Styloid process.

5. External carotid artery.

6 6. Lingual artery.

7. Roots of cervical plexus of nerves.

8. Thyroid axis; 8*, thyroid artery, between which and Q, the trachea, is seen the inferior laryngeal nerve.

9. Omo-hyoid muscle cut.

10. Sternal end of clavicle.

11. Upper rings of trachea, which may with most safety be divided in tracheotomy.

12. Cricoid cartilage.

13. Crico-thyroid interval where laryngotomy is performed.

14. Genio-hyoid muscle.

15. Section of lower maxilla.

16. Parotid duct.

17. Lingual attachment of styloglossus muscle, with part of the gustatory nerve seen above it.

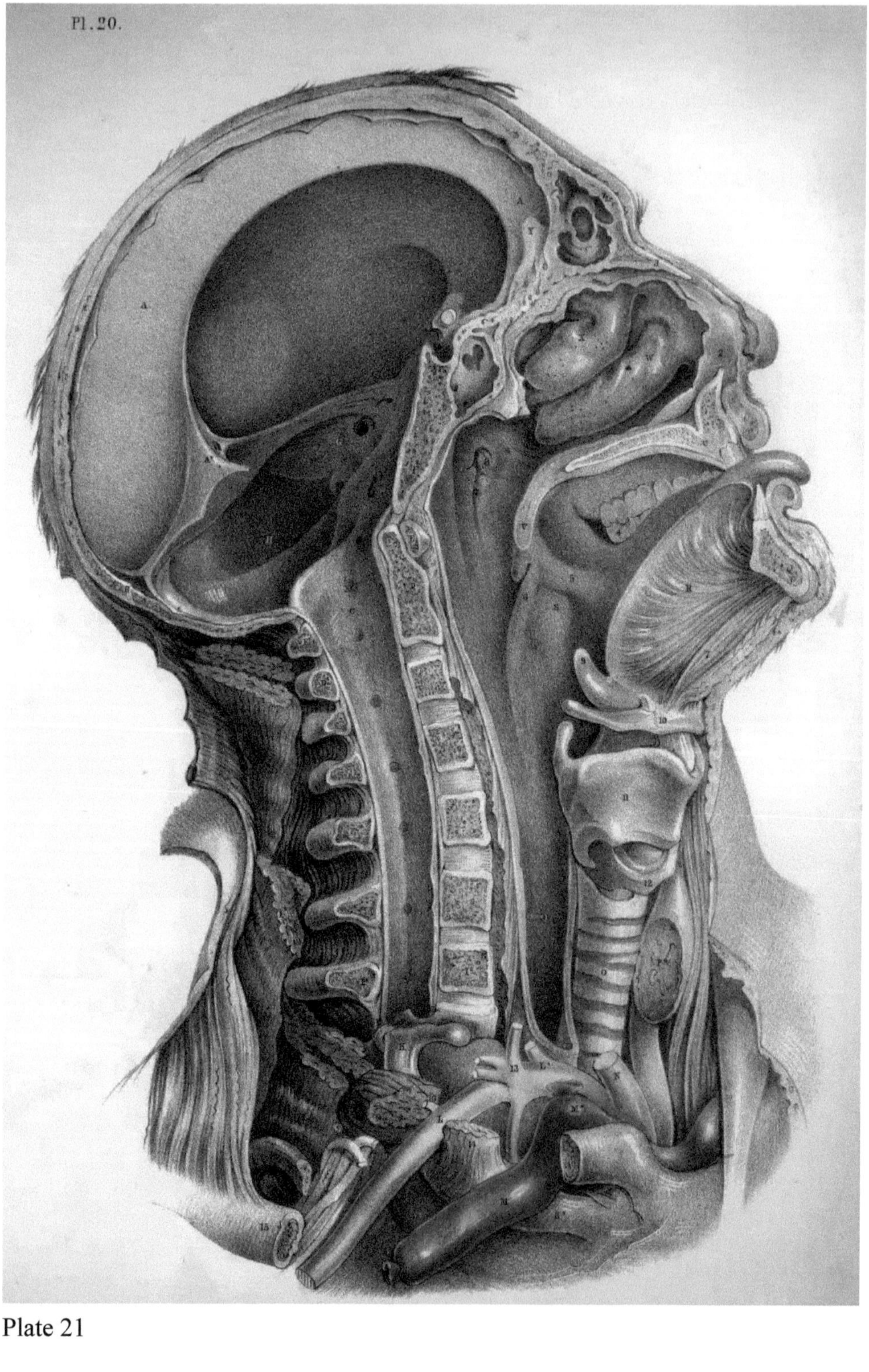

Plate 21

THE RELATIVE POSITION OF THE SUPERFICIAL ORGANS OF THE THORAX AND ABDOMEN.

Application to correct physical diagnosis. Changes in the relative position of the organs during the respiratory motions. Changes effected by disease. Physiological remarks on wounds of the thorax and on pleuritic effusion. Symmetry of the organs, &c.

In the osseous skeleton, the thorax and abdomen constitute a common compartment. We cannot, while we contemplate this skeleton, isolate the one region from the other by fact or fancy. The only difference which I can discover between the regions called thorax and abdomen, in the osseous skeleton, (considering this body morphologically,) results, simply, from the circumstance that the ribs, which enclose thoracic space, have no *osseous* counterparts in the abdomen enclosing abdominal space, and this difference is merely histological. In man and the mammalia the costal arches hold relation with the pulmonary organs, and these costae fail at that region where the ventral organs are located. In birds, and many reptiles, the costal arches enclose the common thoracico-abdominal region, as if it were a common pulmonary region. In fishes the costal arches enclose the thoracico-abdominal region, just as if it were a common abdominal region. I merely mention these general facts to show that costal enclosure does not actually serve to isolate the thorax from the abdomen in the lower classes of animals; and on turning to the human form, I find that this line of separation between the two compartments is so very indefinite, that, as pathologists, we are very liable to err in our diagnosis between the diseased and the healthy organs of either region, as they lie in relation with the moveable diaphragm or septum in the living body. The contents of the whole trunk of the body from the top of the sternum to the perineum are influenced by the respiratory motions; and it is most true that the diaphragmatic line, H F H*, is alternately occupied by those organs situated immediately above and below it during the performance of these motions, even in health.

The organs of the thoracic region hold a certain relation to each other and to the thoracic walls. The organs of the abdomen hold likewise a certain relation to each other and to the abdominal parietes. The organs of both the thorax and the abdomen have a certain relation to each other, as they lie above and below the diaphragm. In dead nature these relations are fixed and readily ascertainable, but in living, moving nature, the organs influence this relative position, not only of each other, but also of that which they bear to the cavities in which they are contained. This change of place among the organs occurs in the normal or healthy state of the living body, and, doubtless, raises some difficulty in the way of our ascertaining, with mathematical precision, the actual state of the parts which we question, by the physical signs of percussion and auscultation. In disease this change of place among these organs is increased, and the difficulty of making a correct diagnosis is increased also in the same ratio. For when an emphysematous lung shall fully occupy the right thoracic side from B to L, then G, the liver, will protrude considerably into the abdomen beneath the right asternal ribs, and yet will not be therefore proof positive that the liver is diseased and abnormally enlarged. Whereas, on the other hand, when G, the liver, is actually diseased, it may occupy a situation in the right side as high as the fifth or sixth ribs, pushing the right lung upwards as high as that level; and, therefore, while percussion elicits a dull sound over this place thus occupied, such sound will not be owing to a hepatized lung, but to the absence of the lung caused by the presence of the liver.

In the healthy adult male body, Plate 22, the two lungs, D D*, whilst in their ordinary expanded state, may be said to range over all that region of the trunk of the body which is marked by the sternal and asternal ribs. The heart, E, occupies the thoracic centre, and part of the left thoracic side. The heart is almost completely enveloped in the two lungs. The only portion of the heart and pericardium, which appears uncovered by the lung on opening the thorax, is the base of the right ventricle, E, situated immediately behind the lower end of the sternum, where this bone is joined by the cartilages of the sixth and seventh ribs. The lungs range perpendicularly from points an inch above B, the first rib, downwards to L, the tenth rib, and obliquely downwards and backwards to the vertebral ends of the last ribs. This space varies in capacity, according to the degree in which the lungs are expanded within it. The increase in thoracic space is attained, laterally, by the expansion of the ribs, C I; and vertically, by the descent of the diaphragm, H, which forces downwards the mass of abdominal viscera. The contraction of thoracic space is caused by the approximation of all the ribs on each side to each other; and by the ascent of the diaphragm. The expansion of the lungs around the heart would compress this organ, were it not that the costal sides yield laterally while the diaphragm itself descends. The heart follows the ascent and descent of the diaphragm, both in ordinary and forced respiration.

But however much the lungs vary in capacity, or the heart as to position in the respiratory motions, still the lungs are always closely applied to the thoracic walls. Between the pleura costalis and pulmonalis there occurs no interval in health. The thoracic parietes expand and contract to a certain degree; and to that same degree, and no further, do the lungs within the thorax expand and contract. By no effort of expiration can the animal expel all the air completely from its lungs, since by no effort of its own, can it contract thoracic space beyond the natural limit. On the other hand, the utmost degree of expansion of which the lungs are capable, exactly equals that degree in which the thoracic walls are dilatable by the muscular effort; and, therefore, between the extremes of inspiration and expiration, the lungs still hold closely applied to the costal parietes. The air within the lungs is separated from the air external to the thorax, by the thoracic parietes. The air within and external to the lungs communicate at the open glottis. When the glottis closes and cuts off the communication, the respiratory act ceases--the lungs become immovable, and the thoracic walls are (so far as the motions of respiration are concerned) rendered immovable also. The muscles of respiration cannot, therefore, produce a vacuum between the pulmonic and costal pleura, either while the external air has or has not access to the lungs. Upon this fact the mechanism of respiration mainly depends; and we may see a still further proof of this in the circumstance that, when the thoracic parietes are pierced, so as to let the external air into the cavity of the pleura, the lung collapses and the thoracic side ceases to exert an expansile influence over the lung. When in cases of fracture of the rib the lung is wounded, and the air of the lung enters the pleura, the same effect is produced as when the external air was admitted through an opening in the side.

When serous or purulent effusion takes place within the cavity of the pleura, the capacity of the lung becomes lessened according to the quantity of the effusion. It is more reasonable to expect that the soft tissue of the lung should yield to the quantity of fluid within the pleural cavity, than that the rigid costal walls should give way outwardly; and, therefore, it seldom happens that the practitioner can discover by the eye any strongly-marked difference between the thoracic walls externally, even when a considerable quantity of either serum, pus, or air, occupies the pleural sacs.

In the healthy state of the thoracic organs, a sound characteristic of the presence of the lung adjacent to the walls of the thorax may be elicited by percussion, or heard during the respiratory act through the stethoscope, over all that costal space ranging anteriorly between B, the first rib, and I K, the eight and ninth ribs. The respiratory murmur can be heard below the level of these ribs posteriorly, for the lung descends behind the arching diaphragm as far as the eleventh rib.

When fluid is effused into the pleural cavity, the ribs are not moved by the intercostal muscles opposite the place occupied by the fluid, for this has separated the lung from the ribs. The fluid has compressed the lung; and in the same ratio as the lung is prevented from expanding, the ribs become less moveable. The presence of fluid in the pleural sac is discoverable by dulness on percussion, and, as might be expected, by the absence of the respiratory murmur at that locality which the fluid occupies. Fluid, when effused into the pleural sac, will of course gravitate; and its position will vary according to the position of the patient. The sitting or standing posture will therefore suit best for the examination of the thorax in reference to the presence of fluid.

Though the lungs are closely applied to the costal sides at all times in the healthy state of these organs, still they slide freely within the thorax during the respiratory motions--forwards and backwards--over the serous pericardium, E, and upwards and downwards along the pleura costalis. The length of the adhesions which supervene upon pleuritis gives evidence of the extent of these motions. When the lung becomes in part solidified and impervious to the inspired air, the motions of the thoracic parietes opposite to the part are impeded. Between a solidified lung and one which happens to be compressed by effused fluid it requires no small experience to distinguish a difference, either by percussion or the use of the stethoscope. It is great experience alone that can diagnose hydro-pericardium from hypertrophy of the substance of the heart by either of these means.

The thoracic viscera gravitate according to the position of the body. The heart in its pericardial envelope sways to either side of the sternal median line according as the body lies on this or that side. The two lungs must, therefore, be alternately affected as to their capacity according as the heart occupies space on either side of the thorax. In expiration, the heart, E, is more uncovered by the shelving edges of the lungs than in inspiration. In pneumothorax of either of the pleural sacs the air compresses the lung, pushes the heart from its normal position, and the space which the air occupies in the pleura yields a clear hollow sound on percussion, whilst, by the ear or stethoscope applied to a corresponding part of the thoracic walls, we discover the absence of the respiratory murmur.

The transverse diameter of the thoracic cavity varies at different levels from above downwards. The diameter which the two first ribs, B B*, measure, is the least. That which is measured by the two eighth ribs, I I*, is the greatest. The perpendicular depth of the thorax, measured anteriorly, ranges from A, the top of the sternum, to F, the xyphoid cartilage. Posteriorly, the perpendicular range of the thoracic cavity measures from the spinous process of the seventh cervical vertebra above, to the last dorsal spinous process below. In full, deep-drawn inspiration in the healthy adult, the ear applied to the thoracic walls discovers the respiratory murmur over all the space included within the above mentioned bounds. After extreme expiration, if the thoracic walls be percussed, this capacity will be found much diminished; and the extreme limits of the thoracic space, which during full inspiration yielded a clear sound, indicative of the presence of the lung, will now, on percussion, manifest a dull sound, in consequence of the absence of the lung, which has receded from the place previously occupied.

Owing to the conical form of the thoracic space, the apex of which is measured by the first ribs, B B*, and the basis by I I*, it will be seen that if percussion be made directly from before, backwards, over the pectoral masses, R R*, the pulmonic resonance will not be elicited. When we raise the arms from the side and percuss the thorax between the folds of the axillae, where the serratus magnus muscle alone intervenes between the ribs and the skin, the pulmonic sound will answer clearly.

At the hypochondriac angles formed between the points F, L, N, on either side the lungs are absent both in inspiration and expiration. Percussion, when made over the surface of the angle of the right side, discovers the presence of the liver, G G*. When made over the median line, and on either side of it above the umbilicus, N, we ascertain the presence of the stomach, M M*. In the left hypochondriac angle, the stomach may also be found to occupy this place wholly.

Beneath the umbilicus, N, and on either side of it as far outwards as the lower asternal ribs, K L, thus ranging the abdominal parietes transversely, percussion discovers the transverse colon, O, P, O*. The small intestines, S S*, covered by the omentum, P*, occupy the hypogastric and iliac regions.

The organs situated within the thorax give evidence that they are developed in accordance to the law of symmetry. The lungs form a pair, one placed on either side of the median line. The heart is a double organ, formed of the right and left heart. The right lung differs from the left, inasmuch as we find the former divided into three lobes, while the latter has only two. That place which the heart now occupies in the left thoracic side is the place where the third or middle lobe of the left lung is wanting. In the abdomen we find that most of its organs are single. The liver, stomach, spleen, colon, and small intestine form a series of single organs: each of these may be cleft symmetrically. The kidneys are a pair.

The extent to which the ribs are bared in the figure Plate 22, marks exactly the form and transverse capacity of the thoracic walls. The diaphragm, H H*, has had a portion of its forepart cut off, to show how it separates the thin edges of both lungs above from the liver, G, and the stomach, M, below. These latter organs, although occupying abdominal space, rise to a considerable height behind K L, the asternal ribs, a fact which should be borne in mind when percussing the walls of the thorax and abdomen at this region.

DESCRIPTION OF PLATE 22.

A. Upper bone of the sternum.

B B*. Two first ribs.

C C*. Second pair of ribs.

D D*. Right and left lungs.

E. Pericardium, enveloping the heart--the right ventricle.

F. Lower end of the sternum.

G G*. Lobes of the liver.

H H*. Right and left halves of the diaphragm in section. The right half separating the right lung from the liver; the left half separating the left lung from the broad cardiac end of the stomach.

I I*. Eighth pair of ribs.

K K*. Ninth pair of ribs.

L L*. Tenth pair of ribs.

M M*. The stomach; M, its cardiac bulge; M*, its pyloric extremity.

N. The umbilicus.

OO*. The transverse colon.

P P*. The omentum, covering the transverse colon and small intestines.

Q. The gall bladder.

R R*. The right and left pectoral prominences.

S S*. Small intestines.

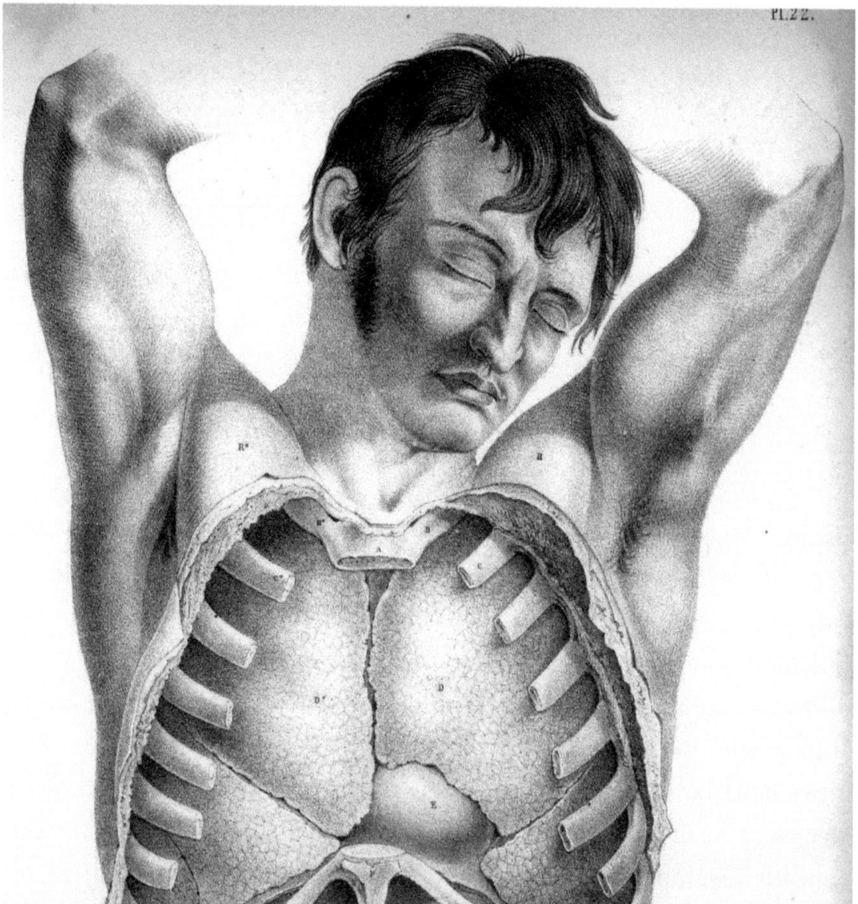

Plate 22

THE RELATIVE POSITION OF THE DEEPER ORGANS OF THE THORAX AND THOSE OF THE ABDOMEN.

Of the heart in reference to auscultation and percussion. Of the lungs, ditto. Relative capacity of the thorax and abdomen as influenced by the motions of the diaphragm. Abdominal respiration. Physical causes of abdominal herniae. Enlarged liver as affecting the capacity of the thorax and abdomen. Physiological remarks on wounds of the lungs. Pneumothorax, emphysema, &c.

The size or capacity of the thorax in relation to that of the abdomen varies in the individual at different periods of life. At an early age, the thorax, compared to the abdomen, is less in proportion than it is at adult age. The digestive organs in early age preponderate considerably over the respiratory organs; whereas, on the contrary, in the healthy and well-formed adult, the thoracic cavity and organs of respiration manifest a greater relative proportion to the ventral cavity and organs. At the adult age, when sexual peculiarities have become fully marked, the thoracic organs of the male body predominate over those of the abdomen, whilst in the female form the ventral organs take precedence as to development and proportions. This diversity in the relative capacity of the thorax and abdomen at different stages of development, and also in persons of different sexes, stamps each individual with characteristic traits of physical conformation; and it is required that we should take into our consideration this normal diversity of character, while conducting our examinations of individuals in reference to the existence of disease.

The heart varies in some measure, not only as to size and weight, but also as to position, even in healthy individuals of the same age and sex. The level at which the heart is in general found to be situated in the thorax is that represented in PLATE 23, where the apex points to the sixth intercostal space on the left side above K, while the arch of the aorta rises to a level with C, the second costal cartilage. In some instances, the heart may be found to occupy a much lower position in the thorax than the one above mentioned, or even a much higher level. The impulse of the right ventricle, F, has been noticed occasionally as corresponding to a point somewhat above the middle of the sternum and the intercostal space between the fourth and fifth left costal cartilages; while in other instances its beating was observable as low down as an inch or more below the xiphoid cartilage, and these variations have existed in a state of health.

Percussion over the region of the heart yields a dull flat sound. The sound is dullest opposite the right ventricle, F; whilst above and to either side of this point, where the heart is overlapped by the anterior shelving edges of both lungs, the sound is modified in consequence of the lung's resonant qualities. The heart-sounds, as heard through the stethoscope, in valvular disease, will, of course, be more distinctly ascertained at the locality of F, the right ventricle, which is immediately substernal. While the body lies supine, the heart recedes from the forepart of the chest; and the lungs during inspiration expanding around the heart will render its sounds less distinct. In the erect posture, the heart inclines forwards and approaches the anterior wall of the thorax. When the heart is hypertrophied, the lungs do not overlap it to the same extent as when it is of its ordinary size. In the latter state, the elastic cushion of the lung muffles the heart's impulse. In the former state, the lung is pushed aside by the overgrown heart, the strong muscular walls of which strike forcibly against the ribs and sternum.

The thorax is separated from the abdomen by the moveable diaphragm. The heart, F E, lies upon the diaphragm, L L*. The liver, M, lies immediately beneath the right side of this muscular septum, L*, while the bulging cardiac end of the stomach, O, is in close contact with it on the left side, L. As these three organs are attached to the diaphragm--the heart by its pericardium, the stomach by the tube of the oesophagus, and the liver by its suspensory ligaments--it must happen that the diaphragm while descending and ascending in the motions of inspiration and expiration will communicate the same alternate motions to the organs which are connected with it.

In ordinary respiration the capacity of the thorax is chiefly affected by the motions of the diaphragm; and the relative position which this septum holds with regard to the thoracic and abdominal chambers will cause its motions of ascent and descent to influence the capacity of both chambers at the same time. When the lungs expand, they follow the descent of the diaphragm, which forces the abdominal contents downwards, and thus what the thorax gains in space the abdomen loses. When the lungs contract, the diaphragm ascends, and by this act the abdomen gains that space which the thorax loses. But the organs of the thoracic cavity perform a different office in the economy from those of the abdomen. The air which fills the lungs is soon again expired, whilst the ingesta of the abdominal viscera are for a longer period retained; and as the space, which by every inspiration the thorax gains from the abdomen, would cause inconvenient pressure on the distended organs of this latter cavity, so we find that to obviate this inconvenience, nature has constructed the anterior parietes of the abdomen of yielding material. The muscular parietes of the abdomen relax during every inspiration, and thus this cavity gains that space which it loses by the encroachment of the dilating lungs.

The mechanical principle upon which the abdominal chamber is constructed, enables it to adjust its capacity to such exigence or pressing necessity as its own visceral organs impose on it, from time to time; and the relation which the abdominal cavity bears to the thoracic chamber, enables it also to be compensatory to this latter. When the inspiratory thorax gains space from the abdomen, or when space is demanded for the increasing bulk of the alimentary canal, or for the enlarging pregnant uterus; or when, in consequence of disease, such as dropsical accumulation, more room is wanted, then the abdominal chamber supplies the demand by the anterior bulge or swell of its expansile muscular parietes.

The position of the heart itself is affected by the expansion of the lungs on either side of it. As the expanding lungs force the diaphragm downwards, the heart follows it, and all the abdominal viscera yield place to the descending thoracic contents. In strong muscular efforts the diaphragm plays an important part, for, previously to making forced efforts, the lungs are distended with air, so as to swell and render fixed the thoracic walls into which so many powerful muscles of the shoulders, the neck, back, and abdomen, are inserted; at the same time the muscular diaphragm L L*, becomes tense and unbent from its arched form, thereby contracting abdominal space, which now has no compensation for this loss of space, since the abdominal parietes are also rendered firm and unyielding. It is at this crisis of muscular effort that the abdominal viscera become impacted together; and, acting by their own elasticity against the muscular force, make an exit for themselves through the weakest parts of the abdominal walls, and thus herniae of various kinds are produced. The most common situations of abdominal herniae are at the inguinal regions, towards which the intestines, T T, naturally gravitate; and at these situations the abdominal parietes are weak and membranous.

The contents of a hernial protrusion through the abdominal parietes, correspond in general with those divisions of the intestinal tube, which naturally lie adjacent to the part where the rupture has taken place. In the umbilical hernia it is either the transverse colon S*, or some part of the small intestine occupying the median line, or both together, with some folds of the omentum, which will be found to form the contents of this swelling. When the diaphragm itself sustains a rupture in its left half, the upper portion of the descending colon, S, protrudes through the opening. A diaphragmatic hernia has not, so far as I am aware, been seen to occur in the right side; and this exemption from rupture of the right half of the diaphragm may be accounted for anatomically, by the fact that the liver, M, defends the diaphragm at this situation. The liver occupies the whole depth of the right hypochondrium; and intervenes between the diaphragm L*, and the right extremity of the transverse colon, S**.

The contents of a right inguinal hernia consist of the small intestine, T. The contents of the right crural hernia are formed by either the small intestine, T, or the intestinum caecum, S***. I have seen a few cases in which the caecum formed the right crural hernia. Examples are recorded in which the intestine caecum formed the contents of a right inguinal hernia. The left inguinal and crural herniae contain most generally the small intestine, T, of the left side.
The right lung, I*, is shorter than the left; for the liver, M, raises

the diaphragm, L, to a higher level within the thorax, on the right side, than it does on the left. When the liver happens to be diseased and enlarged, it encroaches still more on thoracic space; but, doubtless, judging from the anatomical connexions of the liver, we may conclude that when it becomes increased in volume it will accommodate itself as much at the expense of abdominal space. The liver, in its healthy state and normal proportions, protrudes for an inch (more or less) below the margins of the right asternal ribs. The upper or convex surface of the liver rises beneath the diaphragm to a level corresponding with the seventh or sixth rib, but this position will vary according to the descent and ascent of the diaphragm in the respiratory movements. The ligaments by which the liver is suspended do not prevent its full obedience to these motions.

The left lung, I, descends to a lower level than the right; and the left diaphragm upon which it rests is itself supported by the cardiac end of the stomach. When the stomach is distended, it does not even then materially obstruct the expansion of the left lung, or the descent of the left diaphragm, for the abdominal walls relax and allow of the increasing volume of the stomach to accommodate itself. The spleen, R, is occasionally subject to an extraordinary increase of bulk; and this organ, like the enlarged liver and the distended stomach, will, to some extent, obstruct the movements of the diaphragm in the act of respiration, but owing to its free attachments it admits of a change of place. The abdominal viscera, one and all, admit of a change of place; the peculiar forms of those mesenteric bonds by which they are suspended, allow them to glide freely over each other; and this circumstance, together with the yielding nature of the abdominal parietes, allows the thoracic organs to have full and easy play in the respiratory movements performed by agency of the diaphragm.

The muscles of respiration perform with ease so long as the air has access to the lungs through the normal passage, viz., the trachea. While the principle of the thoracic pneumatic apparatus remains underanged, the motor powers perform their functions capably. The physical or pneumatic power acts in obedience to the vital or muscular power, while both stand in equilibrium; but the ascendancy of the one over the other deranges the whole thoracic machine. When the glottis closes by muscular spasm and excludes the external air, the respiratory muscles cease to exert a motor power upon the pulmonary cavity; their united efforts cannot cause a vacuum in thoracic space in opposition to the pressure of the external air. When, in addition to the natural opening of the glottis, a false opening is made in the side at the point K, the air within the lung at I, and external to it in the now open pleural cavity, will stand in equilibrio; the lung will collapse as having no muscular power by which to dilate itself, and the thoracic dilator muscles will cease to affect the capacity of the lung, so long as by their action in expanding the thoracic walls, the air gains access through the side to the pleural sac external to the lung.

Whether the air be admitted into the pleural sac, by an opening made in the side from without, or by an opening in the lung itself, the mechanical principle of the respiratory apparatus will be equally deranged. Pneumo-thorax will be the result of either lesion; and by the accumulation of air in the pleura the lung will suffer pressure. This pressure will be permanent so long as the air has no egress from the cavity of the pleura.

The permanent distention of the thoracic cavity, caused by the accumulation of air in the pleural sac, or by the diffusion of air through the interlobular cellular tissue consequent on a wound of the lung itself, will equally obstruct the breathing; and though the situation of the accumulated air is in fact anatomically different in both cases, yet the effect produced is similar. Interlobular pressure and interpleural pressure result in the same thing, viz., the permanent retention of the air external to the pulmonary cells, which, in the former case, are collapsed individually; and, in the latter case, in the mass. Though the emphysematous lung is distended to a size equal to the healthy lung in deep inspiration, yet we know that emphysematous distention, being produced by extrabronchial air accumulation, is, in fact, obstructive to the respiratory act. The emphysematous lung will, in the same manner as the distended pleural sac, depress the diaphragm and render the thoracic muscles inoperative. The foregoing observations have been made in reference to the effect of wounds of the thorax, the proper treatment of which will be obviously suggested by our knowledge of the state of the contained organs which have suffered lesion.

DESCRIPTION OF PLATE 23.

A. Upper end of the sternum.

B B.* First pair of ribs.

C C.* Second pair of ribs.

D. Aorta, with left vagus and phrenic nerves crossing its transverse arch.

E. Root of pulmonary artery.

F. Right ventricle.

G. Right auricle.

H. Vena cava superior, with right phrenic nerve on its outer border.

I I*. Right and left lungs collapsed, and turned outwards, to show the heart's outline.

K K*. Seventh pair of ribs.

L L*. The diaphragm in section.

M. The liver in section.

N. The gall bladder with its duct joining the hepatic duct to form the common bile duct. The hepatic artery is seen superficial to the common duct; the vena portae is seen beneath it. The patent orifices of the hepatic veins are seen on the cut surface of the liver.

O. The stomach.

P. The coeliac axis dividing into the coronary, splenic and hepatic arteries.

Q. Inferior vena cava.

R. The spleen.

S S* S**. The transverse colon, between which and the lower border of the stomach is seen the gastro-epiploic artery, formed by the splenic and hepatic arteries.

S***. Ascending colon in the right iliac region.

T. Convolutions of the small intestines distended with air.

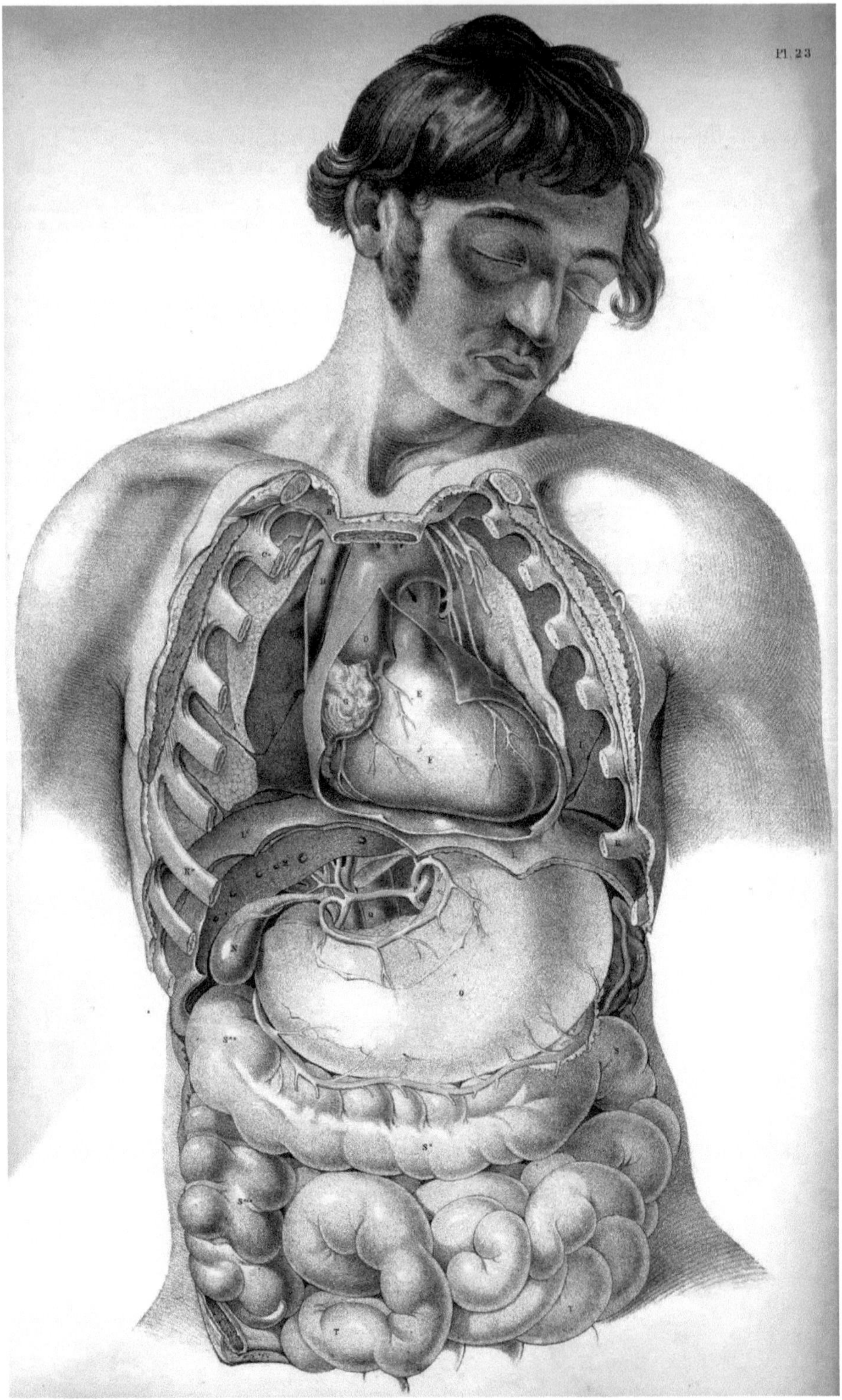

Plate 23

THE RELATIONS OF THE PRINCIPAL BLOODVESSELS TO THE VISCERA OF THE THORACICO-ABDOMINAL CAVITY.

Symmetrical arrangement of the vessels arising from the median thoracico-abdominal aorta, &c. Special relations of the aorta. Aortic sounds. Aortic aneurism and its effects on neighbouring organs. Paracentesis thoracis. Physical causes of dropsy. Hepatic abscess. Chronic enlargements of the liver and spleen as affecting the relative position of other parts. Biliary concretions. Wounds of the intestines. Artificial anus.

The median line of the body is occupied by the centres of the four great systems of organs which serve in the processes of circulation, respiration, innervation, and nutrition. These organs being fashioned in accordance with the law of symmetry, we find them arranged in close connexion with the vertebrate centre of the osseous fabric, which is itself symmetrical. In this symmetrical arrangement of the main organs of the trunk of the body, a mechanical principle is prominently apparent; for as the centre is the least moveable and most protected region of the form, so have these vitally important structures the full benefit of this situation. The aortal trunk, G, of the arterial system is disposed along the median line, as well for its own safety as for the fitting distribution of those branches which spring symmetrically from either side of it to supply the lateral regions of the body.

The visceral system of bloodvessels is moulded upon the organs which they supply. As the thoracic viscera differ in form and functional character from those of the abdomen, so we find that the arterial branches which are supplied by the aorta to each set, differ likewise in some degree. In the accompanying figure, which represents the thoracic and abdominal visceral branches of the aorta taken in their entirety, this difference in their arrangement may be readily recognised. In the thorax, compared with the abdomen, we find that not only do the aortic branches differ in form according to the variety of those organs contained in either region, but that they differ numerically according to the number of organs situated in each. The main vessel itself, however, is common to both regions. It is the one thoracico-abdominal vessel, and this circumstance calls for the comparison, not only of the several parts of the great vessel itself, but of all the branches which spring from it, and of the various organs which lie in its vicinity in the thorax and abdomen, and hence we are invited to the study of these regions themselves connectedly.

In the thorax, the aorta, G G*, is wholly concealed by the lungs in their states both of inspiration and expiration. The first part of the aortic arch, as it springs from the left ventricle of the heart, is the most superficial, being almost immediately sub-sternal, and on a level with the sternal junctions of the fourth ribs. By applying the ear at this locality, the play of the aortic valves may be distinctly heard. From this point the aorta, G, rises and arches from before, backwards, to the left side of the spine, G*. The arch of the vessel lies more deeply between the two lungs than does its ventricular origin. The descending thoracic aorta lies still more deeply situated at the left side of the dorsal spine. At this latter situation it is in immediate contact with the posterior thick part of the left lung; whilst on its right are placed, L, the thoracic duct; I, the oesophagus; K, the vena azygos, and the vertebral column. In Plate 26 may be seen the relation which the superior vena cava, H, bears to the aortic arch, A.

In the span of the aortic arch will be found, H*, the left bronchus, together with the right branch of the pulmonary artery, and the right pulmonary veins. The pneumo-gastric and phrenic nerves descend on either side of the arch. The left pneumo-gastric nerve winds round beneath the arch at the point where the obliterated ductus arteriosus joins it. See Plates 12 & 26.

The pulmonary artery, B, Plates 1 & 2, lies close upon the fore part, and conceals the origin, of the systemic aorta. Whenever, therefore, the semilunar valves of either the pulmonary artery or the systemic aorta become diseased, it must be extremely difficult to distinguish by the sounds alone, during life, in which of the two the derangement exists. The origins of both vessels being at the fore part of the chest, it is in this situation, of course, that the state of their valves is to be examined. The descending part of the thoracic aorta, G*, being at the posterior part of the chest, and lying on the vertebral ends of the left thoracic ribs, will therefore require that we should examine its condition in the living body at the dorsal aspect of the thorax. As the arch of the aorta is directed from before backwards--that is, from the sternum to the spine, it follows that when an aneurism implicates this region of the vessel, the exact situation of the tumour must be determined by antero-posterior examination; and we should recollect, that though on the fore part of the chest the cartilages of the second ribs, where these join the sternum, mark the level of the aortic arch, on the back of the chest its level is to be taken from the vertebral ends of the third or fourth ribs. This difference is caused by the oblique descent of the ribs from the spine to the sternum. The first and second dorsal vertebrae, with which the first and second ribs articulate, are considerably above the level of the first and second pieces of the sternum.

In a practical point of view, the pulmonary artery possesses but small interest for us; and in truth the trunk of the systemic aorta itself may be regarded in the same disheartening consideration, forasmuch as when serious disease attacks either vessel, the "tree of life" may be said to be lopped at its root.

When an aneurism arises from the aortic arch it implicates those important organs which are gathered together in contact with itself. The aneurismal tumour may press upon and obstruct the bronchi, H H*; the thoracic duct, L; the oesophagus, I; the superior vena cava, H, Plate 26, or wholly obliterate either of the vagi nerves. The aneurism of the arch of the aorta may cause suffocation in two ways--viz., either by pressing directly on the tracheal tube, or by compressing and irritating the vagus nerve, whose recurrent branch will convey the stimulus to the laryngeal muscles, and cause spasmodic closure of the glottis. This anatomical fact also fully accounts for the constant cough which attends some forms of aortic aneurism. The pulmonary arteries and veins are also liable to obstruction from the tumour. This will occur the more certainly if the aneurism spring from the right or the inferior side of the arch, and if the tumour should not break at an early period, slow absorption, caused by pressure of the tumour, may destroy even the vertebral column, and endanger the spinal nervous centre. If the tumour spring from the left side or the fore part of the arch, it may in time force a passage through the anterior wall of the thorax.

The principal branches of the thoracic aorta spring from the upper part of its arch. The innominate artery, 2, is the first to arise from it; the left common carotid, 6, and the left subclavian artery, 5, spring in succession. These vessels being destined for the head and upper limbs, we find that the remaining branches of the thoracic aorta are comparatively diminutive, and of little surgical interest. The intercostal arteries occasionally, when wounded, call for the aid of the surgeon; these arteries, like all other branches of the aorta, are largest at their origin. Where these vessels spring from G, the descending thoracic aorta, they present considerable caliber; but at this inaccessible situation, they seldom or never call for surgical interference. As the intercostal arteries pass outwards, traversing the intercostal spaces with their accompanying nerves, they diminish in size. Each vessel divides at a distance of about two inches, more or less, from the spine; and the upper larger branch lies under cover of the inferior border of the adjacent rib. When it is required to perform the operation of paracentesis thoracis, this distribution of the vessel should be borne in mind; and also, that the farther from the spine this operation is performed, the less in size will the vessels be found. The intercostal artery is sometimes wounded by the fractured end of the rib, in which case, if the pleura be lacerated, an effusion of blood takes place within the thorax, compresses the lung, and obstructs respiration.

The thoracic aorta descends along the left side of the spine, as far as the last dorsal vertebra, at which situation the pillars of the diaphragm overarch the vessel. From this place the aorta passes obliquely in front of the five lumbar vertebrae, and on arriving opposite the fourth, it divides into the two common iliac branches. The aorta, for an extent included between these latter boundaries, is named the abdominal aorta, and from its fore part arise those branches, which supply the viscera of the abdomen.

The branches which spring from the abdominal aorta to supply the viscera of this region, are considerable, both as to their number and size. They are, however, of comparatively little interest in practice. To the anatomist they present many peculiarities of distribution and form worthy of notice, as, for example, their frequent anastomosis, their looping arrangement, and their large size and number compared with the actual bulk of the organs which they supply. As to this latter peculiarity, we interpret it according to the fact that here the vessels serve other purposes in the economy besides that of the support and repair of structure. The vessels are large in proportion to the great quantity of fluid matter secreted from the whole extent of the inner surface of this glandular apparatus--the gastro-intestinal canal, the liver, pancreas, and kidneys.

As anatomists, we are enabled, from a knowledge of the relative position of the various organs and bloodvessels of both the thorax and abdomen, to account for certain pathological phenomena which, as practitioners, we possess as yet but little skill to remedy. Thus it would appear most probable that many cases of anasarca of the lower limbs, and of dropsy of the belly, are frequently caused by diseased growths of the liver, P, obstructing the inferior vena cava, R, and vena portae, rather than by what we are taught to be the "want of balance between secreting and absorbing surfaces." The like occurrence may obstruct the gall-ducts, and occasion jaundice. Over-distention of any of those organs situated beneath the right hypochondrium, will obstruct neighbouring organs and vessels. Mechanical obstruction is doubtless so frequent a source of derangement, that we need not on many occasions essay a deeper search for explaining the mystery of disease.

In the right hypochondriac region there exists a greater variety of organs than in the left; and disease is also more frequent on the right side. Affections of the liver will consequently implicate a greater number of organs than affections of the spleen on the left side, for the spleen is comparatively isolated from the more important blood vessels and other organs.

The external surface of the liver, P, lies in contact with the diaphragm, N, the costal cartilages, M, and the upper and lateral parts of the abdominal parietes; and when the liver becomes the seat of abscess, this, according to its situation, will point and burst either into the thorax above, or through the side between or beneath the false ribs, M. The hepatic abscess has been known to discharge itself through the stomach, the duodenum, T, and the transverse colon, facts which are readily explained on seeing the close relationship which these parts hold to the under surface of the liver. When the liver is inflamed, we account for the gastric irritation, either from the inflammation having extended to the neighbouring stomach, or by this latter organ being affected by "reflex action." The hepatic cough is caused by the like phenomena disturbing the diaphragm, N, with which the liver, P, lies in close contact.

When large biliary concretions form in S, the gallbladder, or in the hepatic duct, Nature, failing in her efforts to discharge them through the common bile-duct, into the duodenum, T, sets up inflammation and ulcerative absorption, by aid of which processes they make a passage for themselves through some adjacent part of the intestine, either the duodenum or the transverse colon. In these processes the gall-bladder, which contains the calculus, becomes soldered by effused lymph to the neighbouring part of the intestinal tube, into which the stone is to be discharged, and thus its escape into the peritoneal sac is prevented. When the hepatic abscess points externally towards M, the like process isolates the matter from the cavities of the chest and abdomen.

In wounds of any part of the intestine, whether of X, the caecum, W, the sigmoid flexure of the colon, or Z, the small bowel, if sufficient time be allowed for Nature to establish the adhesive inflammation, she does so, and thus fortifies the peritoneal sac against an escape of the intestinal matter into it by soldering the orifice of the wounded intestine to the external opening. In this mode is formed the artificial anus. The surgeon on principle aids Nature in attaining this result.

DESCRIPTION OF PLATE 24.

A. The thyroid body.

B. The trachea.

C C*. The first ribs.

D D*. The clavicles, cut at their middle.

E. Humeral part of the great pectoral muscle, cut.

F. The coracoid process of the scapula.

G. The arch of the aorta. G*. Descending aorta in the thorax.

H. Right bronchus. H*. Left bronchus.

I. Oesophagus.

K. Vena azygos receiving the intercostal veins.

L. Thoracic duct.

M M*. Seventh ribs.

N N. The diaphragm, in section.

O. The cardiac orifice of the stomach.

P. The liver, in section, showing the patent orifices of the hepatic veins.

Q. The coeliac axis sending off branches to the liver, stomach, and spleen. The stomach has been removed, to show the looping anastomosis of these vessels around the superior and inferior borders of the stomach.

R. The inferior vena cava about to enter its notch in the posterior thick part of the liver, to receive the hepatic veins.

S. The gall-bladder, communicating by its duct with the hepatic duct, which is lying upon the vena portae, and by the side of the hepatic artery.

T. The pyloric end of the stomach, joining T*, the duodenum.

U. The spleen.

V V. The pancreas.

W. The sigmoid flexure of the colon.

X. The caput coli.

Y. The mesentery supporting the numerous looping branches of the superior mesenteric artery.

Z. Some coils of the small intestine.

2. Innominate artery.

3. Right subclavian artery.

4. Right common carotid artery.

5. Left subclavian artery.

6. Left common carotid artery.

7. Left axillary artery.

8. Coracoid attachment of the smaller pectoral muscle.

9. Subscapular muscle.
10. Coracoid head of the biceps muscle.

11. Tendon of the latissimus dorsi muscle.

12. Superior mesenteric artery, with its accompanying vein.

13. Left kidney.

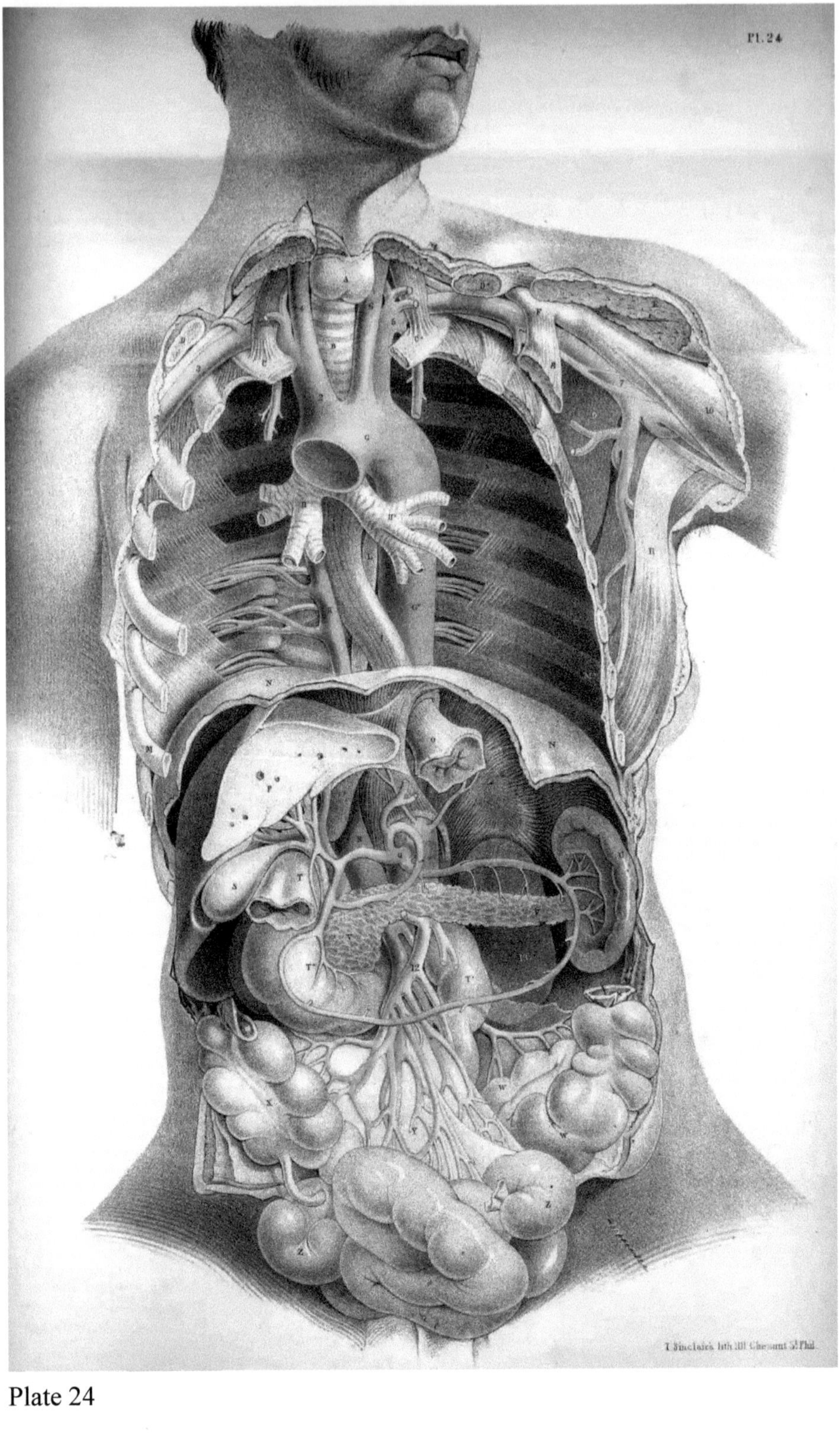

Plate 24

THE RELATION OF THE PRINCIPAL BLOODVESSELS OF THE THORAX AND ABDOMEN TO THE OSSEOUS SKELETON.

The vessels conforming to the shape of the skeleton. Analogy between the branches arising from both ends of the aorta. Their normal and abnormal conditions. Varieties as to the length of these arteries considered surgically. Measurements of the abdomen and thorax compared. Anastomosing branches of the thoracic and abdominal parts of the aorta.

The arterial system of vessels assumes, in all cases, somewhat of the character of the forms upon which they are distributed, or of the organs which they supply. This mode of distribution becomes the more apparent, according as we rise from particulars to take a view of *the whole*. With the same ease that any piece of the osseous fabric, taken separately, may be known, so may any one artery, taken apart from the rest, be distinguished as to the place which it occupied, and the organs which it supplied in the economy. The vascular skeleton, whether taken as a whole or in parts, exhibits characteristics as apparent as are those of the osseous skeleton itself. The main bloodvessel, A B C, of the trunk of the body, possesses character, *sui generis*, just as the vertebral column itself manifests. The main arteries of the head or limbs are as readily distinguishable, the one from the other, as are the osseous fabrics of the head and limbs. The visceral arteries are likewise moulded upon the forms which they supply. But evidently the arterial system of vessels conforms most strictly with the general design of the osseous skeleton.

In Plate 25, viewed as a whole, we find that as the vertebral column stands central to the osseous skeleton, so does the aorta, A B C, take the centre of the arterial skeleton. As the ribs jut symmetrically from either side of the vertebral column, so do the intercostal arteries follow them from their own points of origin in the aorta. The one side of the osseous system is not more like the other than is the system of vessels on one side like that of the other. And in addition to this fact of a similarity of sides in the vascular as in the osseous skeleton, I also remark that both extremities of the aorta divide into branches which are similar to one another above and below, thereby conforming exactly with the upper and lower limbs, which manifest unmistakable points of analogy.

The branches which spring from the aortic arch above are destined to supply the head and upper limbs. They are, H, the innominate artery, and I K, the left common carotid and subclavian arteries. The branches which spring from the other extremity of the aorta are disposed for the support of the pelvis and lower limbs; they are the right and left common iliac arteries, L M. These vessels exhibit, at both ends of the main aortic trunk, a remarkable analogy; and as the knowledge of this fact may serve to lighten the dry and weary detail of descriptive anatomy, at the same time that it points directly to views of practical import, I may be allowed briefly to remark upon it as follows:--

The vessels which spring from both ends of the aorta, as seen in Plate 25, are represented in what is called their normal character--that is, while three vessels, H I K, spring separately from the aortic arch above, only two vessels, L and M, arise from the aorta below. Let the anatomist now recall to mind the "peculiarities" which at times appear amongst the vessels, H I K, above, and he will find that some of them absolutely correspond to the normal arrangement of the vessels, L M, below. And if he will consider the "peculiarities" which occur to the normal order of the vessels, L M, below, he will find that some of these correspond exactly to the normal order of the vessels above. Thus, when I K of the left side join into a common trunk, this resembles the innominate artery, H, of the right side, and then both these vessels perfectly correspond with the two common iliac arteries below. When, on the other hand, L and M, the common iliac arteries, divide, immediately after leaving the aortic trunk, into two pairs of branches, they correspond to the abnormal condition of the vessels, H I K, above; where H, immediately after leaving the aortic arch, divides into two branches, like I K. With this generalization upon the normal and abnormal facts of arrangement, exhibited among the vessels arising from both ends of the aorta, I furnish to the reader the idea that the vessels, H I K, above may present of the same figure as the vessels, L

M, below, and these latter may assume the character of H I K, above. Whenever, therefore, either set of vessels becomes the subject of operation, such as having a ligature applied to them, we must be prepared to meet the "varieties."

The veins assume an arrangement similar to that of the arteries, and the above remarks will therefore equally apply to the veins. In the same way as the arteries, H I K, may present in the condition of two common or brachio-cephalic trunks, and thereby simulate the condition of the common iliac arteries, so we find that the normal forms of the veins above and below actually and permanently exhibit this very type. The brachio-cephalic veins, D B, Plate 26, exactly correspond to each other, and to the common iliac veins, S T; and as these latter correspond precisely with the common iliac arteries, so may we infer that the original or typical condition of the vessels I K, Plate 25, is a brachia-cephalic or common-trunk union corresponding with its brachio-cephalic vein. When the vessels, I K, therefore present of the brachio-cephalic form as the vessel H, we have a perfect correspondence between the two extremes of the aorta, both as regards the arteries arising from it, and the veins which accompany these arteries; and this condition of the vascular skeleton I regard as the typical uniformity. The separate condition of the vessels I K, notwithstanding the frequency of the occurrence of such, may be considered as a special variation from the original type.

The length of the aorta is variable in two or more bodies; and so, likewise, is the length of the trunk of each of those great branches which springs from its arch above, and of those into which it divides below, The modes in which these variations as to length occur, are numerous. The top of the arch of the aorta is described as being in general on a level with the cartilages of the second ribs, from which point it descends on the left side of the spinal column; and after having wound gradually forwards to the forepart of the lumbar spine at C, divides opposite to the fourth lumbar vertebra into the right and left common iliac arteries. The length of that portion of the aorta which is called thoracic, is determined by the position of the pillars of the diaphragm F, which span the vessel; and from this point to where the aorta divides into the two common iliac arteries, the main vessel is named abdominal. The aorta, from its arch to its point of division on the lumbar vertebrae, gradually diminishes in caliber, according to the number and succession of the branches derived from it.

The varieties as to length exhibited by the aorta itself, and by the principal branches which spring from it, occur under the following mentioned conditions:--When the arch of the aorta rises above or sinks below its ordinary position or level,--namely, the cartilages of the second ribs, as seen in Plate 25,--it varies not only its own length, but also that of the vessels H I K; for if the arch of the aorta rises above this level, the vessels H I K become shortened; and as the arch sinks below this level, these vessels become lengthened. Even when the aortic arch holds its proper level in the thorax, still the vessels H I K may vary as to length, according to the height to which they rise in the neck previously to their division. When the aorta sinks below its proper level at the same time that the vessels H I K rise considerably above that point at which they usually arch or divide in the neck, then of course their length becomes greatly increased. When, on the other hand, the aortic arch rises above its usual level, whilst the vessels H I K arch and divide at a low position in the neck, then their length becomes very much diminished. The length of the artery H may be increased even though the arch of the aorta holds its proper level, and though the vessels H I K occupy their usual position in the neck; for it is true that the vessel H may spring from a point of the aortic arch A nearer to the origin of this from the ventricle of the heart, whilst the vessel I may be shortened, owing to the fact of its arising from some part of H, the innominate vessel. All these circumstances are so obvious, that they need no comment, were it not for the necessity of impressing the surgeon with the fact that uncertainty as to a successful result must always attach to his operation of including in a ligature either of the vessels H I K, so as to affect an aneurismal tumour.

Now whilst the length of the aorta and that of the principal branches springing from its arch may be varied according to the above-mentioned conditions, so may the length of the aorta itself, and of the two common iliac vessels, vary according to the place whereat the aorta, C, bifurcates. Or, even when this point of division is opposite the usual vertebra,--viz., the fourth lumbar,--still the common iliac vessels may be short or long, according to the place where they divide into external and internal iliac branches. The aorta may bifurcate almost as high up as where the pillars of the diaphragm overarch it, or as low down as the fifth lumbar vertebra. The occasional existence of a sixth lumbar vertebra also causes a variety in the length, not only of the aorta, but of the two common iliac vessels and their branches.[Footnote]

[Footnote: Whatever may be the number of variations to which the branches arising from both extremes of the aorta are liable, all anatomists admit that the arrangement of these vessels, as exhibited in Plate 25, is by far the most frequent. The surgical anatomist, therefore, when planning his operation, takes this arrangement as the standard type. Haller asserts this order of the vessels to be so constant, that in four hundred bodies which he examined, he found only *one variety*--namely, that in which the left vertebral artery arose from the aorta. Of other varieties described by authors, he observes--"Rara vero haec omnia esse si dixero cum quadringenta nunc cadavera humana dissecuerim, fidem forte inveniam." (Iconum Anatom.) This variety is also stated by J. F. Meckel (Handbuch der Mensch Anat.), Soemmerring (De Corp. Hum Fabrica), Boyer (Tr. d'Anat.), and Mr. Harrison (Surg. Anal. of Art.), to be the most frequent. Tiedemann figures this variety amongst others (Tabulae Arteriarum). Mr. Quain regards as the most frequent change which occurs in the number of the branches of the aortic arch, "that in which the left carotid is derived from the innominate." (Anatomy of the Arteries, &c.) A case is recorded by Petsche (quoted in Haller), in which he states the bifurcation of the aorta to have taken place at the origin of the renal arteries: (query) are we to suppose that the renal arteries occupied their usual position? Cruveilhier records a case (Anal. Descript.) in which the right common iliac was wanting, in consequence of having divided at the aorta into the internal and external iliac branches. Whether the knowledge of these and numerous other varieties of the arterial system be of much practical import to the surgeon, he will determine for himself. To the scientific anatomist, it must appear that the main object in regard to them is to submit them to a strict analogical reasoning, so as to demonstrate the operation of that law which has produced them. To this end I have pointed to that analogy which exists between the vessels arising from both extremities of the aorta. "Itaque convertenda plane est opera ad inquirendas et notandas rerum similitudines et analoga tam integralibus quam partibus; illae enim sunt, quae naturam uniunt, et constituere scientias incipiunt." "Natura enim non nisi parendo vincitur; et quod in contemplatione instar causae est; id in operatione instar regulae est." (Novum Organum Scientiarum, Aph. xxvii-iii, lib. i.)]

The difference between the perpendicular range of the anterior and posterior walls of the thoracic cavity may be estimated on a reference to Plate 25, in which the xyphoid cartilage, E, joined to the seventh pair of ribs, bounds its anterior wall below, while F, the pillars of the diaphragm, bound its posterior wall. The thoracic cavity is therefore considerably deeper in its posterior than in its anterior wall; and this occasions a difference of an opposite kind in the anterior and posterior walls of the abdomen; for while the abdomen ranges perpendicularly from E to W, its posterior range measures only from F to the ventra of the iliac bones, R. The arching form of the diaphragm, and the lower level which the pubic symphysis occupies compared with that of the cristae of the iliac bones, occasion this difference in the measure of both the thorax and abdomen.

The usual position of the kidneys, G G*, is on either side of the lumbar spine, between the last ribs and the cristae of the iliac bones. The kidneys lie on the fore part of the quadratus lumborum and psoae muscles. They are sometimes found to have descended as low as the iliac fossae, R, in consequence of pressure, occasioned by an enlarged liver on the right, or by an enlarged spleen on the left. The length of the abdominal part of the aorta may be estimated as being a third of the entire vessel, measured from the top of its arch to its point of bifurcation. So many and such large vessels arise from the abdominal part of the aorta, and these are set so closely to each other, that it must in all cases be very difficult to choose a proper locality whereat to apply a ligature on this region of the vessel. If other circumstances could fairly justify such an operation, the anatomist believes that the circulation might be maintained through the anastomosis of the internal mammary and intercostal arteries with the epigastric; the branches of the superior mesenteric with those of the inferior; and the branches of this latter with the perineal branches of the pubic. The lumbar, the gluteal, and the circumflex ilii arteries, also communicate around the hip-bone. The same vessels would serve to carryon the circulation if either L, the common iliac, V, the external iliac, or the internal iliac vessel, were the subject of the operation by ligature.

DESCRIPTION OF PLATE 25.

A. The arch of the aorta.

B B. The descending thoracic part of the aorta, giving off *b b*, the intercostal arteries.

C. The abdominal part of the aorta.

D D. First pair of ribs.

E. The xyphoid cartilage.

G G*. The right and left kidneys.

H. The brachio-cephalic artery.

I. Left common carotid artery.

K. Left subclavian artery.

L. Right common iliac artery at its place of division.

M. Left common iliac artery, seen through the meso-rectum.

N. Inferior vena cava.

O O. The sigmoid flexure of the colon.

P. The rectum.

Q. The urinary bladder.

R. The right iliac fossa.

S S. The right and left ureters.

T. The left common iliac vein, joining the right under the right common iliac artery to form the inferior vena cava.

U. Fifth lumbar vertebra.

V. The external iliac artery of right side.

W. The symphysis pubis.

X. An incision made over the locality of the femoral artery.

b b. The dorsal intercostal arteries.

c. The coeliac axis

d. The superior mesenteric artery.

f f. The renal arteries.

g. The inferior mesenteric artery.

h. The vas deferens bending over the epigastric artery and the os pubis, after having passed through the internal abdominal ring.

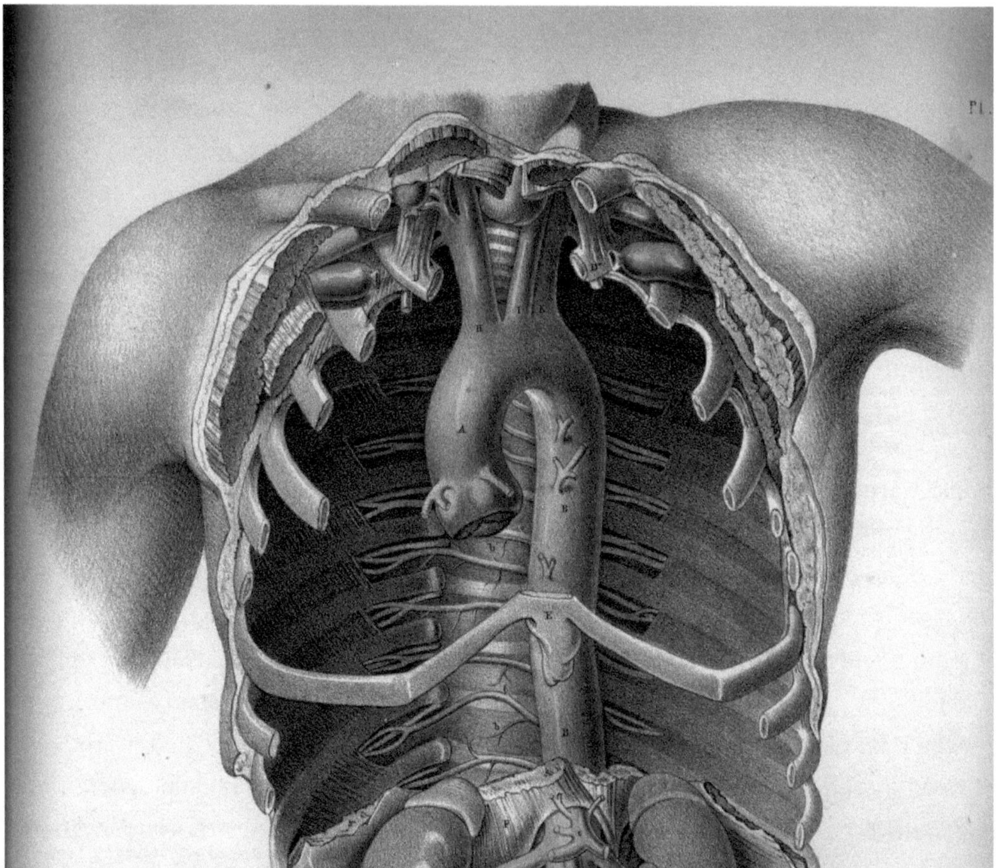

Plate 25

THE RELATION OF THE INTERNAL PARTS TO THE EXTERNAL SURFACE.

In health and disease. Displacement of the lungs from pleuritic effusion. Paracentesis thoracis. Hydrops pericardii. Puncturation. Abdominal and ovarian dropsy as influencing the position of the viscera. Diagnosis of both dropsies. Paracentesis abdominis. Vascular obstructions and their effects.

An exact acquaintance with the normal character of the external form, its natural prominences and depressions, produced by the projecting swell of muscles and points of bone, &c., is of great practical importance to the surgeon. These several marks described on the superficies he takes as certain guides to the precise locality and relations of the more deeply situated organs. And as, by dissection, Nature reveals to him the fact that she holds constant to these relations, so, at least, may all that department of practice which he bases upon this anatomical certainty be accounted as rooted in truth and governed by fixed principles. The same organ bears the same special and general relations in all bodies, not only of the human, but of all other species of vertebrata; and from this evidence we conclude that the same marks on surface indicate the exact situation of the same organs in all similar bodies.

The surface of the well-formed human body presents to our observation certain standard characters with which we compare all its abnormal conditions. Every region of the body exhibits fixed character proper to its surface. The neck, the axilla, the thorax, the abdomen, the groin, have each their special marks, by which we know them; and the eye, well versed in the characters proper to the healthy state of each, will soonest discover the nature of all effects of injury--such as dislocations, fractures, tumours of various kinds, &c. By our acquaintance with the perfect, we discover the imperfect; by a comparison with the geometrically true rectangled triangle, or circle, we estimate the error of these forms when they have become distorted; and in the same way, by a knowledge of what is the healthy normal standard of human form, we diagnose correctly its slightest degree of deformity, produced by any cause whatever, whether by sudden accident, or slowly-approaching disease.

Now, the abnormal conditions of the surface become at once apparent to our senses; but those diseased conditions which concern the internal organs require no ordinary exercise of judgment to discover them. The outward form masks the internal parts, and conceals from our direct view, like the covers of a closed volume, the marvellous history contained within. But still the superficies is so moulded upon the deeper situated structures, that we are induced to study it as a map, which discourses of all which it incloses in the healthy or the diseased state. Thus, the sternum points to A, the aorta; the middle of the clavicles, to C, the subclavian vessels; the localities 9, 10 of the coracoid processes indicate the place of the axillary vessels; the navel, P, points to Q, the bifurcation of the aorta; the pubic symphysis, Z, directs to the urinary bladder, Y. At the points 7, 8, may be felt the anterior superior spinous processes of the iliac bones, between which points and Z, the iliac vessels, V, 6, pass midway to the thigh, and give off the epigastric vessels, 2, 3, to the abdominal parietes. Between these points of general relations, which we trace on the surface of the trunk of the body, the anatomist includes the entire history of the special relations of the organs within contained. And not until he is capable of summing together the whole picture of anatomical analysis, and of viewing this in all its intricate relationary combination--even through and beneath the closed surface of living moving nature, is he prepared to estimate the conditions of disease, or interfere for its removal.

When fluid accumulates on either side of the thoracic compartment to such an excess that an opening is required to be made for its exit from the body, the operator, who is best acquainted with the relations of the parts in a state of health, is enabled to judge with most correctness in how far these parts, when in a state of disease, have swerved from these proper relations. In the normal state of the thoracic viscera, the left thoracic space, G A K N, is occupied by the heart and left lung. The space indicated within the points A N K, in the anterior region of the thorax, is occupied by the heart, which, however, is partially overlapped by the anterior edge of the lung, PLATE 22. If the thorax be deeply penetrated at any part of this region, the instrument will wound either the lung or the heart, according to the situation of the wound. But when fluid becomes effused in any considerable quantity within the pleural sac, it occupies space between the lung and the thoracic walls; and the fluid compresses the lung, or displaces the heart from the left side towards the right. This displacement may take place to such an extent, that the heart, instead of occupying the left thoracic angle, A K N, assumes the position of A K* N on the right side. Therefore, as the fluid, whatever be its quantity, intervenes between the thoracic walls, K K*, and the compressed lung, the operation of paracentesis thoracis should be performed at the point K, or between K and the latissimus dorsi muscle, so as to avoid any possibility of wounding the heart. The intercostal artery at K is not of any considerable size.

In the normal state of the thoracic organs, the pericardial envelope of the heart is at all times more or less uncovered by the anterior edge of the left lung, as seen in PLATE 22. When serous or other fluid accumulates to an excess in the pericardium, so as considerably to distend this sac, it must happen that a greater area of pericardial surface will be exposed and brought into immediate contact with the thoracic walls on the left side of the sternal median line, to the exclusion of the left lung, which now no longer interposes between the heart and the thorax. At this locality, therefore, a puncture may be made through the thoracic walls, directly into the distended pericardium, for the escape of its fluid contents, if such proceeding be in other respects deemed prudent and advisable.

The abdominal cavity being very frequently the seat of dropsical effusion, when this takes place to any great extent, despite the continued and free use of the medicinal diuretic and the hydragogue cathartic, the surgeon is required to make an opening with the instrumental hydragogue--viz., the trocar and cannula. The proper locality whereat the puncture is to be made so as to avoid any large bloodvessel or other important organ, is at the middle third of the median line, between P the umbilicus, and Z the symphysis pubis. The anatomist chooses this median line as the safest place in which to perform paracentesis abdominis, well knowing the situation of 2, 3, the epigastric vessels, and of Y, the urinary bladder.

All kinds of fluid occupying the cavities of the body gravitate towards the most depending part; and therefore, as in the sitting or standing posture, the fluid of ascites falls upon the line P Z, the propriety of giving the patient this position, and of choosing some point within the line P Z, for the place whereat to make the opening, becomes obvious. In the female, the ovary is frequently the seat of dropsical accumulation to such an extent as to distend the abdomen very considerably. Ovarian dropsy is distinguished from ascites by the particular form and situation of the swelling. In ascites, the abdominal swell is symmetrical, when the body stands or sits erect. In ovarian dropsy, the tumour is greatest on either side of the median line, according as the affected ovary happens to be the right or the left one.

The fluid of ascites and that of the ovarian dropsy affect the position of the abdominal viscera variously In ascites, the fluid gravitates to whichever side the body inclines, and it displaces the moveable viscera towards the opposite side. Therefore, to whichever side the abdominal fluid gravitates, we may expect to find it occupying space between the abdominal parietes and the small intestines. The ovarian tumour is, on the contrary, comparatively fixed to either side of the abdominal median line; and whether it be the right or left ovary that is affected, it permanently displaces the intestines on its own side; and the sac lies in contact with the neighbouring abdominal parietes; nor will the intestines and it change position according to the line of gravitation.

Now, though the above-mentioned circumstances be anatomically true respecting dropsical effusion within the general peritonaeal sac and that of the ovary, there are many urgent reasons for preferring to all other localities the line P Z, as the only proper one for puncturing the abdomen so as to give exit to the fluid. For though the peritonaeal ascites does, according to the position of the patient, gravitate to either side of the abdomen, and displace the moveable viscera on that side, we should recollect that some of these are bound fixedly to one place, and cannot be floated aside by the gravitating fluid. The liver is fixed to the right side, 11, by its suspensory ligaments. The spleen occupies the left side, 12. The caecum and the sigmoid flexure of the colon occupy, R R*, the right and left iliac regions. The colon ranges transversely across the abdomen, at P. The stomach lies transversely between the points, 11, 12. The kidneys, O, occupy the lumbar region. All these organs continue to hold their proper places, to whatever extent the dropsical effusion may take place, and notwithstanding the various inclinations of the body in this or that direction. On this account, therefore, we avoid performing the operation of paracentesis abdominis at any part except the median line, P Z; and as to this place, we prefer it to all others, for the following cogent reasons--viz., the absence of any large artery; the absence of any important viscus; the fact that the contained fluid gravitates in large quantity, and in immediate contact with the abdominal walls anteriorly, and interposes itself between these walls and the small intestines, which float free, and cannot approach the parietes of the abdomen nearer than the length which the mesenteric bond allows.

If the ovarian dropsy form a considerable tumour in the abdomen, it may be readily reached by the trocar and cannula penetrating the line P Z. And thus we avoid the situation of the epigastric vessels. The puncture through the linea alba should never be made below the point, midway between P and Z, lest we wound the urinary bladder, which, when distended, rises considerably above the pubic symphysis.

Amongst the many mechanical obstructions which, by impeding the circulation, give rise to dropsical effusion, are the following:--An aneurismal tumour of the aorta, A, or the innominate artery, [Footnote 1] F, may press upon the veins, H or D, and cause an oedematous swelling of the corresponding side of the face and the right arm. In the same way an aneurism of the aorta, Q, by pressing upon the inferior vena cava, T, may cause oedema of the lower limbs. Serum may accumulate in the pericardium, owing to an obstruction of the cardiac veins, caused by hypertrophy of the substance of the heart; and when from this cause the pericardium becomes much distended with fluid, the pressure of this upon the flaccid auricles and large venous trunks may give rise to general anasarca, to hydrothorax or ascites, either separate or co-existing.

Tuberculous deposits in the lungs and scrofulous bronchial glands may cause obstructive pressure on the pulmonary veins, followed by effusion of either pus or serum into the pleural sac. [Footnote 2] An abscess or other tumour of the liver may, by pressing on the vena portae, cause serous effusion into the peritonaeal sac; or by pressure on the inferior vena cava, which is connected with the posterior thick border of the liver, may cause anasarca of the lower limbs. Matter accumulating habitually in the sigmoid flexure of the colon may cause a hydrocele, or a varicocele, by pressing on the spermatic veins of the left side. It is quite true that these two last-named affections appear more frequently on the left side than on the right; and it seems to me much more rational to attribute them to the above-mentioned circumstance than to the fact that the left spermatic veins open, at a disadvantageous right angle, into the left renal vein.

[Footnote 1: The situation of this vessel, its close relation to the pleura, the aorta, the large venous trunks, the vagus and phrenic nerves, and the uncertainty as to its length, or as to whether or not a thyroid or vertebral branch arises from it, are circumstances which render the operation of tying the vessel in cases of aneurism very doubtful as to a successful issue. The operation (so far as I know) has hitherto failed. Anatomical relations, nearly similar to these, prevent, in like manner, an easy access to the iliac arteries, and cause the operator much anxiety as to the issue.]

[Footnote 2: The effusion of fluid into the pleural sac (from whatever cause it may arise) sometimes takes place to a very remarkable extent. I have had opportunities of examining patients, in whom the heart appeared to be completely dislocated, from the left to the right side, owing to the large collection of serous fluid in the left pleural sac. The heart's pulsations could be felt distinctly under the right nipple. Paracentesis thoracis was performed at the point indicated in PLATE 26. In these cases, and another observed at the Hotel Dieu, the heart and lung, in consequence of the extensive adhesions which they contracted in their abnormal position, did not immediately resume their proper situation when the fluid was withdrawn from the chest. Nor is it to be expected that they should ever return to their normal character and position, when the disease which caused their displacement has been of long standing.]

DESCRIPTION OF PLATE 26.

A. The systemic aorta. Owing to the body being inclined forwards, the root of the aorta appears to approach too near the lower boundary (N) of the thorax.

B. The left brachio-cephalic vein.

C. Left subclavian vein.

D. Right brachia-cephalic vein.

E. Left common carotid artery.

F. Brachio-cephalic artery.

G G*. The first pair of ribs.

H. Superior vena cava.

I. Left bronchus.

K K*. Fourth pair of ribs.

L. Descending thoracic aorta.

M. Oesophagus.

N. Epigastrium.

O. Left kidney.

P. Umbilicus.

Q. Abdominal aorta, at its bifurcation.

R R*. Right and left iliac fossae.

S. Left common iliac vein.

T. Inferior vena cava.

U. Psoas muscle, supporting the right spermatic vessels.

V. Left external iliac artery crossed by the left ureter.

W. Right external iliac artery crossed by the right ureter.

X. The rectum.

Y. The urinary bladder, which being fully distended, and viewed from above, gives it the appearance of being higher than usual above the pubic symphysis.

Z. Pubic symphysis.

2. The left internal abdominal ring complicated with the epigastric vessels, the vas deferens, and the spermatic vessels.

3. The right internal abdominal ring in connection with the like vessels and duct as that of left side.

4. Superior mesenteric artery.

5, 6. Right and left external iliac veins.

7, 8. Situations of the anterior superior iliac spinous processes.

9, 10. Situations of the coracoid processes.

11, 12. Right and left hypochondriac regions.

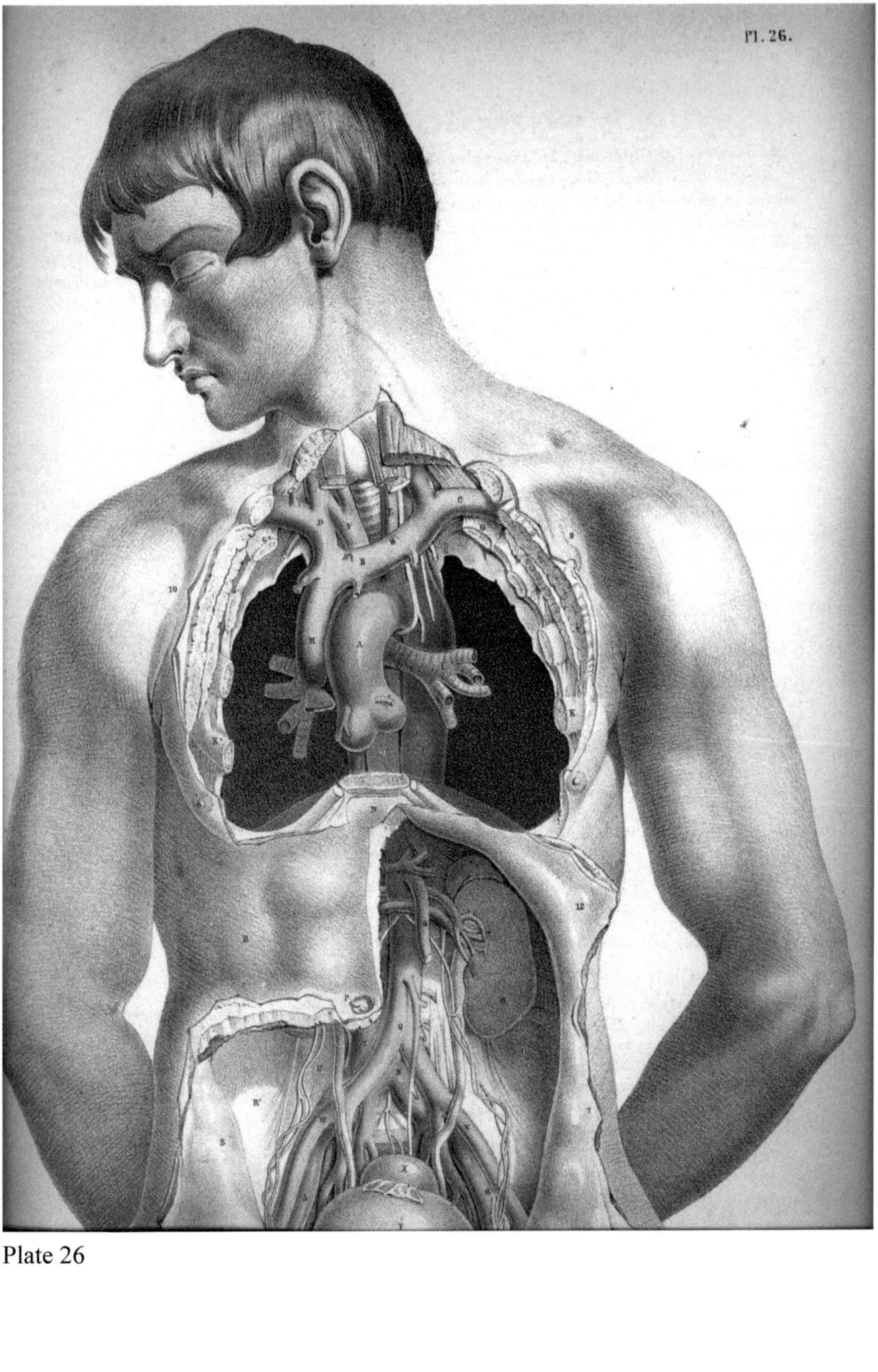

Plate 26

THE SURGICAL DISSECTION OF THE SUPERFICIAL PARTS AND BLOODVESSELS OF THE INGUINO-FEMORAL REGION.

Physical causes of the greater frequency of inguinal and femoral herniae.

The surface considered in reference to the subjacent parts.

Hernial protrusions are very liable to occur at the inguino-femoral region; and this fact has led the surgeon to study the anatomical relations of this part with more than ordinary care and patience. So minutely has he dissected every structure proper to this locality, and so closely has he investigated every possible condition of it as being the seat of hernial, that the only novelty which now remains to be sought for is that of a simplification of the facts, already known to be far too much obscured by an unwieldy nomenclature, and a useless detail of trifling evidence. And it would seem that nothing can more directly tend to this simplification, than that of viewing the inguinal and femoral regions, not separately, but as a relationary whole. For as both regions are blended together by structures which are common to both, so do the herniae which are described as being proper to either region, occur in such close connexion as at times to render it very difficult to distinguish between them.

The human species is, of all others, most subject to hernial in the groin. The erect attitude of the human form, and the fact that many of its more powerful muscular efforts are performed in this posture, cause its more frequent liability to the accidents called abdominal herniae or ruptures.

The viscera of the abdomen occupy this cavity completely, and indeed they naturally, at all times, subject the abdominal parietes to a state of constant pressure, as may be proved by their escape from the abdomen in cases of large wounds of this region. In the erect posture of the body this pressure is increased, for the viscera now gravitate and force downwards and forwards against the abdominal parietes. In addition to this gravitating force, another power impels the viscera from above downwards--namely, that of the muscles of the trunk, and the principal agent amongst these is the diaphragm. The lungs, again, expanding above the diaphragm, add also to the gravitation of the abdominal contents, and these, under the pressure thus accumulated, occasionally make an exit for themselves at the groins, which are the weakest and most depending parts of the abdomen.

Herniae are variously named in accordance with the following circumstances--viz., the precise locality at which they occur--the size and form of the tumour--the time of life at which they happen. Sexual peculiarities do not serve to distinguish herniae, though it is true that the inguinal form, at the part D F, occurs more commonly in the male, whilst the crural form, at the opening E, happens more frequently in the female.

The most common forms of herniae happen at those localities where the abdominal walls are traversed by the bloodvessels on their way to the outstanding organs, and where, in consequence, the walls of the abdomen have become weakened. It also happens, that at these very situations the visceral pressure is greatest whilst the body stands erect. These localities are, A, the umbilicus, a point characterized as having given passage (in the foetal state) to the umbilical vessels; D, the place where the spermatic vessels and duct pass from the abdomen to the testicle; and immediately beneath this, the crural arch, which gives exit to the crural vessels. Herniae may happen at other localities, such as at the thyroid aperture, which transmits the thyroid vessels; and at the greater sacrosciatic notch, through which the gluteal vessels pass; and all regions of the abdominal walls may give exit to intestinal protrusion in consequence of malformations, disease, or injury. But as the more frequent varieties of herniae are those which traverse the localities, A, D, E, and as these, fortunately, are the most manageable under the care of the surgical anatomist, we proceed to examine the structures concerned in their occurrence.

A direct opening from within outwards does not exist in the walls of the abdomen; and anatomy demonstrates to us the fact, that where the spermatic cord, D F, and the femoral vessels, pass from the abdomen to the external parts, they carry with them a covering of the several layers of structures, both muscular and membranous, which they encounter in their passage. The inguinal and crural forms of herniae which follow the passages made by the spermatic cord, and the crural vessels, must necessarily carry with them the like investments, and these are what constitute the coverings of the herniae themselves.

The groin in its undissected state is marked by certain elevations and depressions which indicate the general relations of the subcutaneous parts. The abdomen is separated from the thigh by an undulating grooved line, extending from C*, the point of the iliac bone, to B, the symphysis pubis This line or fold of the groin coincides exactly with the situation of that fibrous band of the external oblique muscle named Poupart's ligament. From below the middle of this abdomino-femoral groove, C B, another curved line, D, *b*, springs, and courses obliquely, inwards and downwards, between the upper part of the thigh and the pubis, to terminate in the scrotum. The external border of this line indicates the course of the spermatic cord, D F, which can be readily felt beneath the skin. In all subjects, however gross or emaciated they may happen to be, these two lines are readily distinguishable, and as they bear relations to the several kinds of rupture taking place in these parts, the surgeon should consider them with keen regard. A comparison of the two sides of the figure, PLATE 27, will show that the spermatic cord, D F, and Poupart's ligament, C B, determine the shape of the inguino-femoral region. When the integument with the subcutaneous adipose tissue is removed from the inguino-femoral region, we expose that common investing membrane called the superficial fascia. This fascia, *a a a*, stretches over the lower part of the abdomen and the upper part of the thigh. It becomes intimately attached to Poupart's ligament along the ilio-pubic line, C B; it invests the spermatic cord, as shown at *b*, and descends into the scrotum, so as to encase this part. Where this superficial fascia overlies the saphenous opening, E, of the fascia lata, it assumes a "cribriform" character, owing to its being pierced by numerous lymphatic vessels and some veins. As this superficial fascia invests all parts of the inguino-femoral region, as it forms an envelope for the spermatic cord, D F, and sheathes over the saphenous opening, E, it must follow of course that wherever the hernial protrusion takes place in this region, whether at D, or F, or E, or adjacent parts, this membrane forms the external subcutaneous covering of the bowel.

There is another circumstance respecting the form and attachments of the superficial fascia, which, in a pathological point of view, is worthy of notice--viz., that owing to the fact of its enveloping the scrotum, penis, spermatic cord, and abdominal parietes, whilst it becomes firmly attached to Poupart's ligament along the abdomino-femoral fold, B C, it isolates these parts, in some degree, from the thigh; and when urine happens to be from any cause extravasated through this abdominal-scrotal bag of the superficial fascia, the thighs do not in general participate in the inflammation superinduced upon such accident.

The spermatic cord, D, emerges from the abdomen and becomes definable through the fibres of the sheathing tendon of the external oblique muscle, H, at a point midway between the extremities of the ilio-pubic line or fold. In some cases, this place, whereat the cord first manifests itself in the groin, lies nearer the pubic symphysis; but however much it may vary in this particular, we may safely regard the femoro-pubic fold, D, *b*, as containing the cord, and also that the place where this fold meets the iliopubic line, C B, at the point D, marks the exit of the cord from the abdomen.

The spermatic cord does not actually pierce the sheathing tendon of the external oblique muscle at the point D, and there does not, in fact, exist naturally such an opening as the "external abdominal ring," for the cord carries with it a production of the tendon of the external oblique muscle, and this has been named by surgical anatomists the "intercolumnar fascia," [Footnote] the "spermatic fascia." The fibres of this spermatic fascia are seen at D F, crossing the cord obliquely, and encasing it. This covering of the cord lies beneath the spermatic envelope formed by, *a b,* the superficial fascia; and when a hernial protrusion descends through the cord, both these investing membranes form the two outermost envelopes for the intestine in its new and abnormal situation.

[Footnote: On referring to the works of Sir Astley Cooper, Hesselbach, Scarpa, and, others, I find attempts made to establish a distinction between what is called the "intercolumnar fascia" and the "spermatic fascia," and just as if these were structures separable from each other or from the aponeurotic sheath of the external oblique muscle. I find, in like manner, in these and other works, a tediously-laboured account of the superficial fascia, as being divisible into two layers of membrane, and that this has given rise to considerable difference of opinion as to whether or not we should regard the deeper layer as being a production of the fascia lata, ascending from the thigh to the abdomen, or rather of the membrane of the abdomen descending to the thigh, &c. These and such like considerations I omit to discuss here; for, with all proper deference to the high authority of the authors cited, I dare to maintain, that, in a practical point of view, they arc absolutely of no moment, and in a purely scientific view, they are, so far as regards the substance of the truth which they would reveal, wholly beneath the notice of the rational mind. The practitioner who would arm his judgment with the knowledge of a broad fact or principle, should not allow his serious attention to be diverted by a pursuit after any such useless and trifling details, for not only are they unallied to the stern requirements of surgical skill, but they serve to degrade it from the rank and roll of the sciences. Whilst operating for the reduction of inguinal hernia by the "taxis" or the bistoury, who is there that feels anxiety concerning the origin or the distinctiveness of the "spermatic fascia?" Or, knowing it to be present, who concerns himself about the better propriety of naming it "tunica vaginalis communis," "tunique fibreuse du cordon spermatique," "fascia cremasterica," or "tunica aponeurotica?"]

The close relations which the cord, D F, bears to the saphenous opening, E, of the fascia lata, should be closely considered, forasmuch as when an oblique inguinal hernia descends from D to F, it approaches the situation of the saphenous opening, E, which is the seat of the femoral or crural hernia, and both varieties of hernia may hence be confounded. But with a moderate degree of judgment, based upon the habit of referring the anatomy to the surface, such error may always be avoided. This important subject shall be more fully treated of further on.

The superficial bloodvessels of the inguino-femoral region are, *e e*, the saphenous vein, which, ascending from the inner side of the leg and thigh, pierces the saphenous opening, E, to unite with the femoral vein. The saphenous vein, previously to entering the saphenous opening, receives the epigastric vein, *i*, the external circumflex ilii vein, *h*, and another venous branch, *d*, coming from the fore part of the thigh. In the living body the course of the distended saphenous vein may be traced beneath the skin, and easily avoided in surgical operations upon the parts contained in this region. Small branches of the femoral artery pierce the fascia lata, and accompany these superficial veins. Both these orders of vessels are generally divided in the operation required for the reduction of either the inguinal or the femoral strangulated hernia; but they are, for the most part, unimportant in size. Some branches of nerves, such as, *k*, the external cutaneous, which is given off from the lumbar nerves, and, *f*, the middle cutaneous, which is derived from the crural nerve, pierce the fascia lata, and appear upon the external side and middle of the thigh.

Numerous lymphatic glands occupy the inguino-femoral region; these can be felt, lying subcutaneous, even in the undissected state of the parts. These glands form two principal groups, one of which, *c*, lies along the middle of the inguinal fold, C B; the other, G *g*, lies scattered in the neighbourhood of the saphenous opening. The former group receive the lymphatic vessels of the generative organs; and the glands of which it is composed are those which suppurate in, syphilitic or other affections of these parts.

The general relations which the larger vessels of the inguino-femoral region bear to each other and to the superficies, may be referred to in PLATE 27, with practical advantage. The umbilicus, A, indicates pretty generally the level at which the aorta bifurcates on the forepart of the lumbar vertebrae. In the erect, and even in the recumbent posture, the aorta may (especially in emaciated subjects) be felt pulsating under the pressure of the hand; for the vertebrae bear forward the vessel to a level nearly equal with, C C, the anterior superior spinous processes of the iliac bones. If a gunshot were to pass through the abdomen, transversely, from these points, and through B, it would penetrate the aorta at its bifurcation. The line A B coincides with the linea alba. The oblique lines, A D, A D,* indicate the course of the iliac vessels. The point D marks the situation where the spermatic vessels enter the abdomen; and also where the epigastric artery is given off from the external iliac. The most convenient line of incision that can be made for reaching the situation of either of the iliac arteries, is that which ranges from C, the iliac spine, to D, the point where the spermatic cord enters the abdomen. The direct line drawn between D and G marks the course of the femoral artery, and this ranges along the outer border, E, of the saphenous opening.

DESCRIPTION OF PLATE 27.

A. The umbilicus.

B. The upper margin of the pubic symphysis.

C. The anterior superior spine of the left iliac bone. C*, the situation of the corresponding part on the right side.

D. The point where, in this subject, the cord manifested itself beneath the fibres of the external oblique muscle. D*, a corresponding part on the opposite side.

E. The saphenous opening in the fascia lata, receiving *e*, the saphenous vein.

F. The lax and pendulous cord, which in this case, overlies the upper part of the saphenous opening.

G. Lymphatic glands lying on the fascia lata in the neighbourhood of the saphenous opening.

H. The fleshy part of the external oblique muscle.

a a a. The superficial fascia of the abdomen.

b. The same fascia forming an envelope for the spermatic cord and scrotum.

c. Inguinal glands lying near Poupart's ligament.

d. A common venous trunk, formed by branches from the thigh and abdomen, and joining--

e e. The saphenous vein.

f. The middle cutaneous nerve, derived from the anterior crural nerve.

g. Femoral lymphatic glands.

h. Superficial external iliac vein.

i. Superficial epigastric vein.

k. External cutaneous branches of nerves from the lumbar plexus.

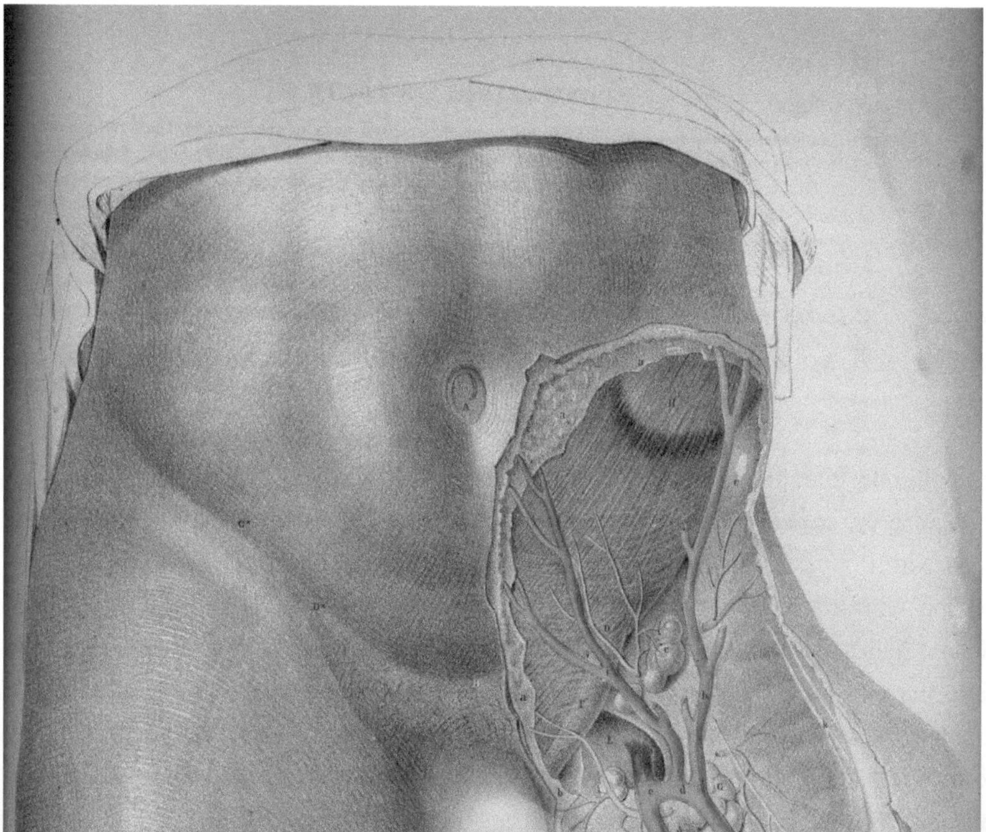

PLATE 27

THE SURGICAL DISSECTION OF THE FIRST, SECOND, THIRD, AND FOURTH LAYERS OF THE INGUINAL REGION, IN CONNEXION WITH THOSE OF THE THIGH.

The external abdominal ring and spermatic cord. Cremaster muscle—how formed. The parts considered in reference to inguinal hernia. The saphenous opening, spermatic cord, and femoral vessels in relation to femoral hernia.

The common integument or *first layer* of the inguino-femoral region being removed, we expose the superficial fascia constituting the *second layer*. The connexion of this fascia with Poupart's ligament along the line C D, together with the facts, that corresponding with this line the fascia is devoid of adipous substance, and the integument thin and delicate, whilst above over the abdomen, and below over the upper part of the thigh, the meshes of the fascia are generally loaded with a considerable quantity of adipous tissue, will account for the permanency and distinctness of the fold of the groin. As this fold corresponds with Poupart's ligament, it is taken as a guide to distinguish between the inguinal and femoral forms of herniae.

The general relations of the superficial fascia are well described by Camper in the following sentence: "Musculus obliquus igitur externus abdominis, qua parte carneus est, membrana quadam propria, quali omnes musculi, tegitur, quae sensim in aponeurosin mutata, ac cum tendineis hujus musculi partibus unita, externe ac anteriore parte abdomen tegit; finem vero nullibi habere perspicuum est, ad pubem enim miscet cellulosa membrana, cum ligamento penis in viris ac clitoridis in feminis, involucrum dat musculo cremasteri, ac aponeuroseos speciem musculis anterioribus femoris, qua glandulae inguinales, ac cruris vasa majora obteguntur." (Icones Herniarum.)

Owing to the varied thickness of the adipous tissue contained in the superficial fascia at several regions of the same body, and at some corresponding regions of different individuals, it will be evident that the depth of the incision required to divide it, so as to expose subjacent structures, must vary accordingly. Where the superficial fascia, after encasing the cord, descends into the scrotum, it is also devoid of the fatty tissu.

By the removal of the superficial fascia and glands we expose the aponeurosis of the external oblique muscle, A *a,* Pl. 28, (constituting t*he third layer* of the groin,) and also the fascia of the thigh, H L. These strong fibrous structures will be observed to hold still *in situ* the other parts, and to be the chief agents in determining the normal form of this region.

The inguino-femoral region, as being the seat of hernial protrusions, may in this stage of the dissection be conveniently described as a space formed of two triangles--the one inguinal, the other femoral, placed base to base. The inguinal triangle may be drawn between the points, B C D, Pl. 28, while the femoral triangle may be marked by the points, C D N. The conjoined bases of these triangles correspond to Poupart's ligament along the line, C D. The inguinal varieties of herniae occur immediately above the line, C D, while the femoral varieties of herniae take place below this line. The herniae of the inguinal triangle are, therefore, distinguishable from those of the femoral triangle by a reference to the line, C D, or Poupart's ligament.

The aponeurosis of the external oblique muscle occupies the whole of that space which I have marked as the inguinal triangle, B C D, Pl. 28. The fleshy fibres of the muscle, A, after forming the lateral wall of the abdomen, descend to the level of C, the iliac spinous process, and here give off the inguinal part of their broad tendon, *a.* The fibres of this part of the tendon descend obliquely downwards and forwards to become inserted at the median line of the abdomen into the linea alba, B D, as also into the symphysis and crista of the os pubis. The lower band of the fibres of this tendinous sheath--viz., that which is stretched between C, the iliac spine, and D, the crista pubis, is named Poupart's ligament; and this is strongly connected with H, the iliac portion of the fascia lata of the thigh.

Poupart's ligament is not stretched tensely in a right line, like the string of a bow, between the points, C and D. With regard to these points it is lax, and curves down towards the thigh like the arc of a circle. The degree of tension which it manifests when the thigh is in the extended position is chiefly owing to its connexion with the fascia lata. If in this position of the limb we sever the connexion between the ligament and fascia, the former becomes relaxed in the same degree as it does when we flex the thigh upon the abdomen. The utmost degree of relaxation which can be given to Poupart's ligament is effected by flexing the thigh towards the abdomen, at the same time that we support the body forwards. This fact has its practical application in connexion with the reduction of herniae.

Immediately above the middle of Poupart's ligament, at the point E, Pl. 28, we observe the commencement of a separation taking place among the fibres of the aponeurosis. These divide into two bands, which, gradually widening from each other as they proceed inwards, become inserted, the upper one into the symphysis pubis, the lower into the spine and pectineal ridge of this bone. The lower band identifies itself with Poupart's ligament. The interval which is thus formed by the separation of these fibres assumes the appearance of an acute triangle, the apex of which is at E, and the base at D. But the outer end of this interval is rounded off by certain fibres which cross those of the bands at varying angles. At this place, the aponeurosis, thus constituted of fibres disposed crossways, is elongated into a canal, forming an envelope for the cord, K. This elongation is named the "external spermatic fascia," and is continued over the cord as far as the testicle. In the female, a similar canal encloses the round ligament of the uterus. From the above-mentioned facts, it will appear that the so-called "external abdominal ring" does not exist as an aperture with defined margins formed in the tendon of the external oblique muscle. It is only when we divide the spermatic fascia upon the cord at K, that we form the external ring, and then it must be regarded as an artificial opening, as at D, Pl. 29.

The part of the groin where the spermatic fascia is first derived from the aponeurosis, so as to envelope the cord, varies in several individuals; and thereupon depends, in great measure, the strength or weakness of the groin. In some cases, the cord becomes pendulous as far outwards as the point E, Pl. 28, which corresponds to the internal ring, thereby offering a direct passage for the hernial protrusion. In other instances, the two bands of the aponeurosis, known as the "pillars of the ring," together with the transverse fibres, or "intercolumnar fascia," firmly embrace and support the cord as far inwards as the point K, and by the oblique direction thus given to the cord in traversing the inguinal parietes, these parts are fortified against the occurrence of hernia. In Pl. 28, the cord, K, will be observed to drop over the lower band of fibres, ("external pillar of the ring,") and to have D, the crista pubis, on its inner side. In Pl. 29, the upper band of fibres ("internal pillar of the ring") may be seen proceeding to its insertion into the symphysis pubis. When a hernial tumour protrudes at the situation K, it is invested, in the same manner as the cord, by the spermatic fascia, and holds in respect to the fibrous bands or pillars the same relations also as this part.

After removing the tendon of the external oblique muscle, A *a*, Pl. 28, together with its spermatic elongation, E, we expose the internal oblique, F E, Pl. 29, and the cremaster, constituting the *fourth inguinal layer*. The fleshy part of this muscle, F E, occupies a much greater extent of the inguinal region than does that of the external oblique. Whilst the fleshy fibres of the latter terminate on a level with C, the iliac spine, those of the internal oblique are continued down as far as the external abdominal ring, E D *h*, and even protrude through this place in the form of a cremaster. The muscular fibres of the internal oblique terminate internally at the linea semilunaris, *g*; while Poupart's ligament, the spinous process and crest of the ilium, give origin to them externally. At the linea semilunaris, the tendon of the internal oblique is described as dividing into two layers, which passing, one before and the other behind the rectus abdominis, thus enclose this muscle in a sheath, after which they are inserted into the linea alba, G. The direction of the fibres of the inguinal portion of the muscle, F E, is obliquely downwards and forwards, and here they are firmly overlaid by the aponeurosis of the external oblique.

The cremaster muscle manifests itself as being a part of the internal oblique, viewing this in its totality. Cloquet (Recherches anatomiques sur les Hernies de l'Abdomen) first demonstrated the correctness of this idea.

The oblique and serial arrangement of the muscular fibres of the internal oblique, F, Pl. 29, is seen to be continued upon the spermatic cord by the fibres of the cremaster, E e. These fibres, like those of the lower border of the internal oblique, arise from the middle of Poupart's ligament, and after descending over the cord as far as the testicle in the form of a series of inverted loops, *e*, again ascend to join the tendon of the internal oblique, by which they become inserted into the crest and pectineal ridge of the os pubis. The peculiar looping arrangement exhibited by the cremasteric fibres indicates the fact that the testicle, during its descent from the loins to the scrotum, carried with it a muscular covering, at the expense of the internal oblique muscle. The cremaster, therefore, is to be interpreted as a production of the internal oblique, just as the spermatic fascia is an elongation of the external oblique. The hernia, which follows the course of the spermatic vessels, must therefore necessarily become invested by cremasteric fibres.

The fascia lata, H, Pl. 28, being strongly connected and continuous with Poupart's ligament along its inferior border, the boundary line, which Poupart's ligament is described as drawing between the abdomen and thigh, must be considered as merely an artificial one.

In the upper region of the thigh the fascia lata is divided into two parts--viz., H, the iliac part, and L, the pubic. The iliac part, H, which is external, and occupying a higher plane than the pubic part, is attached to Poupart's ligament along its whole extent, from C to D, Pl. 28; that is, from the anterior iliac spinous process to the crista pubis. From this latter point over the upper and inner part of the thigh, the iliac division of the fascia appears to terminate in an edge of crescentic shape, *h;* but this appearance is only given to it by our separating the superficial fascia with which it is, in the natural state of the parts, blended. The pubic part of the fascia, L, Pl. 28, which is much thinner than the iliac part, covers the pectineus muscle, and is attached to the crest and pectineal ridge of the os pubis, occupying a plane, therefore, below the iliac part, and in this way passes outwards beneath the sheath of the femoral vessels, K I, Pl. 29. These two divisions of the fascia lata, although separated above, are united and continuous on the same plane below. An interval is thus formed between them for the space of about two inches below the inner third of Poupart's ligament; and this interval is known as the "saphenous opening," L *h*, Pl. 28. Through this opening, the saphena vein, O, Pl. 29, enters the femoral vein, I.

From the foregoing remarks it will appear that no such aperture as that which is named "saphenous," and described as being shaped in the manner of L *h*, Pl. 28, with its "upper and lower cornua," and its "falciform process," or edge, *h*, exists naturally. Nor need we be surprised, therefore, that so accurate an observer as Soemmering (de Corporis Humani Fabrica) appears to have taken no notice of it.

Whilst the pubic part of the fascia lata passes beneath the sheath of the femoral vessels, K I, Pl. 29, the iliac part, H *h*, blends by its falciform margin with the superficial fascia, and also with N *n,* the sheath of the femoral vessels. The so-called saphenous opening, therefore, is naturally masked by the superficial fascia; and this membrane being here perforated for the passage of the saphena vein, and its tributary branches, as also the efferent vessels of the lymphatic glands, is termed "cribriform."

The femoral vessels, K I, contained in their proper sheath, lie immediately beneath the iliac part of the fascia lata, in that angle which is expressed by Poupart's ligament, along the line C D above; by the sartorius muscle in the line C M externally; and by a line drawn from D to N, corresponding to the pectineus muscle internally. The femoral vein, I, lies close to the outer margin of the saphenous opening. The artery, K, lies close to the outer side of the vein; and external to the artery is seen, L, the anterior crural nerve, sending off its superficial and deep branches.

When a femoral hernia protrudes at the saphenous space L *h*, Pl. 28, the dense falciform process, *h,* embraces its outer side, while the pubic portion of the fascia, L, lies beneath it. The cord, K, is placed on the inner side of the hernia; the cribriform fascia covers it; and the upper end of the saphena vein, M, passes beneath its lower border. The upper cornu, *h,* Pl. 29, of the falciform process would seem, by its situation, to be one of the parts which constrict a crural hernia. An inguinal hernia, which descends the cord, K, Pl. 28, provided it passes no further than the point indicated at K, and a crural hernia turning upwards from the saphenous interval over the cord at K, are very likely to present some difficulty in distinctive diagnosis.

DESCRIPTION OF THE FIGURES OF PLATES 28 & 29.

PLATE 28.

A. The fleshy part of the external oblique muscle; *a,* its tendon covering the rectus muscle.

B. The umbilicus.

C. The anterior superior spinous process of the ilium.

D. The spinous process of the os pubis.

E. The point where in this instance the fibres of the aponeurotic tendon of the external oblique muscle begin to separate and form the pillars of the external ring.

F G. See Plate 29.

H. The fascia lata--its iliac portion. The letter indicates the situation of the common femoral artery; *h,* the falciform edge of the saphenous opening.

I. The sartorius muscle covered by a process of the fascia lata.
K. The spermatic fascia derived from the external oblique tendon.

L. The pubic part of the fascia lata forming the inner and posterior boundary of the saphenous opening.

M. The saphenous vein.

N. A tributary vein coming from the fore part of the thigh.

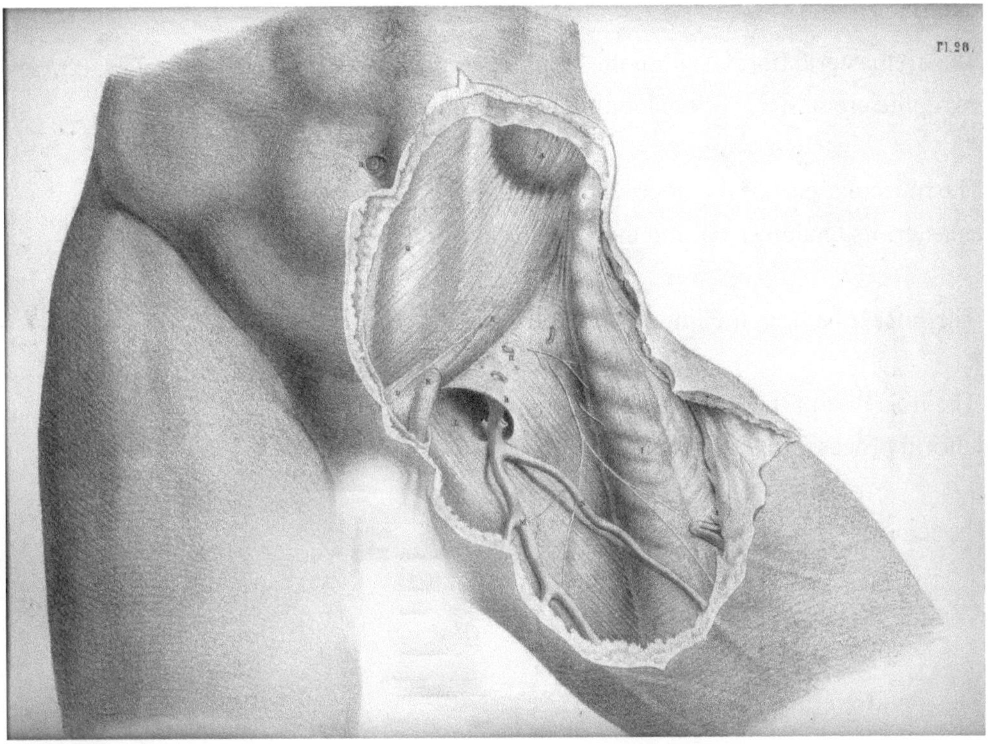

Plate 28

PLATE 29.

A. The muscular part of the external oblique; *a,* its tendon.

B. The umbilicus.

C. The anterior superior iliac spine.

D. The spine of the os pubis.

E. The cremasteric fibres, within the external ring, surrounding the cord; *e*, the cremasteric fibres looping over the cord outside the ring.

F. The muscular part of the internal oblique giving off, E, the cremaster; its tendon sheathing the rectus muscle.

G. The linea alba; f, *g*, the linea semilunaris.

H. The iliac part of the fascia lata; *h,* the upper cornu of its falciform process.

I. The femoral vein.

K. The femoral artery.

L. The anterior crural nerve.

M. The sartorius muscle.

N. The sheath of the femoral vessels; *n,* its upper part.

O. The saphena vein.

P. The pubic part of the fascia lata.

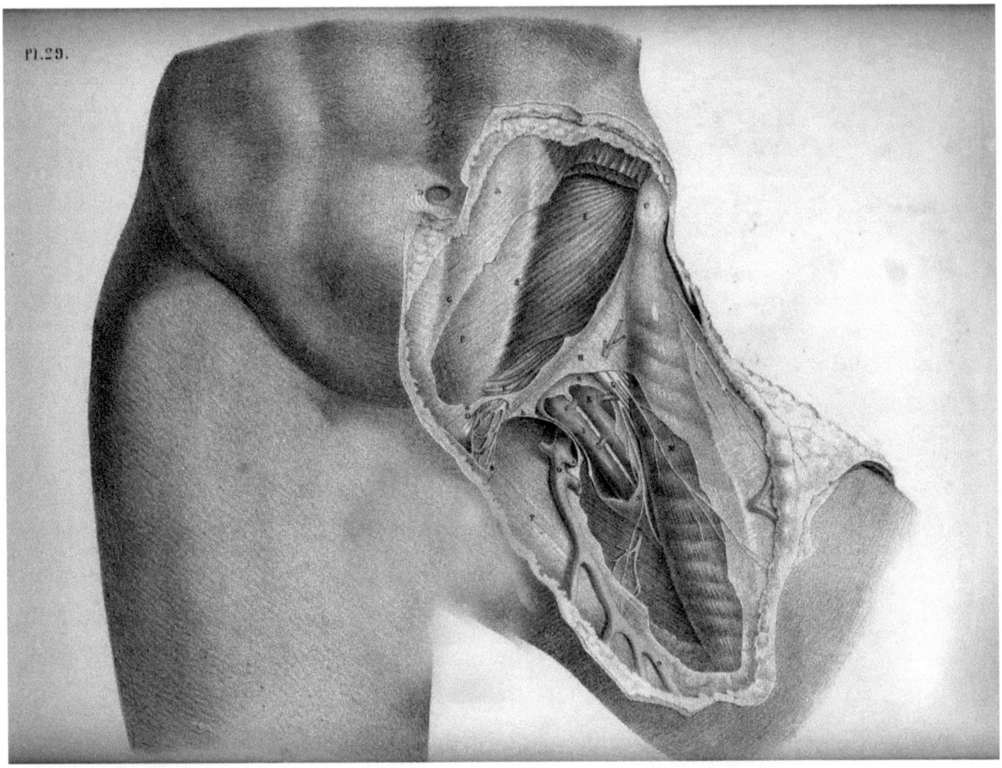

Plate 29

THE SURGICAL DISSECTION OF THE FIFTH, SIXTH, SEVENTH, AND EIGHTH LAYERS OF THE INGUINAL REGION, AND THEIR CONNEXION WITH THOSE OF THE THIGH.

The conjoined tendon, internal inguinal ring, and cremaster muscle, considered in reference to the descent of the testicle and of the hernia. The structure and direction of the inguinal canal.

When we remove the internal oblique and cremaster muscles, we expose the transverse muscle, which may be regarded as the *fifth inguinal layer*, F, Pl. 30. This muscle is similar in shape and dimensions to the internal oblique. The connexions of both are also similar, inasmuch as they arise from the inner edge of the crista ilii, and from the outer half of, V, Poupart's ligament. The fleshy fibres of these two muscles vary but little in direction, and terminate at the same place--viz., the linea semilunaris, which marks the outer border of the rectus muscle. But whilst the fleshy parts of these three abdominal muscles, D E F, form successive strata in the groin, their aponeurotic tendons present the following peculiarities of arrangement in respect to the rectus muscle. The tendon of the external oblique, *d*, passes altogether in front of the rectus; that of the internal oblique is split opposite the linea semilunaris into two layers, which enclose the rectus between them as they pass to be inserted into the linea alba. But midway between the navel and pubes, at the point marked G, both layers of the tendon are found to pass in front of the rectus. The tendon of the transverse muscle passes behind the rectus; but opposite the point G, it joins both layers of the internal oblique tendon, and with this passes in front of the rectus. The fibrous structure thus constituted by the union of the tendons of the internal oblique and transverse muscles, *e f*, is named the "conjoined tendon."

The conjoined tendon, *f*, Plates 30 and 31, appears as a continuation of the linea semilunaris, for this latter is in itself a result of the union of the tendons of the abdominal muscles at the external border of the rectus. As the conjoined tendon curves so far outwards to its insertion into the pectineal ridge of the pubic bone, as to occupy a situation immediately behind the external ring, it thereby fortifies this part against the occurrence of a direct protrusion of the bowel. But the breadth, as well as the density, of this tendon varies in several individuals, and these will accordingly be more or less liable to the occurrence of hernia.

The arched inferior border of the transverse muscle, F, Plate 30, expresses by its abrupt termination that some part is wanting to it; and this appearance, together with the fact that the fibres of this part of the muscle blend with those of the internal oblique and cremaster, and cannot be separated except by severing the connexion, at once suggests the idea that the cremaster is a derivation from both these muscles.

Assuming this to be the case, therefore, it follows that when the dissector removes the cremaster from the space L *h*, he himself causes this vacancy in the muscular parietes of the groin to occur, and at the same time gives unnatural definition to the lower border of the transverse and oblique muscles. In a dissection so conducted, the cord is made to assume the variable positions which anatomists report it to have in respect to the neighbouring muscles. But when we view nature as she is, and not as fashioned by the scalpel, we never fail to find an easy explanation of her form.

In the foetus, prior to the descent of the testicle, the cremaster muscle does not exist. (Cloquet, op cit.) From this we infer, that those parts of the muscles, E F, Plate 30, which at a subsequent period are converted into a cremaster, entirely occupy the space L *h*. In the adult body, where one of the testicles has been arrested in the inguinal canal, the muscles, E F, do not present a defined arched margin, above the vacant space L *h*, but are continued (as in the foetus) as low down as the external abdominal ring. In the adult, where the testicle has descended to the scrotum, the cremaster exists, and is serially continuous with the muscles, E F, covering the space L *h*; the meaning of which is, that the cremasteric parts of the muscles, E F, cover this space. The name cremaster therefore must not cancel the fact that the fibres so named are parts of the muscles, E F. Again, in the female devoid of a cremaster, the muscles, E F, present of their full quantities, having sustained no diminution of their bulk by the formation of a cremaster. But when an external inguinal hernia occurs in the female body, the bowel during its descent carries before it a cremasteric covering at the expense of the muscles E F, just in the same way as the testicle does in the foetus. (Cloquet.)

From the above-mentioned facts, viewed comparatively, it seems that the following inferences may be legitimately drawn:--1st, that the space L *h* does not naturally exist devoid of a muscular covering; for, in fact, the cremaster overlies this situation; 2nd, that the name cremaster is one given to the lower fibres of the internal oblique and transverse muscles which cover this space; and 3rd, that to separate the cremasteric elongation of these muscles, and then describe them as presenting a defined arched margin, an inch or two above Poupart's ligament, is an act as arbitrary on the part of the dissector as if he were to subdivide these muscles still more, and, while regarding the subdivisions as different structures, to give them names of different signification. When once we consent to regard the cremaster as constituted of the fibres originally proper to the muscles, E F, we then are led to the discovery of the true relations of the cord in respect to these muscles.

On removing the transverse muscle, we expose the inguinal part of the transversalis fascia--the *sixth inguinal layer*, L *h*, Plate 30--K *k*, Plate 31. This fascia or membrane affords a general lining to the abdominal walls, in some parts of which it presents of a denser and stronger texture than in others. It is stretched over the abdomen between the muscles and the peritonaeum. The fascia iliaca, the fascia pelvica, and the fascia transversalis, are only regional divisions of the one general membrane. On viewing this fascia in its totality, I find it to exhibit many features in common with those other fibrous structures which envelope serous cavities. The transversalis fascia supports externally the peritonaeum, in the same way as the dura mater supports the arachnoid membrane, or as the pleural fascia supports the serous pleura. While the serous membranes form completely shut sacs, the fibrous membranes which lie external to those sacs are pierced by the vessels which course between them and the serous membranes, and afford sheaths or envelopes for these vessels in their passage from the interior to the external parts. The sheath, H *h*, Plates 30 and 31, which surrounds the spermatic vessels, and the sheath, R, Plate 31, which envelopes the crural vessels, are elongations of the fascia transversalis.

In the groin, the transversalis fascia, K *k*, Plate 31, presents, in general, so dense a texture as to offer considerable resistance to visceral pressure. Here it is stretched between the transverse muscle, F, Plate 31, and the peritonaeum, I. It adheres to the external surface of the peritonaeum, and to the internal surface of the transverse muscle, by means of an intervening cellular tissue. It is connected below to Poupart's ligament, along the line of which it joins the fascia iliaca. It lines the lower posterior aspect of the rectus muscle, where this is devoid of its sheath; and it is incorporated with *f,* the conjoined tendon, thereby fencing the external abdominal ring. Immediately above the middle of Poupart's ligament, this membrane, at the point marked *h,* Plate 30, is pouched into a canal-shaped elongation, which invests the spermatic vessels as far as the testicle in the scrotum; and to this elongation is given the names "fascia spermatica interna" (Cooper), "fascia infundibuliform" (Cloquet). The same part, when it encloses an external oblique hernia, is named "fascia propria." The neck or inlet of this funnel-shaped canal is oval, and named the "internal abdominal ring." As this ring looks towards the interior of the abdomen, and forms the entrance of the funnel-shaped canal, it cannot of course be seen from before until we slit open this canal. Compare the parts marked H *h* in Plates 30 and 31.

The inguinal and iliac portions of the fascia transversalis join along the line of Poupart's ligament, A C. The iliac vessels, in their passage to the thigh, encounter the fascia at the middle third of the crural arch formed by the ligament, and take an investment (the sheath, R) from the fascia. The fore part of this sheath is mentioned as formed by the fascia transversalis--the back part by the fascia iliaca; but these distinctions are merely nominal, and it is therefore unnecessary to dwell upon them. The sheath of the femoral vessels is also funnel-shaped, and surrounds them on all sides. Its broad entrance lies beneath the middle of Poupart's ligament. Several septa are met with in its interior. These serve to separate the femoral vessels from each other. The femoral vein, O, Plate 30, is separated from the falciform margin, S *s*, of the saphenous opening by one of these septa. Between this septum and the falx an interval occurs, and through it the crural hernia usually descends. These parts will be more particularly noticed when considering the anatomy of crural hernia.

Beneath the fascia transversalis is found the subserous cellular membrane, which serves as a connecting medium between the fascia and the peritonaeum. This cellular membrane may be considered as the *seventh inguinal layer.* It is described by Scarpa (sull' Ernie) as forming an investment for the spermatic vessels inside the sheath, where it is copious, especially in old inguinal hernia. It is also sometimes mixed with fatty tissue. In it is found embedded the infantile cord--the remains of the upper part of the peritoneal tunica vaginalis--a structure which will be considered in connexion with congenital herniae.

By removing the subserous cellular tissue, we lay bare the peritonaeum, which forms the *eighth layer of the inguinal region.* Upon it the epigastric and spermatic vessels are seen to rest. These vessels course between the fascia transversalis and the peritonaeum. The internal ring which is formed in the fascia, K *h*, may be now seen to be closed by the peritonaeum, I. The inguinal canal, therefore, does not, in the normal state of these parts, communicate with the general serous cavity; and here it must be evident that before the bowel, which is situated immediately behind the peritonaeum, I, can be received into the canal, H *h,* it must either rupture that membrane, or elongate it in the form of a sac.

The exact position which the epigastric, L, Plate 31, and spermatic vessels, M, bear in respect to the internal ring, is a point of chief importance in the surgical anatomy of the groin; for the various forms of herniae which protrude through this part have an intimate relation to these vessels. The epigastric artery, in general, arises from the external iliac, close above the middle of Poupart's ligament, and ascends the inguinal wall in an oblique course towards the navel. It applies itself to the inner border of the internal ring, and here it is crossed on its outer side by the spermatic vessels, as these are about to enter the inguinal canal.

The inguinal canal is the natural channel through which the spermatic vessels traverse the groin on their way to the testicle in the scrotum. In the remarks which have been already made respecting the several layers of structures found in the groin, I endeavoured to realize the idea of an inguinal canal as consisting of elongations of these layers invaginated the one within the other, the outermost layer being the integument of the groin elongated into the scrotal skin, whilst the innermost layer consisted of the transversalis fascia elongated into the fascia spermatica interna, or sheath. The peritonaeum, which forms the eighth layer of the groin, was seen to be drawn across the internal ring of this canal above in such a way as to close it completely, whilst all the other layers, seven in number, were described as being continued over the spermatic vessels in the form of funnel-shaped investments, as far down as the testicle.

With the ideas of an inguinal canal thus naturally constituted, I need not hesitate to assert that the form, the extent, and the boundaries of the inguinal canal, as given by the descriptive anatomist, are purely conventional, and do not exist until after dissection; for which reason, and also because the form and condition of these parts so described and dissected do not appear absolutely to correspond in any two individuals, I omit to mention the scale of measurements drawn up by some eminent surgeons, with the object of determining the precise relative position of the several parts of the inguinal region.

The existence of an inguinal canal consisting, as I have described it, of funnel-shaped elongations from the several inguinal layers continued over the cord as far as the testicle, renders the adult male especially liable to hernial protrusions at this part. The oblique direction of the canal is, in some measure, a safeguard against these accidents; but this obliquity is not of the same degree in all bodies, and hence some are naturally more prone to herniae than others.

DESCRIPTION OF THE FIGURES OF PLATES 30 & 31.

PLATE 30.

A. The anterior superior iliac spine.

B. The umbilicus.

C. The spine of the pubis.

D. The external oblique muscle; *d*, its tendon. .

E. The internal oblique muscle; *e*, its tendon.

F. The transverse muscle; *f*, its tendon, forming, with *e*, the conjoined tendon.

G. The rectus muscle enclosed in its sheath.

H. The fascia spermatica interna covering the cord; *h*, its funnel-shaped extremity.

I, K, L, M. See Plate 31.

N. The femoral artery; *n*, its profunda branch.

O. The femoral vein.

P. The saphena vein.

Q. The sartorius muscle.

R. The sheath of the femoral vessels.

S. The falciform margin of the saphenous opening.

T. The anterior crural nerve.

U. The pubic portion of the fascia lata.

V. The iliac portion attached to Poupart's ligament.

W. The lower part of the iliacus muscle.

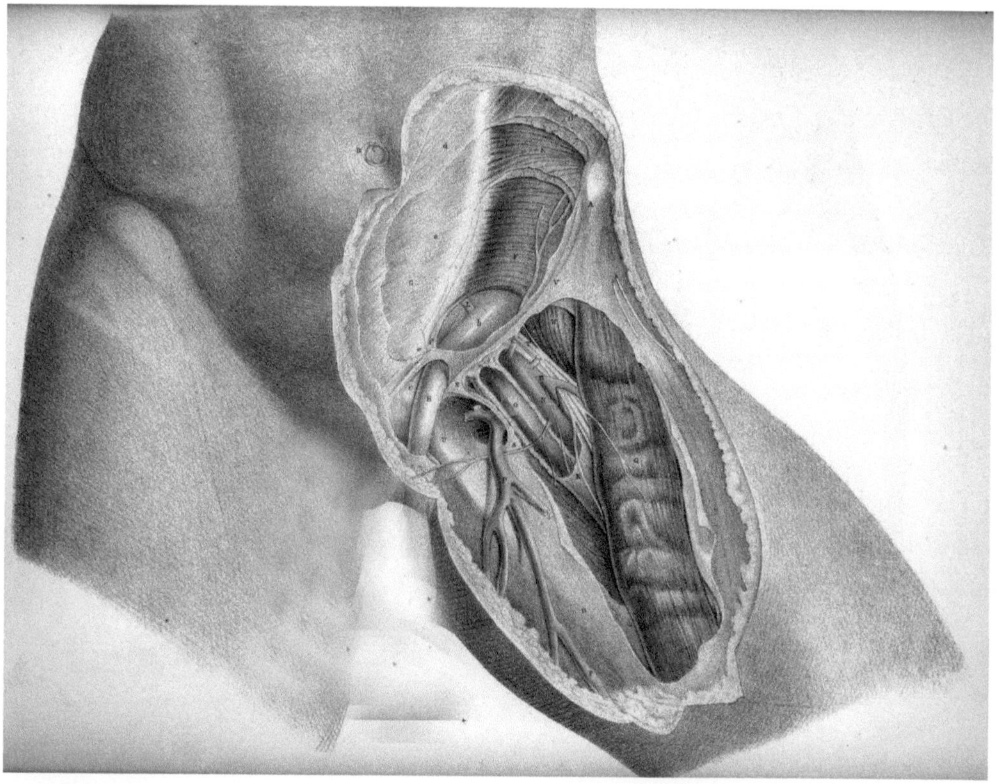

PLATE 30

PLATE 31.

A. The anterior superior iliac spine.

B. The umbilicus.

C. The spine of the pubis.

D. The external oblique muscle; *d,* its tendon; *d**, the external ring.

E. The internal oblique muscle.

F. The transverse muscle; *f,* its tendon; forming, with *e*, the conjoined tendon.

G. The rectus muscle laid bare.

H *h.* The fascia spermatica interna laid open above and below *d**, the external ring.

I. The peritonaeum closing the internal ring.

K. The fascia transversalis; *k*, its pubic part.

L. The epigastric artery and veins.

M. The spermatic artery, veins, and vas deferens bending round the epigastric artery at the internal ring; m, the same vessels below the external ring.

N. The femoral artery; *n*, its profunda branch.

O. The femoral vein, joined by--

P. The saphena vein.

Q. The sartorius muscle.

R. The sheath of the femoral vessels.

S S. The falciform margin of the saphenous opening,

T. The anterior crural nerve.

U. The pubic part of the fascia lata.

V. The iliac part of the fascia lata.

W. The lower part of the iliacus muscle.

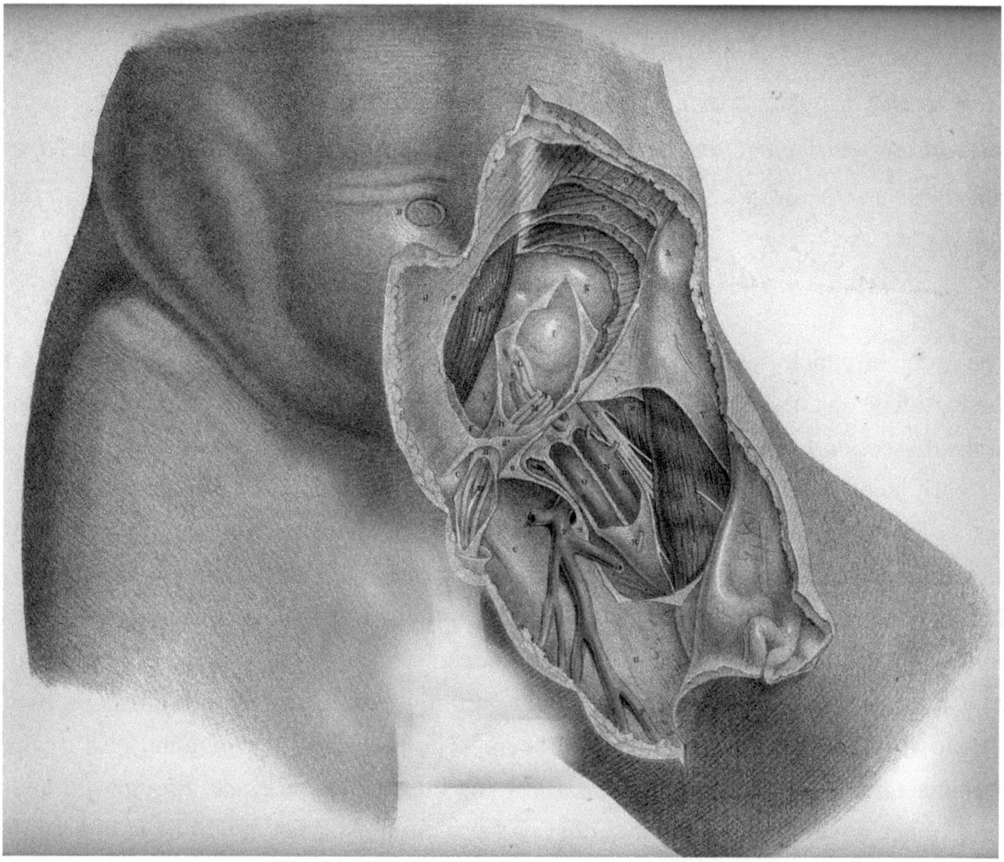

Plate 31

THE DISSECTION OF THE OBLIQUE OR EXTERNAL, AND OF THE DIRECT OR INTERNAL INGUINAL HERNIA.

Their points of origin and their relations to the inguinal rings. The triangle of Hesselbach. Investments and varieties of the external inguinal hernia, its relations to the epigastric artery, and its position in the canal. Bubonocele, complete and scrotal varieties in the male. Internal inguinal hernia considered in reference to the same points. Corresponding varieties of both herniae in the female.

The order in which the herniary bowel takes its investments from the eight layers of the inguinal region, is precisely the reverse of that order in which these layers present in the dissection from before backwards. The innermost layer of the inguinal region is the peritonaeum, and from this membrane the intestine, when about to protrude, derives its first covering. This covering constitutes the hernial sac. Almost all varieties of inguinal herniae are said to be enveloped in a sac, or elongation of the peritonaeum. This is accounted as the general rule. The exceptions to the rule are mentioned as occurring in the following modes: 1st, the caecum and sigmoid flexure of the colon, which are devoid of mesenteries, and only partially covered by the peritonaeum, may slip down behind this membrane, and become hernial; 2nd, the inguinal part of the peritonaeum may suffer rupture, and allow the intestine to protrude through the opening. When a hernia occurs under either of these circumstances, it will be found deprived of a sac.

All the blood vessels and nerves of the abdomen lie external to the peritonaeum. Those vessels which traverse the abdomen on their way to the external organs course outside the peritonaeum; and at the places where they enter the abdominal parietes, the membrane is reflected from them. This disposition of the peritonaeum in respect to the spermatic and iliac vessels is exhibited in Plate 32.

The part of the peritonaeum which lines the inguinal parietes does not (in the normal state of the adult body) exhibit any aperture corresponding to that named the internal ring. The membrane is in this place, as elsewhere, continuous throughout, being extended over the ring, as also over other localities, where subjacent structures may be in part wanting. It is in these places, where the membrane happens to be unsupported, that herniae are most liable to occur. And it must be added, that the natural form of the internal surface of the groin is such as to guide the viscera under pressure directly against those parts which are the weakest.

The inner surface of the groin is divided into two pouches or fossae, by an intervening crescentic fold of the peritonaeum, which corresponds with the situation of the epigastric vessels. This fold is formed by the epigastric vessels and the umbilical ligament, which, being tenser and shorter than the peritonaeum, thereby cause this membrane to project. The outer fossa represents a triangular space, the apex of which is below, at P; the base being formed by the fibres of the transverse muscle above; the inner side by the epigastric artery; and the outer side by Poupart's ligament. The apex of this inverted triangle is opposite the internal ring. The inner fossa is bounded by the epigastric artery externally; by the margin of the rectus muscle internally; and by the os pubis and inner end of Poupart's ligament inferiorly. The inner fossa is opposite the external abdominal ring, and is known as the triangle of Hesselbach.

The two peritonaeal fossae being named external and internal, in reference to the situation of the epigastric vessels, we find that the two varieties of inguinal herniae which occur in these fossae are named external and internal also, in reference to the same part.

The *external* inguinal hernia, so called from its commencing in the outer peritonaeal fossa, on the outer side of the epigastric artery, takes a covering from the peritonaeum of this place, and pushes forward into the internal abdominal ring at the point marked P, Plate 32. In this place, the incipient hernia or bubonocele, covered by its sac, lies on the forepart of the spermatic vessels, and becomes invested by those same coverings which constitute the inguinal canal, through which these vessels pass. In this stage of the hernia, its situation in respect to the epigastric artery is truly external, and in respect to the spermatic vessels, anterior, while the protruded intestine itself is separated from actual contact with either of these vessels by its proper sac. The bubonocele, projecting through the internal ring at the situation marked F, (Plate 33,) midway between A, the anterior iliac spine, and I, the pubic spine, continues to increase in size; but as its further progress from behind directly forwards becomes arrested by the tense resisting aponeurosis of the external oblique muscle, *h,* it changes its course obliquely inwards along the canal, traversing this canal with the spermatic vessels, which still lie behind it, and, lastly, makes its exit at the external ring, H. The obliquity of this course, pursued by the hernia, from the internal to the external ring, has gained for it the name of *oblique* hernia. In this stage of the hernial protrusion, the only part of it which may be truly named external is the neck of its sac, F, for the elongated body, G, of the hernia lies now actually in front of the epigastric artery, P, and this vessel is separated from the anterior wall of the canal, H *h,* by an interval equal to the bulk of the hernia. While the hernia occupies the canal, F H, without projecting through the external ring, H, it is named "incomplete." When it has passed the external ring, H, so as to form a tumour of the size and in the situation of *f g*, it is named "complete." When, lastly, the hernia has extended itself so far as to occupy the whole length of the cord, and reach the scrotum, it is termed "scrotal hernia." These names, it will be seen, are given only to characterise the several stages of the one kind of hernia--viz., that which commences to form at a situation external to the epigastric artery, and, after following the course of the spermatic vessels through the inguinal canal, at length terminates in the scrotum.

The external inguinal hernia having entered the canal, P, (Plate 32,) at a situation immediately in front of the spermatic vessels, continues, in the several stages of its descent, to hold the same relation to these vessels through the whole length of the canal, even as far as the testicle in the scrotum. This hernia, however, when of long standing and large size, is known to separate the spermatic vessels from each other in such a way, that some are found to lie on its fore part--others to its outer side. However great may be the size of this hernia, even when it becomes scrotal, still the testicle is invariably found below it. This fact is accounted for by the circumstance, that the lower end of the spermatic envelopes is attached so firmly to the coats of the testicle as to prevent the hernia from either distending and elongating them to a level below this organ, or from entering the cavity of the tunica vaginalis.

The external form of inguinal hernia is, comparatively speaking, but rarely seen in the female. When it does occur in this sex, its position, investments, and course through the inguinal canal, where it accompanies the round ligament of the uterus, are the same as in the male. When the hernia escapes through the external abdominal ring of the female groin, it is found to lodge in the labium pudendi. In the male body, the testicle and spermatic cord, which have carried before them investments derived from all the layers of the inguinal region, have, as it were, already marked out the track to be followed by the hernia, and prepared for it its several coverings. The muscular parietes of the male inguinal region, from which the loose cremaster muscle has been derived, have by this circumstance become weakened, and hence the more frequent occurrence of external inguinal hernia in the male. But in the female, where no such process has taken place, and where a cremaster does not exist at the expense of the internal oblique and transverse muscles, the inguinal parietes remain more compact, and are less liable to suffer distention in the course of the uterine ligament.

The *internal* inguinal hernia takes its peritonaeal covering (the sac) from the inner fossa, Q R, Plate 32, internal to the epigastric artery, and forces directly forwards through the external abdominal ring, carrying investments from each of such structures as it meets with in this locality of the groin. As the external ring, H, Plate 34, is opposite the inner peritonaeal fossa, Q R, Plate 32, this hernia, which protrudes thus immediately from behind forwards, is also named *direct.* In this way these two varieties of hernia, (the external, Plate 33, and the internal, Plate 34,) though commencing in different situations, P and R, Plate 32, within the abdomen, arrive at the same place--viz., the external ring, H, Plates 33 and 34. The coverings of the internal hernia, Plate 34, though not derived exactly from the same locality as those which invest the cord and the external variety, are, nevertheless, but different parts of the same structures; these are, 1st, the peritonaeum, G, which forms its sac; 2nd, the pubic part of the fascia transversalis; 3rd, the conjoined tendon itself, or (according as the hernia may occur further from the mesial line) the cremaster, which, in common with the internal oblique and transverse muscles, terminates in this tendon; 4th, the external spermatic fascia, derived from the margins of the external ring; 5th, the superficial fascia and integuments.

The coverings of the internal inguinal hernia are (as to number) variously described by authors. Thus with respect to the conjoined tendon, the hernia is said, in some instances, to take an investment of this structure; in others, to pass through a cleft in its fibres; in others, to escape by its outer margin. Again, the cremaster muscle is stated by some to cover this hernia; by others, to be rarely met with, as forming one of its coverings; and by others, never. Lastly, it is doubted by some whether this hernia is even covered by a protrusion of the fascia transversalis in all instances. [Footnote]

[Footnote: Mr. Lawrence (Treatise on Ruptures) remarks, "How often it may be invested by a protrusion of the fascia transversalis, I cannot hitherto determine." Mr. Stanley has presented to St. Bartholomew's Hospital several specimens of this hernia invested by the fascia. Hesselbach speaks of the fascia as being always present. Cloquet mentions it as being present always, except in such cases as where, by being ruptured, the sac protrudes through it. Langenbeck states that the fascia is constantly protruded as a covering to this hernia: "Quia hernia inguinalis interna non in canalis abdominalis aperturam internam transit, tunicam vaginalem communem intrare nequit; parietem autem canalis abdominalis internum aponeuroticum, in quo fovea inguinalis interna, et qui ex adverso annulo abdominali est, ante se per annulum trudit." (Comment, ad illust. Herniarum, &c.) Perhaps the readiest and surest explanation which can be given to these differences of opinion may be had from the following remark:--"Culter enim semper has partes extricat, quae involucro adeo inhaerent, ut pro lubitu musculum (membranam) efformare queas unde magnam illam inter anatomicos discrepantiam ortam conjicio." (Camper. Icones Herniarum.)]

The variety in the number of investments of the internal inguinal hernia (especially as regards the presence or absence of the conjoined tendon and cremaster) appears to me to be dependent, 1st, upon the position whereat this hernia occurs; 2nd, upon the state of the parts through which it passes; and 3rd, upon the manner in which the dissection happens to be conducted.

The precise relations which the internal hernia holds in respect to the epigastric and spermatic vessels are also mainly dependent (as in the external variety) upon the situation where it traverses the groin. The epigastric artery courses outside the neck of its sac, sometimes in close connexion with this part--at other times, at some distance from it, according as the neck may happen to be wide and near the vessel, or narrow, and removed from it nearer to the median line. At the external ring, H, (Plate 34,) the sac of this hernia, *g*, protrudes on the inner side of the spermatic vessels, *f;* and the size of the hernia distending the ring, removes these vessels at a considerable interval from, I, the crista pubis. At the ring, H, (Plate 34,) the investments, *g f,* of the direct hernia are not *always* distinct from those of the oblique hernia, *g f,* (Plate 33); for whilst in both varieties the intestine and the spermatic vessels are separated from actual contact by the sac, yet it is true that the direct hernia, as well as the oblique, may occupy the inguinal canal. It is in relation to the epigastric artery alone that the direct hernia differs essentially from the oblique variety; for I find that both may be enclosed in the same structures as invest the spermatic vessels.

The external ring of the male groin is larger than that of the female; and this circumstance, with others of a like nature, may account for the fact, that the female is very rarely the subject of the direct hernia. In the male, the direct hernia is found to occur much less frequently than the oblique, and this we might, *a priori,* expect, from the anatomical disposition of the parts. But it is true, nevertheless, that the part where the direct hernia occurs is not defended so completely in some male bodies as it is in others. The conjoined tendon, which is described as shielding the external ring, is in some cases very weak, and in others so narrow, as to offer but little support to this part of the groin.

DESCRIPTION OF THE FIGURES OF PLATES 32, 33, & 34.

PLATE 32.

A. That part of the ilium which abuts against the sacrum.

B. The spine of the ischium.

C. The tuberosity of the ischium.

D. The symphysis pubis.

E. Situation of the anterior superior iliac spine.

F. Crest of the ilium.

G. Iliacus muscle.

H. Psoas magnus muscle supporting the spermatic vessels.

I. Transversalis muscle.

K. Termination of the sheath of the rectus muscle.

L1 L2 L3. The iliac, transverse and pelvic portions of the transversalis fascia.

M M. The peritonaeum lining the groin.

N. The epigastric vessels lying between the peritonaeum, M, and the transversalis fascia, L2. O. The umbilical ligament.

P. The neck of the sac of an external inguinal hernia formed before the spermatic vessels.

Q. An interval which occasionally occurs between the umbilical ligament and the epigastric artery.

R and Q. Situations where the direct inguinal hernia occurs when, as in this case, the umbilical ligament crosses the space named the internal fossa--the triangle of Hesselbach.

S. Lower part of the right spermatic cord.

T. The bulb of the urethra.

U. External iliac vein covered by the peritonaeum.

V. External iliac artery covered by the peritonaeum.

W. Internal iliac artery.

X. Common iliac artery.

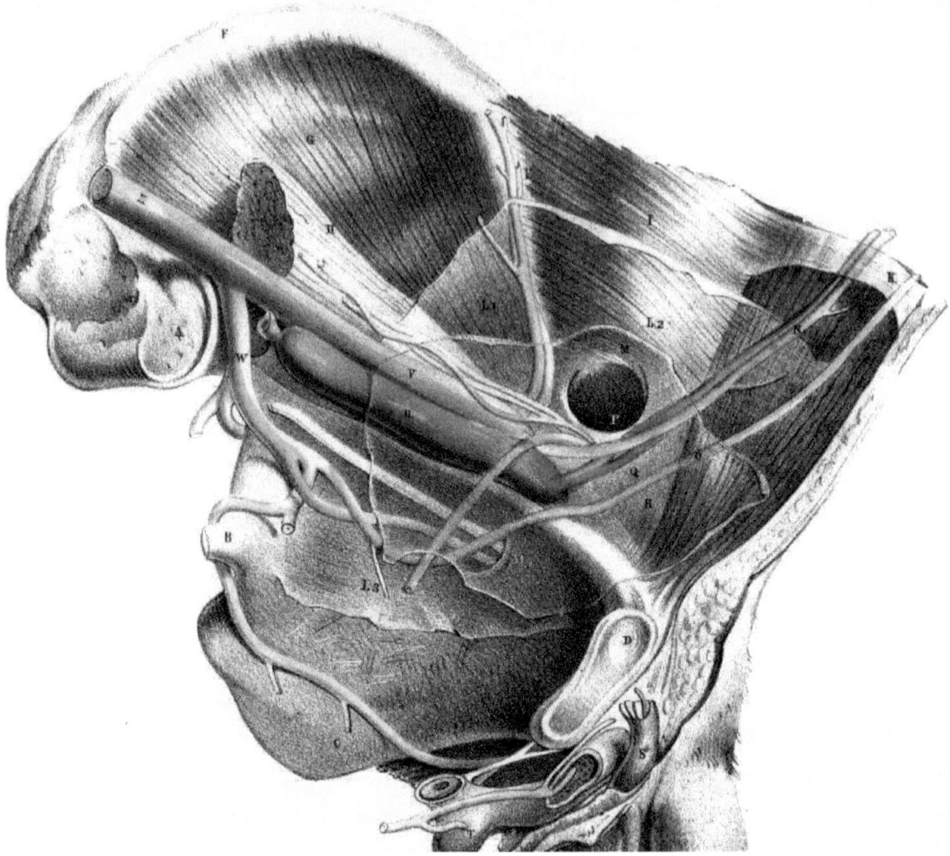

Plate 32

PLATE 33.--*The External Inguinal Hernia.*

A. Anterior iliac spinous process.

B. The umbilicus.

C. Fleshy part of the external oblique muscle; *c*, its tendon.

D. Fleshy part of the internal oblique muscle; *d,* its tendon.

E. Transversalis muscle; *e,* the conjoined tendon.

F *f.* The funnel-shaped sheath of the spermatic vessels covering the external hernia; upon it are seen the cremasteric fibres.

G *g*. The peritonaeal covering or sac of the external hernia within the sheath.

H. The external abdominal ring.

I. The crista pubis.

K *k*. The saphenous opening.

L. The saphena vein.

M. The femoral vein.

N. The femoral artery; *n*, its profunda branch.

O. The anterior crural nerve.

P. The epigastric vessels overlaid by the neck of the hernia.

Q Q. The sheath of the femoral vessels.

R. The sartorius muscle.

S. The iliacus muscle.

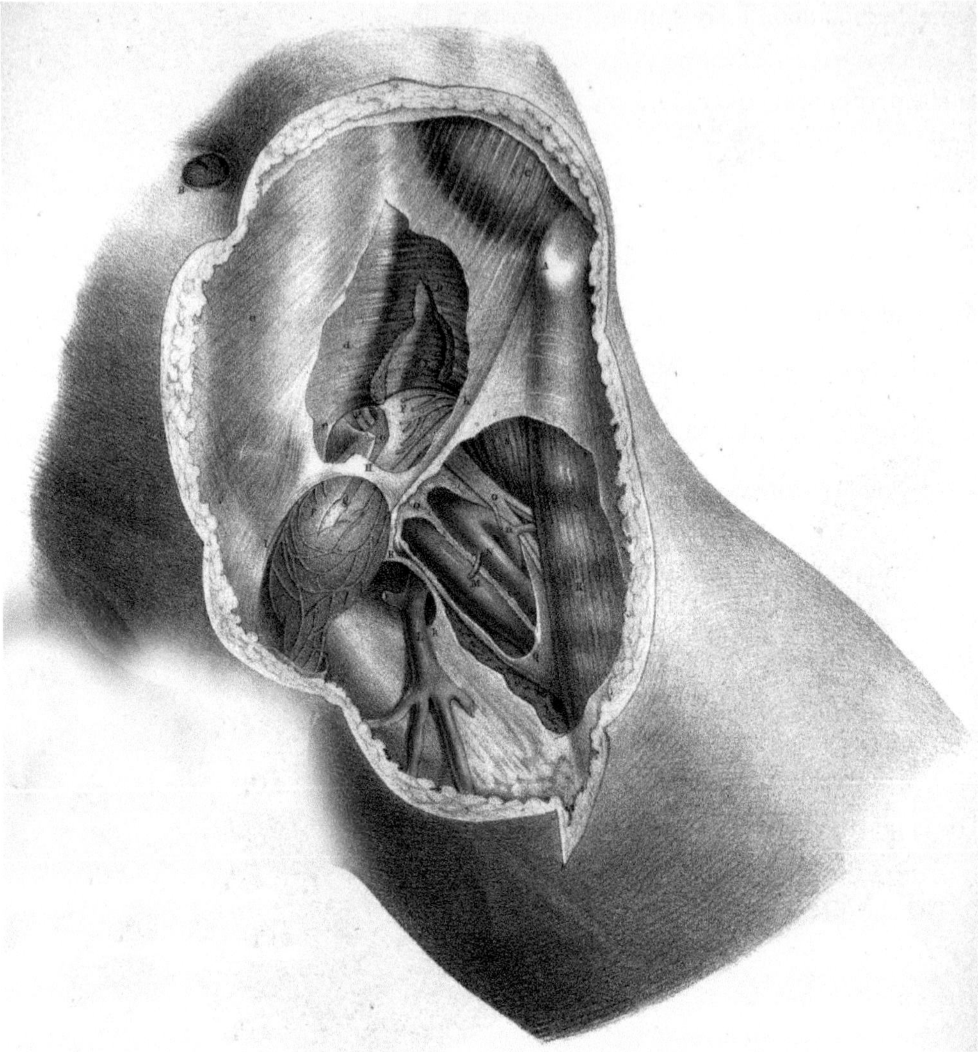

Plate 33

PLATE 34.--*The Internal Inguinal Hernia.*

The letters indicate the same parts as in Plate 33

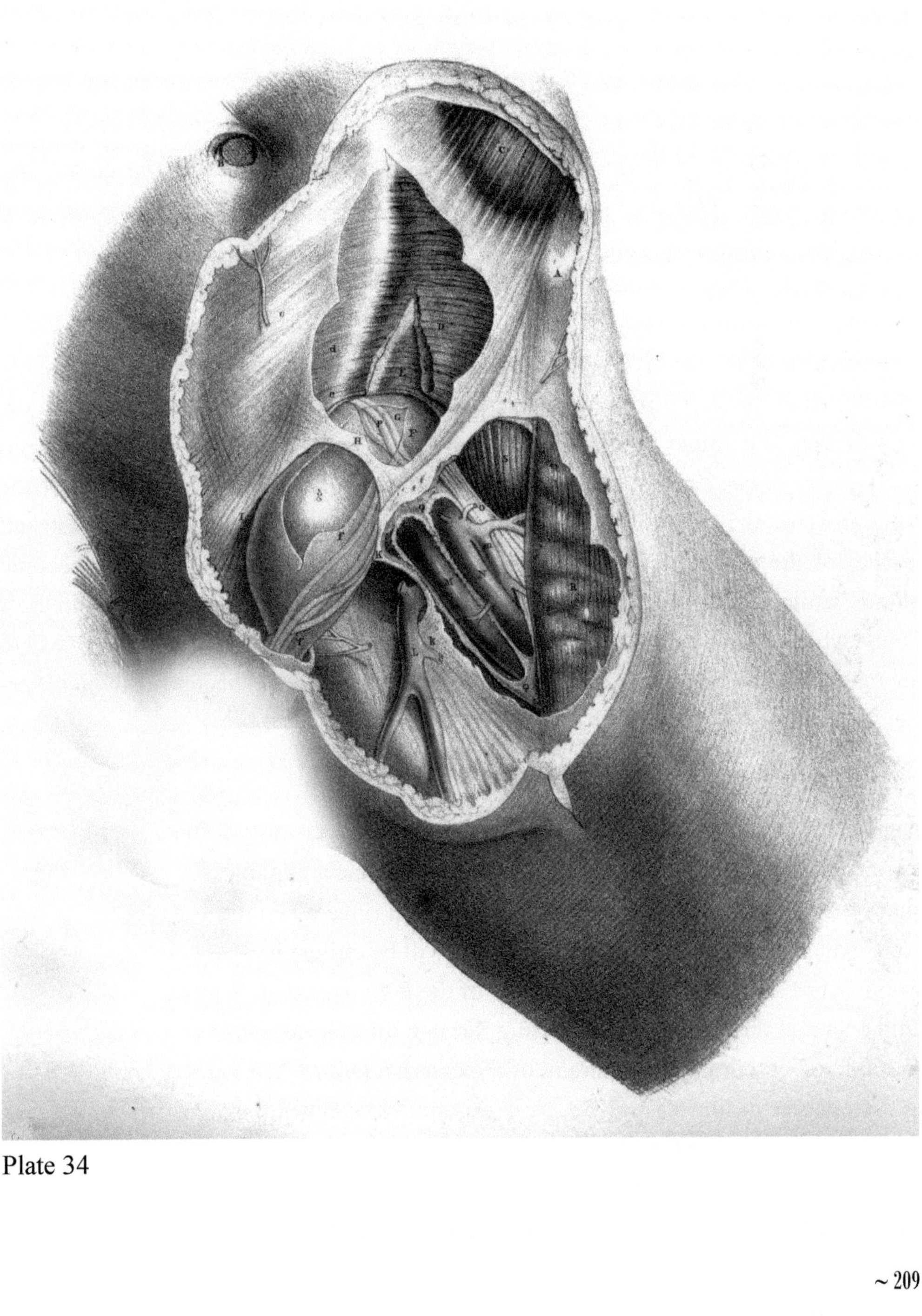

Plate 34

THE DISTINCTIVE DIAGNOSIS BETWEEN EXTERNAL AND INTERNAL INGUINAL HERNIAE, THE TAXIS, SEAT OF STRICTURE, AND THE OPERATION.

Both herniae compared as to position and structural characters. The co-existence of both rendering diagnosis difficult. The oblique changing to the direct hernia as to position, but not in relation to the epigastric artery. The taxis performed in reference to the position of both as regards the canal and abdominal rings. The seat of stricture varying. The sac. The lines of incision required to avoid the epigastric artery. Necessity for opening the sac.

A comparison of the relative position of these two varieties of herniae is in ordinary cases the chief means by which we can determine their distinctive diagnosis; but oftentimes they are found to exhibit such an interchange of characters, that the name *direct* or *oblique* can no longer serve to distinguish between them. The nearer the one approaches the usual place of the other, the more likely are they to be mistaken the one for the other. An internal hernia may enter the inguinal canal, and become oblique; while an external hernia, though occupying the canal, may become direct. It is only when these herniae occur at the situations commonly described, and where they manifest their broadest contrast, that the following diagnostic signs can be observed.

The external bubonocele, H, Plate 37, G, Plate 38, when recently formed, may be detected at a situation midway between the iliac and pubic spinous processes, where it has entered the internal ring. When the hernia extends itself from this part, its course will be obliquely inwards, corresponding with the direction of the inguinal canal. While it still occupies the canal without passing through the external ring, it is rendered obscure by the restraint of the external oblique tendon; but yet a degree of fulness may be felt in this situation. When the hernia has passed the external ring, T, Plate 36, it dilates considerably, and assumes the form of an oblong swelling, H, Plate 36, behind which the spermatic vessels are situated. When it has become scrotal, the cord will be found still on its posterior aspect, while the testicle itself occupies a situation directly below the swelling.

The internal hernia, H, Plate 38, also traverses the external ring, T, where it assumes a globular shape, and sometimes projects so far inwards, over the pubes, C, as to conceal the crista of this bone. As the direction of this hernia is immediately from behind forwards, the inguinal canal near the internal ring is found empty, unswollen. The cord, Q, lies external to and somewhat over the fore part of this hernia; and the testicle does not occupy a situation exactly beneath the fundus of the sac, (as it does in the external hernia,) but is found to be placed either at its fore part or its outer side. This difference as to the relative position of the cord and testicle in both these forms of herniae, is accounted for under the supposition that whilst the external variety descends inside the sheaths of the inguinal canal, the internal variety does not. But this statement cannot apply to all cases of internal hernia, for this also occasionally enters the canal. Both forms of inguinal herniae may exist at the same time on the same side: the external, G, Plate 38, being a bubonocele, still occupying the inguinal canal; while the internal, H, protrudes through the external ring, T, in the usual way. In this form of hernia--a compound of the oblique and direct--while the parts remain still covered by the integuments, it must be difficult to tell its nature, or to distinguish any mark by which to diagnose the case from one of the external variety, H, Plate 36, which, on entering the canal at the internal ring, protrudes at the external ring. In both cases, the swelling produced in the groin must be exactly of the same size and shape. The epigastric artery in the case where the two herniae co-exist lies between them, holding in its usual position with respect to each when occurring separately--that is, on the outer side of the internal hernia, H, and on the inner side of the external one, G; and the external hernia, G, not having descended the canal as far as the external ring, T, allows the internal hernia, H, to assume its usual position with respect to the cord, Q. [Footnote]

[Footnote: Cases of this double hernia (external and internal) have been met with by Wilmer, Arnaud, Sandifort, Richter, and others. A plurality of the same variety of hernia may also occur on the same side. A complete and incomplete external inguinal hernia existing in the one groin, is recorded by Mr. Aston Key in his edition of Sir Astley Cooper's work on Hernia. Sir Astley Cooper states his having met with three internal inguinal herniae in each inguinal region. (Ing. et Congenit. Hernire.)]

Returning, however, to the more frequent conditions of inguinal hernia--viz., those in which either the direct or the oblique variety occurs alone--it should be remembered that a hernia originally oblique, H, Plates 35 and 37, may, when of long standing, and having attained a large size, destroy, by its gravitation, the obliquity of the inguinal canal to such a degree as to bring the internal, H, Plate 35, opposite to the external ring, as at I, and thereby exhibit all the appearance of a hernia originally direct, I, Plate 37. In such a case, the epigastric artery, F, which lies on the outer side of the neck of a truly direct hernia, I, Plate 37, will be found to course on the inner side, G, of the neck of this false-seeming direct hernia, I, Plate 35.

In the trial made for replacing the protruded bowel by the *taxis*, two circumstances should be remembered in order to facilitate this object: 1st, the abdominal parietes should be relaxed by supporting the trunk forward, and at the same time flexing the thigh on the trunk; 2nd, as every complete hernial protrusion becomes distended more or less beyond the seat of stricture--wherever this may happen to be--its reduction by the taxis should be attempted, with gradual, gentle, equable pressure, so that the sac may be first emptied of its fluid. That part of the hernia which protruded last should be replaced first. The direction in which the hernia protrudes must always determine the direction in which it is to be reduced. If it be the external or oblique variety, the viscus is to be pushed upwards, outwards, and backwards; if it be the internal or direct variety, it is to be reduced by pressure, made upwards and backwards. Pressure made in this latter direction will serve for the reduction of that hernia which, from being originally external and oblique, has assumed the usual position of the internal or direct variety.

The seat of the stricture in an external inguinal hernia is found to be situated either at the internal ring, corresponding to the neck of the sac, or at the external ring. Between these two points, which "bound the canal," and which are to be regarded merely as passive agents in causing stricture of the protruding bowel, the lower parts of the transversalis and internal oblique muscles embrace the herniary sac, and are known at times to be the cause of its active strangulation or spasm.

The seat of stricture in an internal hernia may be either at the neck of its sac, I, Plate 37, or at the external ring, T, Plate 38; and according to the locality where this hernia enters the inguinal wall, the nature of its stricture will vary. If the hernia pass through a cleft in the conjoined tendon, *f,* Plate 38, this structure will constrict its neck all around. If it pass on the outer margin of this tendon, then the neck of the sac, bending inwards in order to gain the external ring, will be constricted against the sharp resisting edge of the tendon. Again, if the hernia enter the inguinal wall close to the epigastric artery, it will find its way into the inguinal canal, become invested by the structures forming this part, and here it may suffer active constriction from the muscular fibres of the transverse and internal oblique or their cremasteric parts. The external ring may be considered as always causing some degree of pressure on the hernia which passes through it.

In both kinds of inguinal herniae, the neck of the sac is described as being occasionally the seat of stricture, and it certainly is so; but never from a cause originating in itself *per se*, or independently of adjacent structures. The form of the sac of a hernia is influenced by the parts through which it passes, or which it pushes and elongates before itself. Its neck, H, Plate 37, is narrow at the internal ring of the fascia transversalis, because this ring is itself narrowed; it is again narrowed at the external ring, T, Plate 36, from the same cause. The neck of the sac of a direct hernia, I, Plate 37, being formed in the space of the separated fibres of the conjoined tendon, or the pubic part of the transversalis fascia, while the sac itself passes through the resisting tendinous external ring, is equal to the capacities of these outlets. But if these constricting outlets did not exist, the neck of the sac would be also wanting. When, however, the neck of the sac has existed in the embrace of these constricting parts for a considerable period--when it suffers inflammation and undergoes chronic thickening--then, even though we liberate the stricture of the internal ring or the external, the neck of the sac will be found to maintain its narrow diameter, and to have become itself a real seat of stricture. It is in cases of this latter kind of stricture that experience has demonstrated the necessity of opening the sac (a proceeding otherwise not only needless, but objectionable) and dividing its constricted neck.

The fact that the stricture may be seated in the neck of the sac independent of the internal ring, and also that the duplicature of the contained bowel may be adherent to the neck or other part of the interior, or that firm bands of false membrane may exist so as to constrict the bowel within the sac, are circumstances which require that this should be opened, and the state of its contained parts examined, prior to the replacement of the bowel in the abdomen. If the bowel were adherent to the neck of the sac, we might, when trying to reduce it by the taxis, produce visceral invagination; or while the stricture is in the neck of the sac, if we were to return this and its contents *en masse* (the "reduction en bloc") into the abdomen, it is obvious that the bowel would be still in a state of strangulation, though free of the internal ring or other opening in the inguinal wall.

The operation for the division of the stricture by the knife is conducted in the following way: an incision is to be made through the integuments, adipous membrane, and superficial fascia, of a length and depth sufficient to expose the tendon of the external oblique muscle for an inch or so above the external ring; and the hernia for the same extent below the ring. The length of the incision will require to be varied according to circumstances, but its direction should be oblique with that of the hernia itself, and also over the centre of its longitudinal axis, so as to avoid injuring the spermatic vessels. If the constriction of the hernia be caused by the external ring, a director is to be inserted beneath this part, and a few of its fibres divided. But when the stricture is produced by either of the muscles which lie beneath the aponeurosis of the external oblique, it will be necessary to divide this part in order to expose and incise them.

When the thickened and indurated neck of the sac is felt to be the cause of the strangulation, or when the bowel cannot be replaced, in consequence of adhesions which it may have contracted with some part of the sac, it then becomes necessary to open this envelope. And now the position of the epigastric artery is to be remembered, so as to avoid wounding it in the incision about to be made through the constricted neck of the sac. The artery being situated on the *inner side of the neck* of the sac of an oblique hernia, requires the incision to be made *outwards* from the *external side of the neck*; whereas in the direct hernia, the artery being on its *outer side*, the incision should be conducted *inwards* from *the inner side of the neck*. But as the external or oblique hernia may by its weight, in process of time, gravitate so far inwards as to assume the position and appearance of a hernia originally direct and internal, and as by this change of place the oblique hernia, becoming direct as to position, does not at the same time become internal in respect to the epigastric artery,--for this vessel, F, Plate 35, has been borne inwards to the place, G, where it still lies, internal to the neck of the sac, and since, moreover, it is very difficult to diagnose a case of this kind with positive certainty, it is therefore recommended to incise the stricture at the neck of the sac in a line carried *directly upwards*. (Sir Astley Cooper.) It will be seen, however, on referring to Plates 32, 33, 34, 35, 36, 37, & 38, that an incision carried *obliquely upwards* towards the umbilicus would be much more likely to avoid the epigastric artery through all its varying relations.

DESCRIPTION OF THE FIGURES OF PLATES 35, 36, 37, & 38.

PLATE 35.

A. Anterior superior spine of the ilium; *a*, indicates the situation of the middle of Poupart's ligament.

B. Symphysis pubis.

C. Rectus abdominis muscle covered by the fascia transversalis.

D. The peritonaeum lining the groin.

E. The situation of the conjoined tendon resisting the further progress of the external hernia gravitating inwards.

F. A dotted line indicating the original situation of the epigastric artery in the external hernia.

G. The new position assumed by the epigastric artery borne inwards by the weight of the old external hernia.

H. The original situation of the neck of the sac of the external hernia.

I. The new situation assumed by the neck of the sac of an old external hernia which has gravitated inwards from its original place at H.

K. The external iliac vein covered by the peritonaeum.

L. The external iliac artery covered by the peritonaeum and crossed by the spermatic vessels.

M. The psoas muscle supporting the spermatic vessels and the genito-crural nerve.

N. The iliacus muscle.

O. The transversalis fascia lining the transverse muscle.

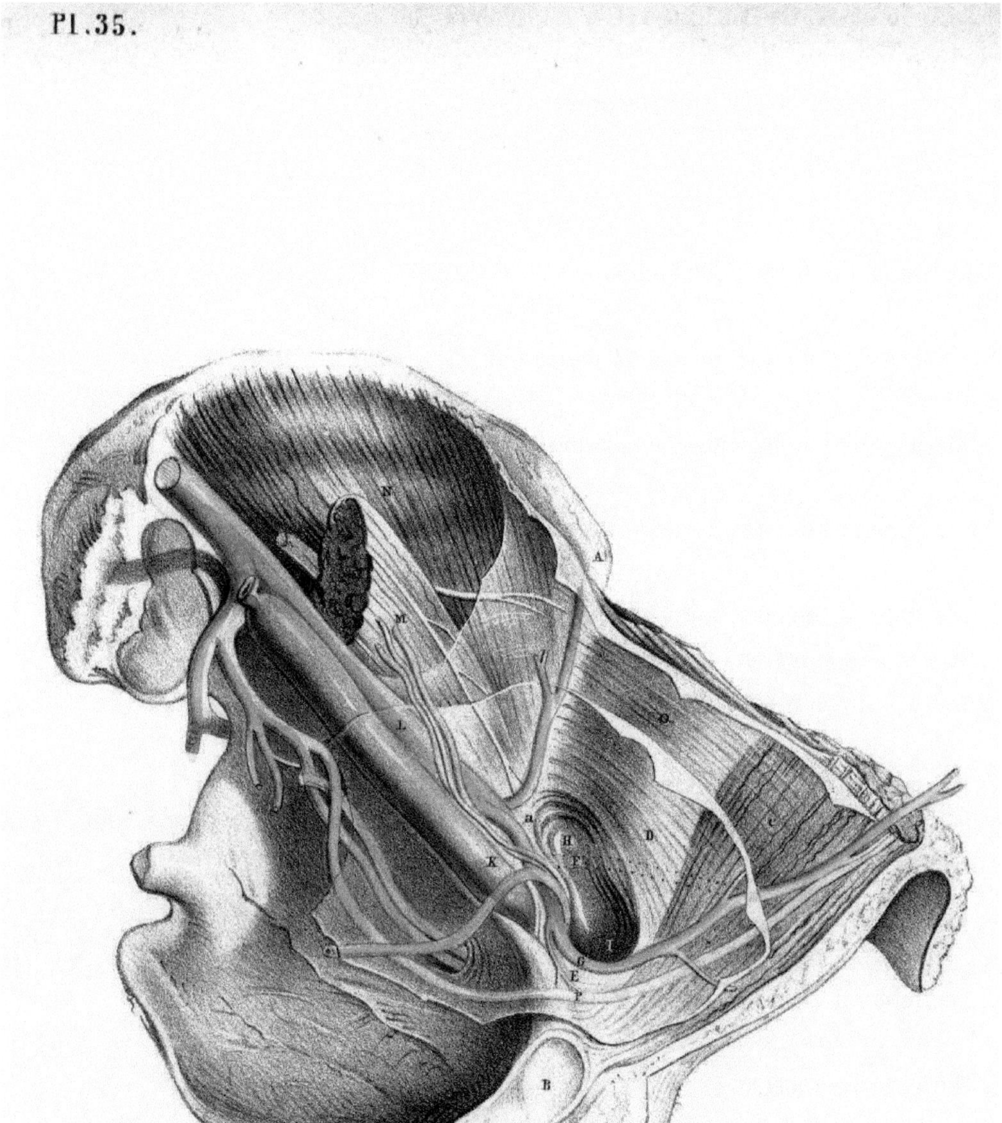

Plate 35

PLATE 36.--AN ANTERIOR VIEW OF PLATE 35.

A. Anterior superior iliac spinous process.

B. The navel.

C. The situation of the crista pubis.

D. The external oblique muscle; *d*, its tendon.

E. Internal oblique muscle; *e*, its tendon, covering the rectus muscle.

F. Lower part of the transverse muscle; *f*, the conjoined tendon.

G. The transversalis fascia investing the upper part of the hernial sac; *g*, the original situation of the epigastric artery internal to this hernia; *g**, the new situation of the artery pushed inwards.

H. The hernial sac, invested by *h*, the elongation of the fascia transversalis, or funnel-shaped sheath.

I. The femoral artery.

K. The femoral vein.

L. The sartorius muscle.

M. Iliac part of the fascia lata joining Poupart's ligament.

N. Pubic part of the fascia lata.

O. Saphena vein.

P P. Falciform margin of the saphenous opening.

Q. See Plate 38.

R. Sheath of the femoral vessels.

S. Anterior crural nerve.

T. The external ring.

Pl. 36.

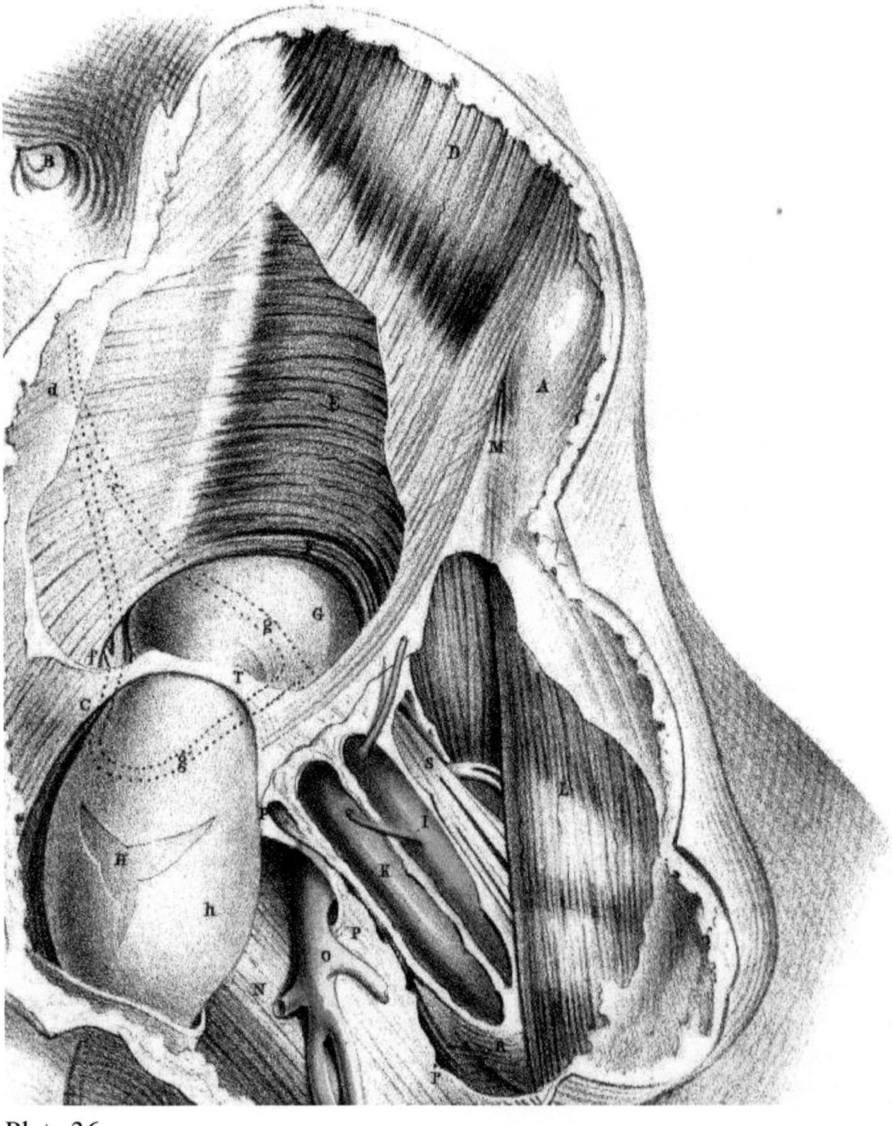

Plate 36

PLATE 37.

All the letters except the following indicate the same parts as in Plate 35.

F. The epigastric artery passing between the two hernial sacs

G. The umbilical ligament.

H. The neck of the sac of the external hernia.

I. The neck of the sac of the internal hernia.

Pl.37.

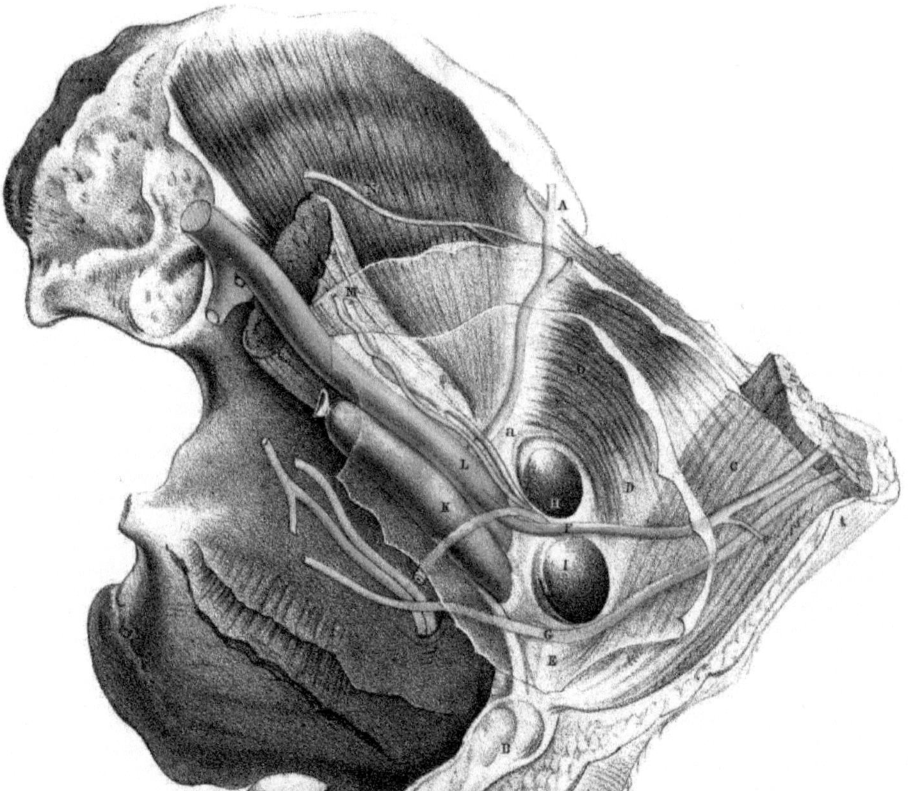

Plate 37

PLATE 38.--AN ANTERIOR VIEW OF PLATE 37.

All the letters, with the exception of the following, refer to the same parts as in Plate 36.

G. The funnel-shaped elongation of the fascia transversalis receiving *g*, the sac of the external bubonocele.

H. The sac of the internal inguinal hernia invested by *h,* the transversalis fascia.

Q. The spermatic vessels lying on the outer side of H, the direct inguinal hernia.

Pl.38.

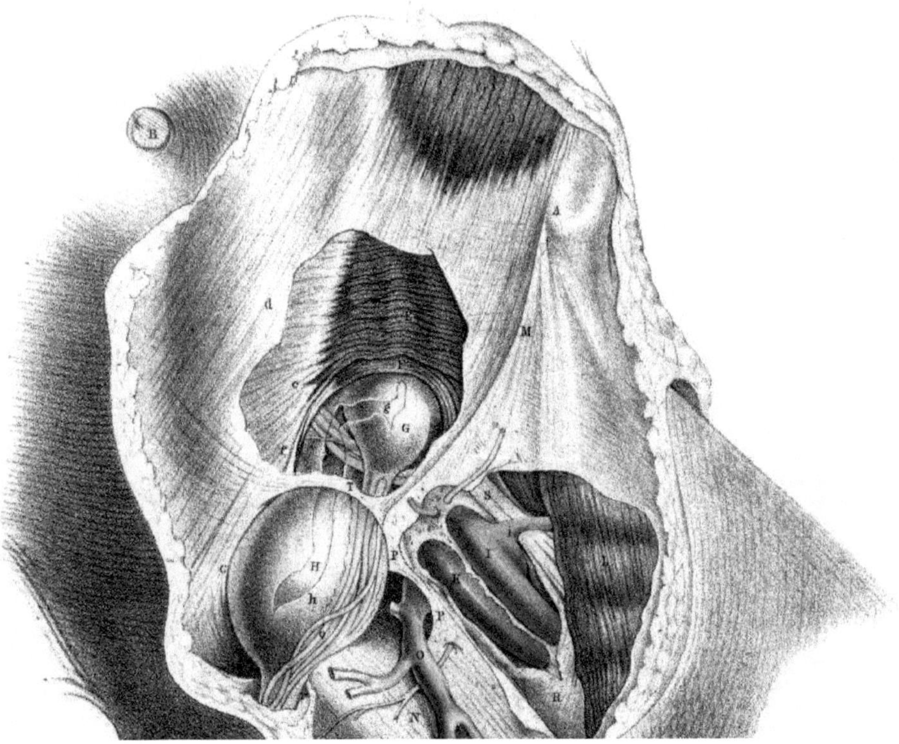

Plate 38

DEMONSTRATIONS OF THE NATURE OF CONGENITAL AND INFANTILE INGUINAL HERNIAE, AND OF HYDROCELE.

Descent of the testicle. The testicle in the scrotum. Isolation of its tunica vaginalis. The tunica vaginalis communicating with the abdomen. Sacculated serous spermatic canal. Hydrocele of the isolated tunica vaginalis. Congenital hernia and hydrocele. Infantile hernia. Oblique inguinal hernia. How formed and characterized.

PLATE 39. Fig. 1--*The descent of the testicle from the loins to the scrotum.*--The foetal abdomen and scrotum form one general cavity, and are composed of parts which are structurally identical. The cutaneous, fascial, muscular, and membranous layers of the abdominal parietes are continued into those of the scrotum. At the fifth month of foetal life, the testicle, 3, is situated in the loins beneath the kidney, 2. The testicle is then numbered amongst the abdominal viscera, and, like these, it is developed external to the peritonaeal membrane, which forms an envelope for it. At the back and sides of the testicle, where the peritonaeum is reflected from it, a small membranous fold or mesentery (mesorchium, *Seiler)* is formed, and between the layers of this the nerves and vessels enter the organ, the nerves being derived from the neighbouring sympathetic ganglia (aortic plexus), while the arteries and veins spring directly from the main abdominal bloodvessels. It being predetermined that the testicle, 3, should migrate from the loins to the scrotum, 6 *a,* 7, at a period included between the sixth and ninth month, certain structural changes are at this time already effected for its sure and easy passage. By the time that the testis, 5, is about to enter the internal inguinal ring, 6 *a*, (seventh or eighth month,) a process or pouch of the peritonaeal membrane (processus vaginalis) has already descended through this aperture into the scrotum, and the testicle follows it.

The descent of the testis is effected by a very slow and gradual process of change. (Tout va par degres dans la nature, et rien par sauts.--*Bonnet.*) But how, or by what distinct and active structural agent, this descent is effected, or whether there does exist, in fact, any such agent as that which anatomists name "gubernaculum testis," are questions which appear to me by no means settled.[Footnote]

[Footnote: Dr. Carpenter (Principles of Human Physiology) remarks, that "the cause of this descent is not very clear. It can scarcely be due merely, as some have supposed, to the contraction of the gubernaculum, since that does not contain any fibrous structure until after the lowering of the testis has commenced." Dr. Sharpey (Quain's Anatomy, 5th edition) observes, that "the office of the gubernaculum is yet imperfectly understood." The opinions of these two distinguished physiologists will doubtless be regarded as an impartial estimate of the results of the researches prosecuted in reference to these questions by Haller, Camper, Hunter, Arnaud, Lobstein, Meckel, Paletta, Wrisberg, Vicq d'Azyr, Brugnone, Tumiati, Seiler, Girardi, Cooper, Bell, Weber, Carus, Cloquet, Curling, and others. From my own observations, I am led to believe that no such muscular structure as a gubernaculum exists, and therefore that the descent of the testis is the effect of another cause. Leaving these matters, however, to the consideration of the physiologist, it is sufficient for the surgeon to know that the testis in its transition derives certain coverings from the parietes of the groin, and that a communication is thereby established between the scrotal and abdominal cavities.]

The general lining membrane of the foetal abdomen is composed of two layers--an outer one of fibrous, and an inner one of serous structure. Of these two layers, the abdominal viscera form for themselves a double envelope. [Footnote] The testis in the loins has a covering from both membranes, and is still found to be enclosed by both, even when it has descended to the scrotum. The two coverings of fibro-serous structure which surrounded the testis in the loins become respectively the tunica albuginea and tunica vaginalis when the gland occupies the scrotal cavity.

[Footnote: Langenbeck describes the peritonaeum as consisting of two layers; one external and fibrous, another internal and serous. By the first, he means, I presume, that membrane of which the transversalis and iliac fasciae are parts. (See Comment. de Periton. Structura, &c.)]

Fig. 1

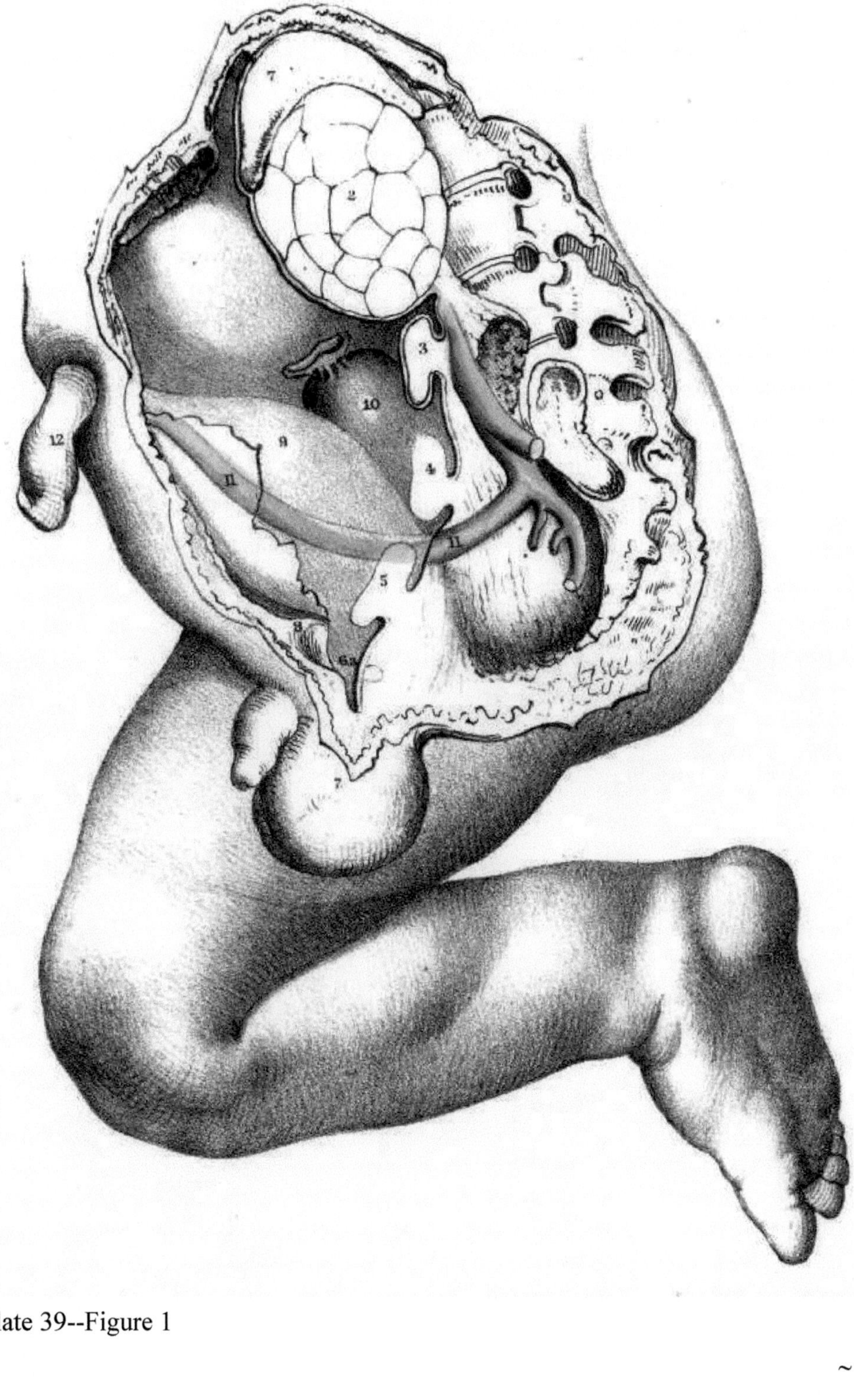

Plate 39--Figure 1

PLATE 39, Fig. 2.--*The testicle in the scrotum*.--When the testicle, 5, descends into the scrotum, 7, which happens in general at the time of birth, the abdomino-scrotal fibro-serous membrane, 6 *a*, 6 *d*, is still continuous at the internal ring, 6 *b*. From this point downwards, to a level with the upper border of the testicle, the canal of communication between the scrotal cavity and the abdomen becomes elongated and somewhat constricted. At this part, the canal itself consists, like the abdominal membrane above and the scrotal membrane below, of a fibrous and serous layer, the latter enclosed within the former. The serous lining of this canal is destined to be obliterated, while the outer fibrous membrane is designed to remain in its primitive condition. When the serous canal contracts and degenerates to the form of a simple cord, it leaves the fibrous canal still continuous above with the fibrous membrane (transversalis fascia) of the abdomen, and below with the fibrous envelope (tunica albuginea) of the testis; and at the adult period, this fibrous canal is known as the internal spermatic sheath, or infundibuliform fascia enclosing the remains of the serous canal, together with the spermatic vessels, &c.

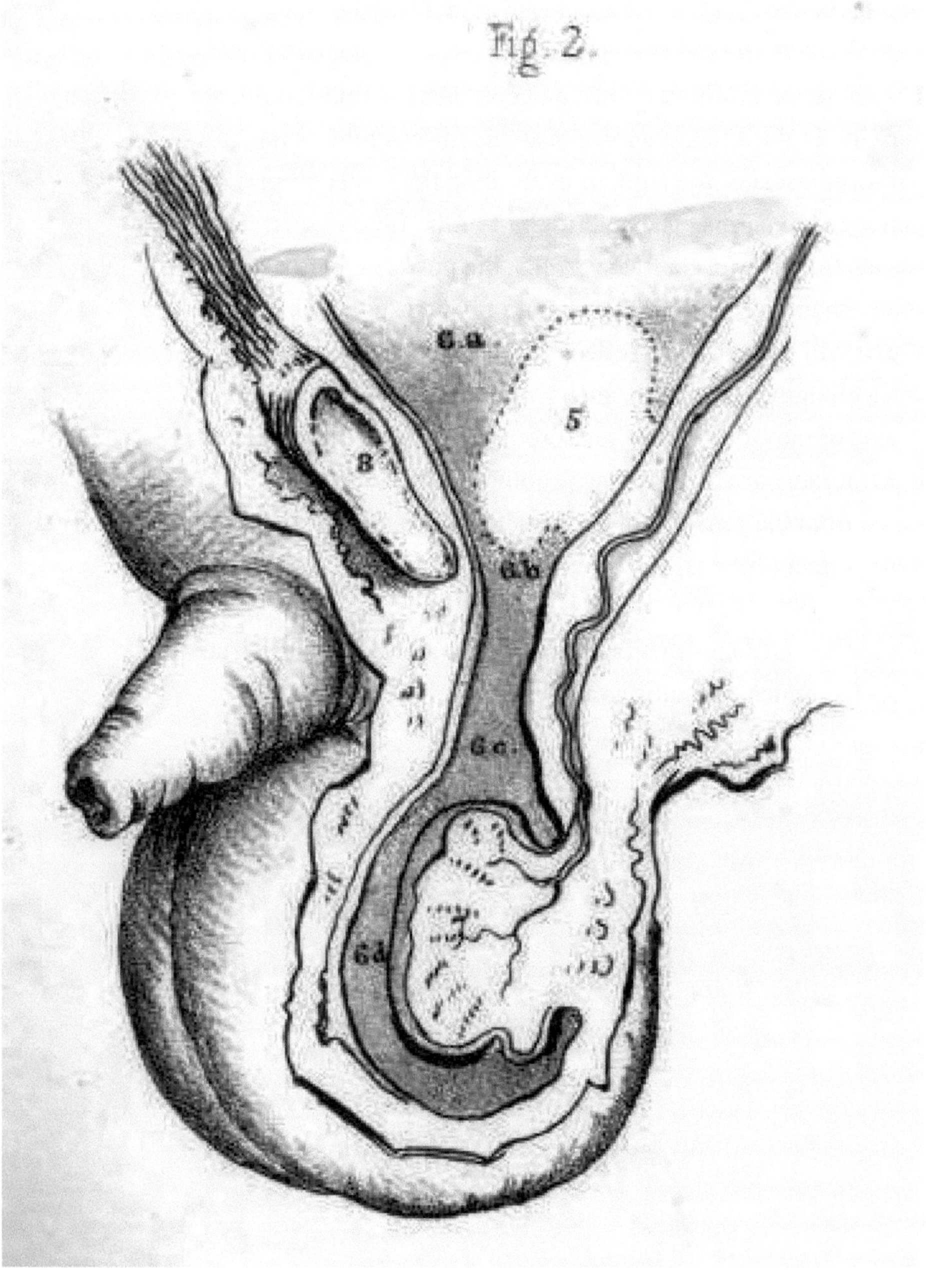

Plate 39--Figure 2

PLATE 39, Fig. 3.--*The serous tunica vaginalis is separated from the peritonaeum.*--When the testicle, 7, has descended to the scrotum, the serous tube or lining of the inguinal canal and cord, 6 *b*, 6 *c*, closes and degenerates into a simple cord, (infantile spermatic cord,) and thereby the peritonaeal sac, 6 *a*, becomes distinct from the serous tunica vaginalis, 6 *d*. But the fibrous tube, or outer envelope of the inguinal canal, remains still pervious, and continues in this condition throughout life. In the adult, we recognise this fibrous tube as the infundibuliform fascia of the cord, or as forming the fascia propria of an external inguinal hernia. The anterior part of the fibrous spermatic tube descends from the fascia transversalis; the posterior part is continuous with the fascia iliaca. In relation to the testicle, the posterior part will be seen to be reflected over the body of the gland as the tunica albuginea, while the anterior part blends with the cellular tissue of the front wall of the scrotum. The tunica vaginalis, 6 *d*, is now traceable as a distinct sac,[Footnote] closed on all sides, and reflected from the fore part of the testicle, above and below, to the posterior aspect of the front wall of the scrotum.

[Footnote: Mr. Owen states that the Chimpanzee alone, amongst brute animals, has the tunica vaginalis as a distinct sac.]

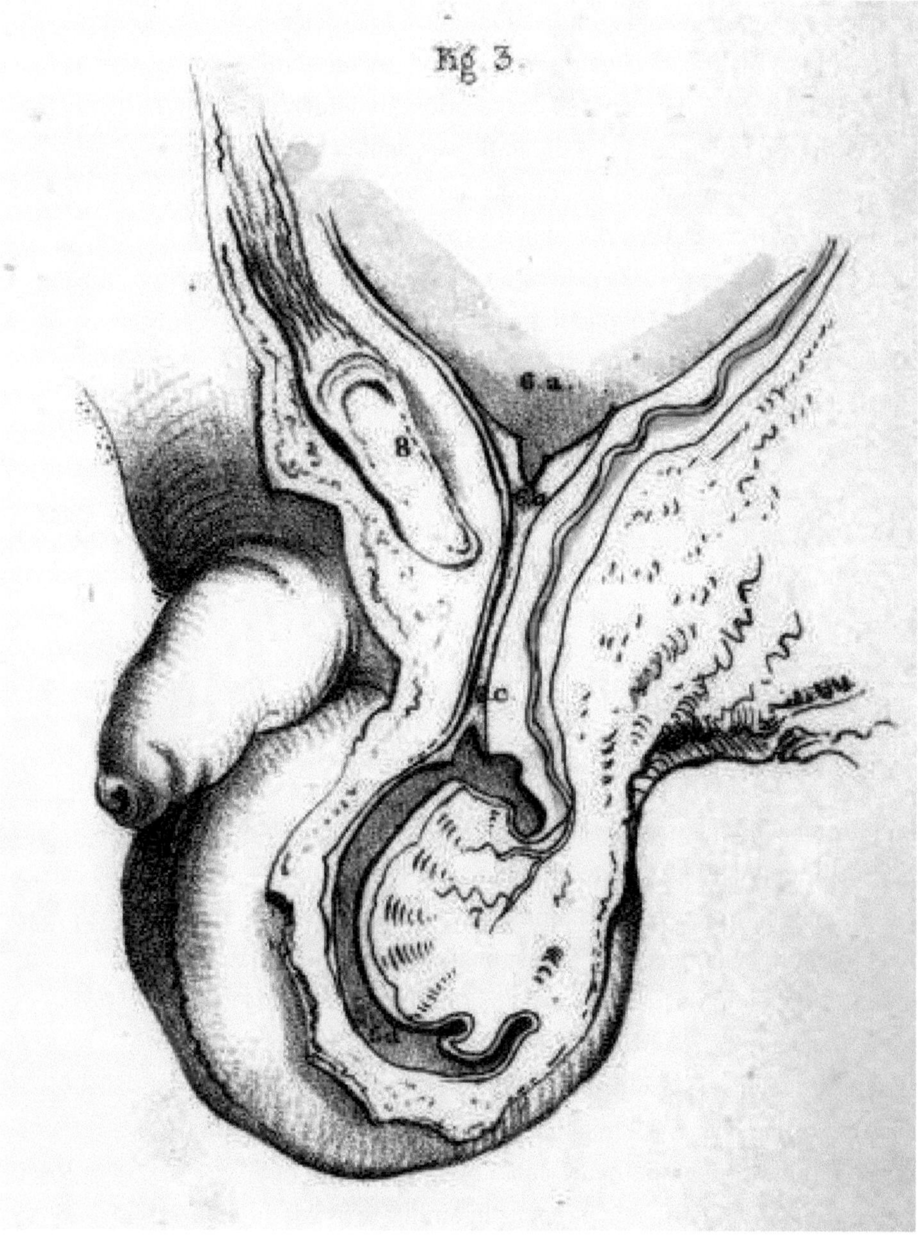

Plate 39--Figure 3

PLATE 40, Fig. 1.--*The abdomino-scrotal serous lining remains continuous at the internal ring, and a congenital hydrocele is formed.*--When the serous spermatic tube, 6 *b*, 6 *c*, remains pervious and continuous above with the peritonaeum, 6 *a*, and below with the serous tunica vaginalis, 6 *d*, the serous fluid of the abdomen will naturally gravitate to the most depending part--viz., the tunica vaginalis; and thus a hydrocele is formed. This kind of hydrocele is named congenital, owing to the circumstance that the natural process of obliteration, by which the peritonaeum becomes separated from the tunica vaginalis, has been, from some cause, arrested. [Footnote 1] As long as the canal of communication, 6 *b*, 6 *c*, between the tunica vaginalis, 6 *d*, and the peritonaeum 6 *a*, remains pervious, which it may be throughout life, this form of hydrocele is, of course, liable to occur. It may be diagnosed from diseased enlargements of the testicle, by its transparency, its fluctuation, and its smooth, uniform fulness and shape, besides its being of less weight than a diseased testis of the same size would be. It may be distinguished from the common form of hydrocele of the isolated tunica vaginalis by the fact, that pressure made on the scrotum will cause the fluid to pass freely into the general cavity of the peritonaeum. As the fluid distends the tunica vaginalis, 6 *c*, 6 *d*, in front of the testis, this organ will of course lie towards the back of the scrotum, and therefore, if it be found necessary to evacuate the fluid, the puncture may be made with most safety in front of the scrotum. If ascites should form in an adult in whom the tunica vaginalis still communicates with the peritonaeal sac, the fluid which accumulates in the latter membrane will also distend the former, and all the collected fluid may be evacuated by tapping the scrotum. When a hydrocele is found to be congenital, it must be at once obvious that to inject irritating fluids into the tunica vaginalis (the radical cure) is inadmissible. In an adult, free from all structural disease, and in whom a congenital hydrocele is occasioned by the gravitation of the ordinary serous secretion of the peritonaeum, a cure may be effected by causing the obliteration of the serous spermatic canal by the pressure of a truss. When a congenital hydrocele happens in an infant in whom the testicle, 5, Fig. 1, Plate 39, is arrested in the inguinal canal, [Footnote 2] if pressure be made on this passage with a view of causing its closure, the testicle will be prevented from descending.

[Footnote 1: The serous spermatic tube remains open in all quadrupeds; but their natural prone position renders them secure against hydrocele or hernial protrusion. It is interesting to notice how in man, and the most anthropo-morphous animals, where the erect position would subject these to the frequent accident of hydrocele or hernia, nature causes the serous spermatic tube to close.]

[Footnote 2: In many quadrupeds (the Rodentia and Monotremes) the testes remain within the abdomen. In the Elephant, the testes always occupy their original position beneath the kidneys, in the loins. Human adults are occasionally found to be "testi-conde;" the testes being situated below the kidneys, or at some part between this position and the internal inguinal ring. Sometimes only one of the testes descends to the scrotum.]

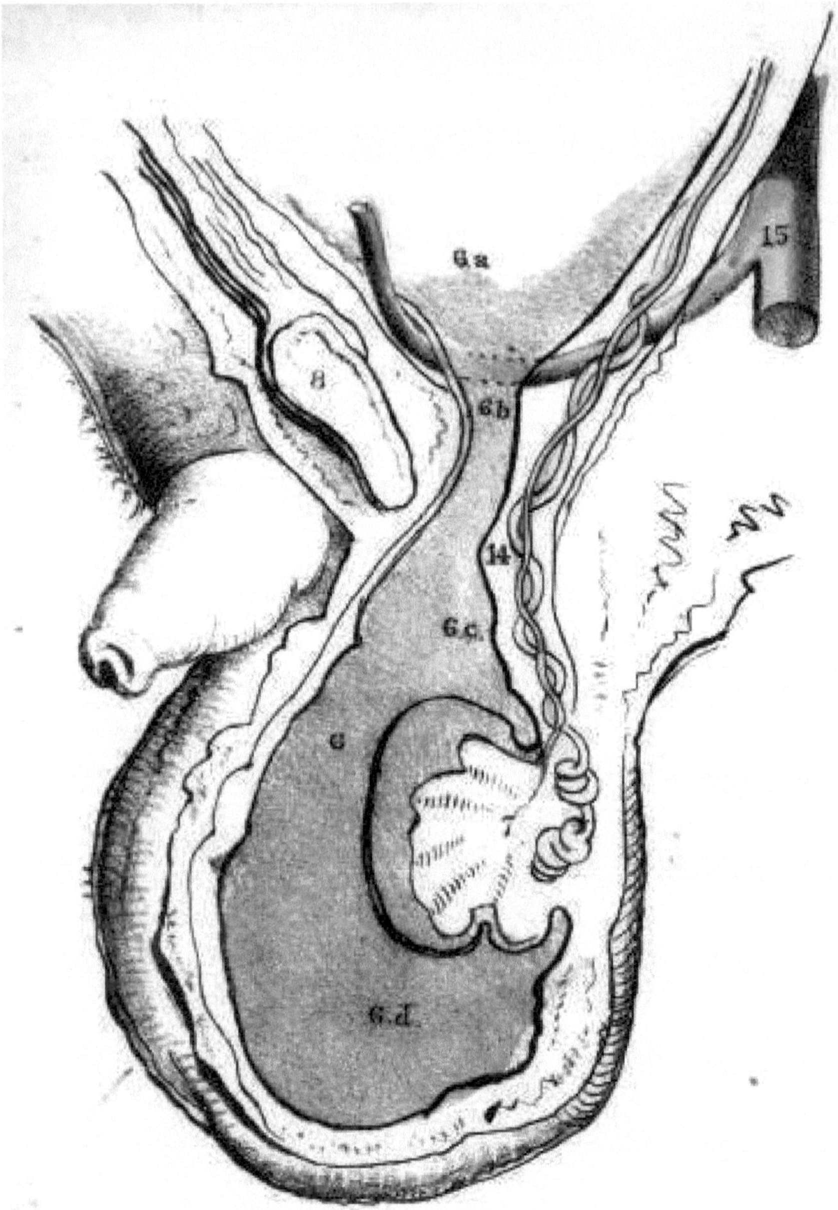

Plate 40--Figure 1.

PLATE 40, Fig. 2.--*The serous spermatic canal closes imperfectly, so as to become sacculated, and thus a hydrocele of the cord is formed.*--After the testicle, 7, has descended to the scrotum, the sides of the serous tube, or lining of the inguinal canal and cord, 6 *b*, 6 *c,* may become adherent at intervals; and the intervening sacs of serous membrane continuing to secrete their proper fluid, will occasion a hydrocele of the cord. This form of hydrocele will differ according to the varieties in the manner of closure; and these may take place in the following modes:--1st, if the serous tube close only at the internal ring, 6 *a*, while the lower part of it, 6 *b*, 6 *c*, remains pervious, and communicating with the tunica vaginalis, 6 *d*, a hydrocele will be formed of a corresponding shape; 2nd, if the tube close at the upper part of the testicle, 6 *c*, thus isolating the tunica vaginalis, 6 *d*, while the upper part, 6 *b*, remains pervious, and the internal ring, 6 *a*, open, and communicating with the peritonaeal sac, a hydrocele of the cord will happen distinct from the tunica vaginalis; or this latter may be, at the same time, distended with fluid, if the disposition of the subject be favourable to the formation of dropsy; 3rd, the serous tube may close at the internal ring, form sacculi along the cord, and close again at the top of the testicle, thus separating the tunica vaginalis from the abdomen, and thereby several isolated hydroceles may be formed. If in this condition of the parts we puncture one of the sacs for the evacuation of its contents, the others, owing to their separation, will remain distended.

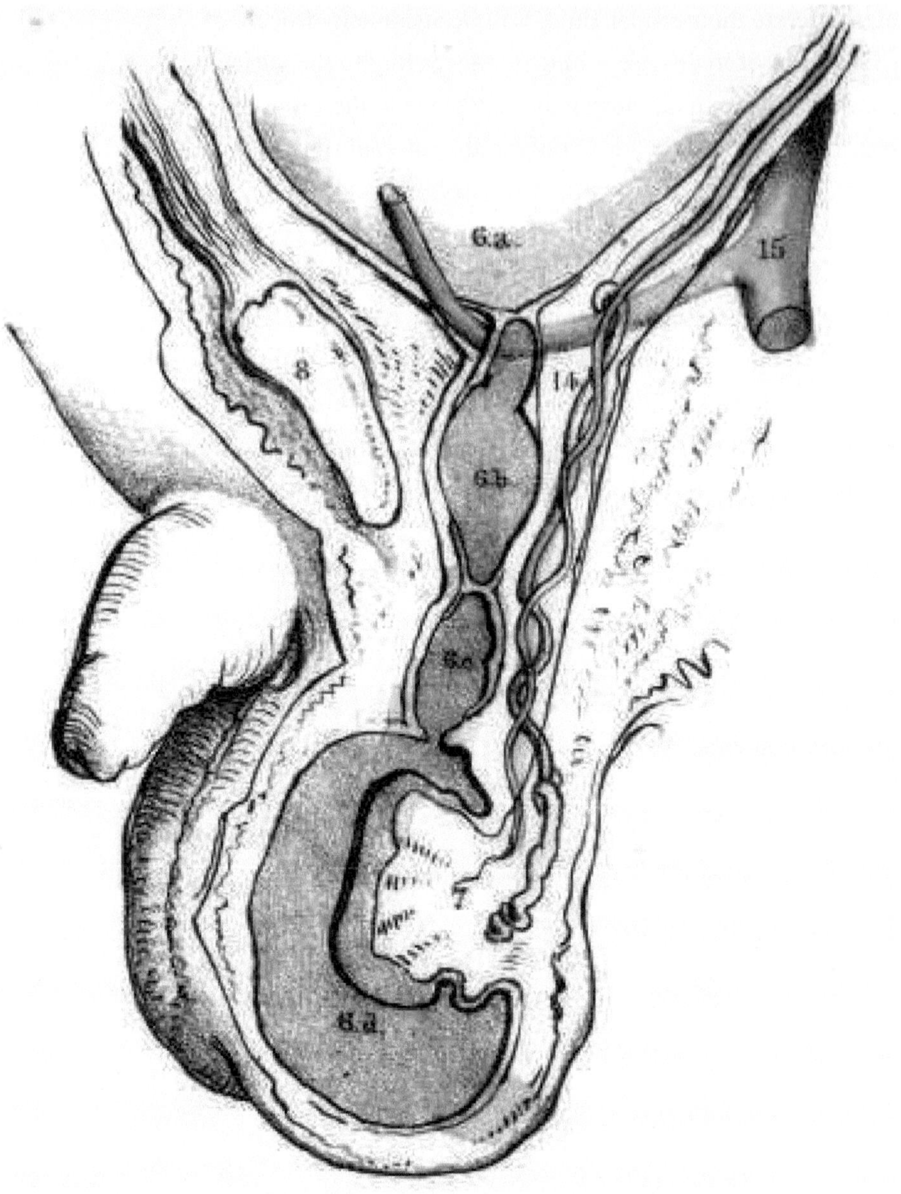

Plate 40--Figure 2.

PLATE 40, Fig. 3.--*Hydrocele of the isolated tunica vaginalis*.--When the serous spermatic tube, 6 *b*, 6 *c*, becomes obliterated, according to the normal rule, after the descent of the testicle, 7, the tunica vaginalis, 6 *d*, is then a distinct serous sac. If a hydrocele form in this sac, it may be distinguished from the congenital variety by its remaining undiminished in bulk when the subject assumes the horizontal position, or when pressure is made on the tumour, for its contents cannot now be forced into the abdomen. The testicle, 7, holds the same position in this as it does in the congenital hydrocele. [Footnote] The radical cure may be performed here without endangering the peritonaeal sac. Congenital hydrocele is of a cylindrical shape; and this is mentioned as distinguishing it from isolated hydrocele of the tunica vaginalis, which is pyriform; but this mark will fail when the cord is at the same time distended, as it may be, in the latter form of the complaint.

[Footnote: When a hydrocele is interposed between the eye and a strong light, the testis appears as an opaque body at the back of the tunica vaginalis. But this position of the organ is, from several causes, liable to vary. The testis may have become morbidly adherent to the front wall of the serous sac, in which case the hydrocele will distend the sac laterally. Or the testis may be so transposed in the scrotum, that, whilst the gland occupies its front part, the distended tunica vaginalis is turned behind. The tunica vaginalis, like the serous spermatic tube, may, in consequence of inflammatory fibrinous effusion, become sacculated-multilocular, in which case, if a hydrocele form, the position of the testis will vary accordingly.--See Sir Astley Cooper's work, ("Anatomy and Diseases of the Testis;") Morton's "Surgical Anatomy;" Mr. Curling's "Treatise on Diseases of the Testis;" and also his article "Testicle," in the Cyclopaedia of Anatomy and Physiology.]

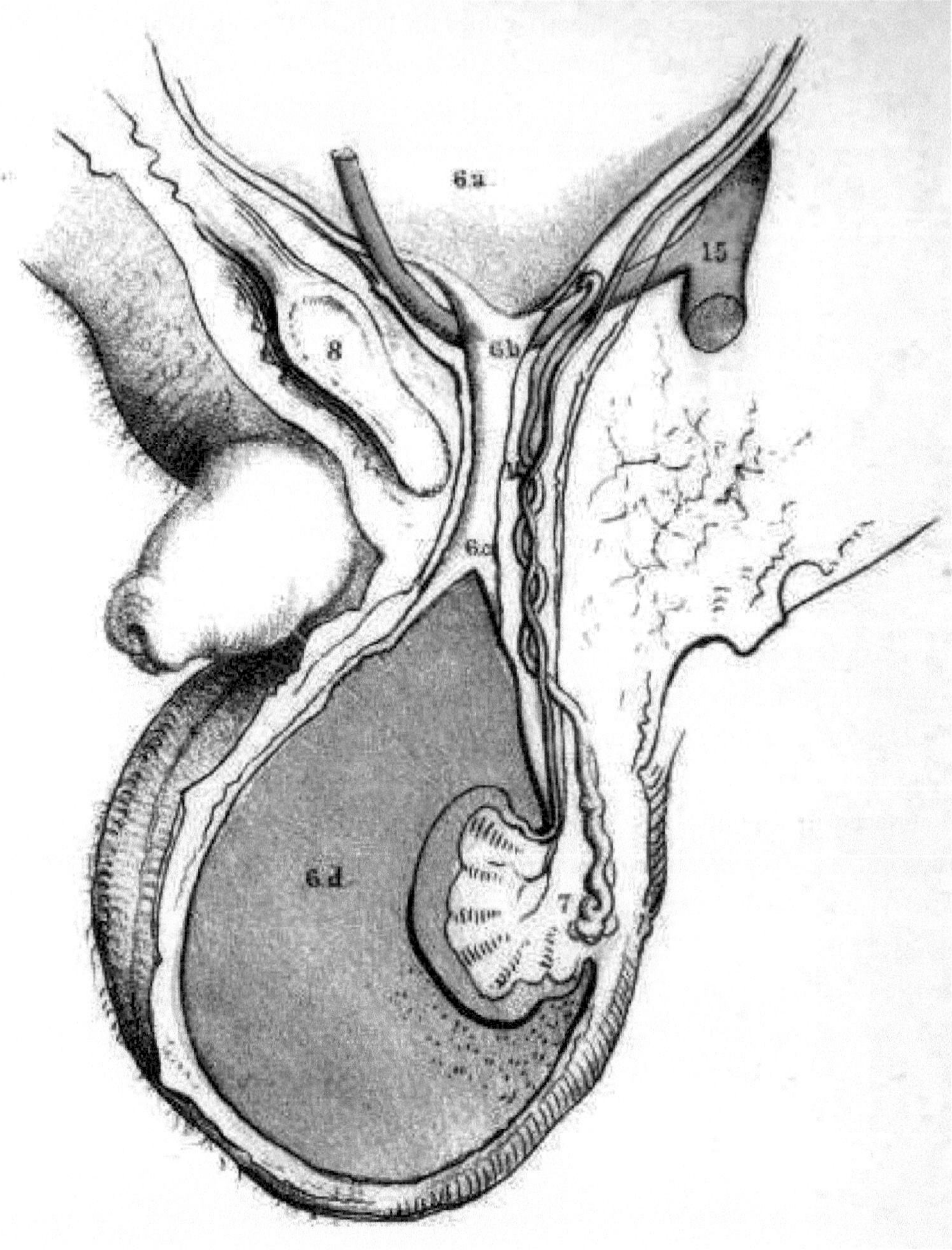

Plate 40--Figure 3.

PLATE 40, Fig. 4.--*The serous spermatic tube remaining pervious, a congenital hernia is formed.*--When the testicle, 7, has descended to the scrotum, if the communication between the peritonaeum, 6 *a*, and the tunica vaginalis, 6 *c*, be not obliterated, a fold of the intestine, 13, will follow the testicle, and occupy the cavity of the tunica vaginalis, 6 *d*. In this form of hernia (hernia tunicae vaginalis, *Cooper*), the intestine is in front of, and in immediate contact with, the testicle. The intestine may descend lower than the testicle, and envelope this organ so completely as to render its position very obscure to the touch. This form of hernia is named congenital, since it occurs in the same condition of the parts as is found in congenital hydrocele--viz., the inguinal ring remaining unclosed. It may occur at any period of life, so long as the original congenital defect remains. It may be distinguished from hydrocele by its want of transparency and fluctuation. The impulse which is communicated to the hand applied to the scrotum of a person affected with scrotal hernia, when he is made to cough, is also felt in the case of congenital hydrocele. But in hydrocele of the separate tunica vaginalis, such impulse is not perceived. Congenital hernia and hydrocele may co-exist; and, in this case, the diagnostic signs which are proper to each, when occurring separately, will be so mingled as to render the precise nature of the case obscure.

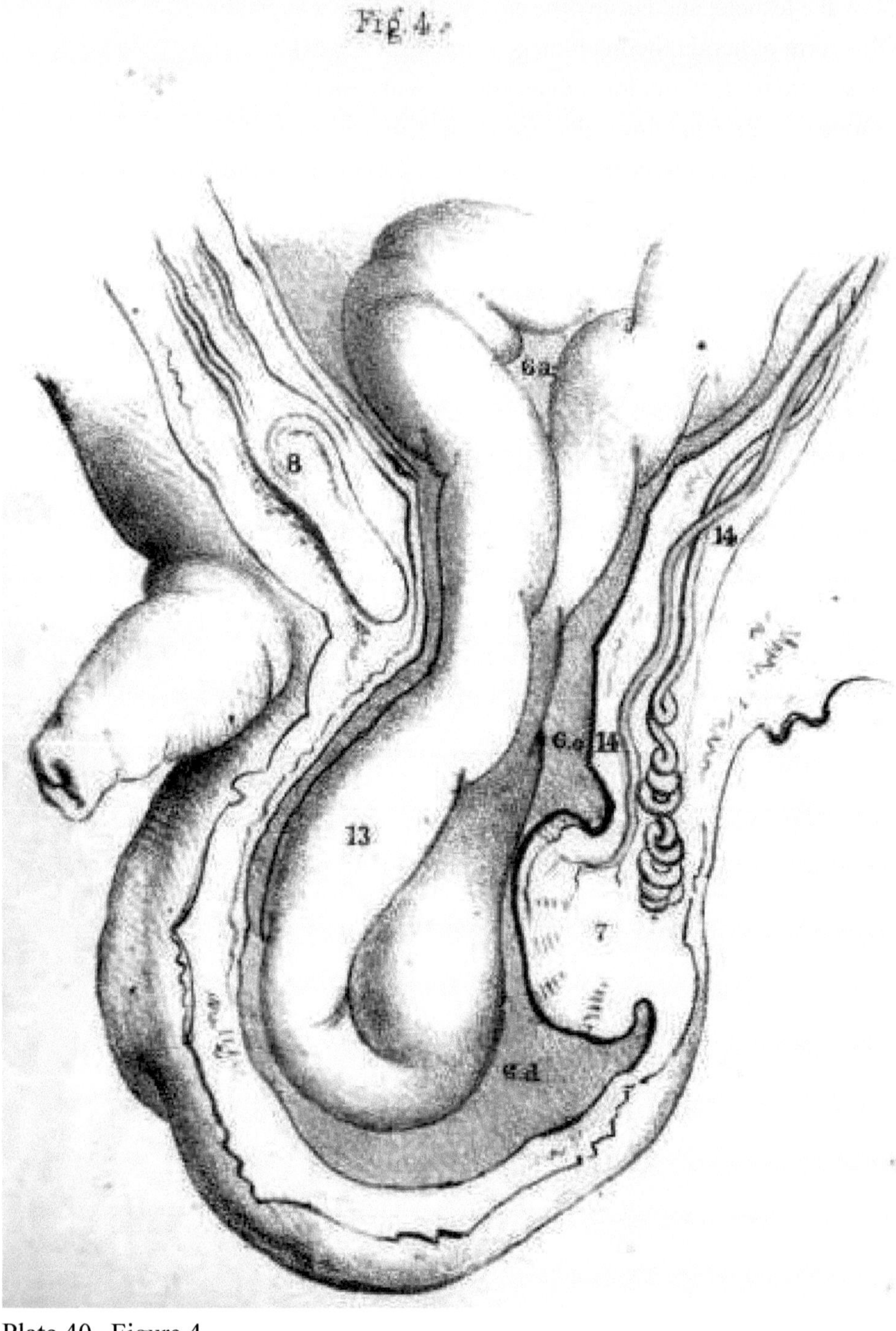

Plate 40--Figure 4.

PLATE 40, Fig. 5.--*Infantile hernia.*--When the serous spermatic tube becomes merely closed, or obliterated at the inguinal ring, 6 *b*, the lower part of it, 6 *c,* is pervious, and communicating with the tunica vaginalis, 6 *d.* In consequence of the closure of the tube at the inguinal ring, if a hernia now occur, it cannot enter the tunica vaginalis, and come into actual contact with the testicle. The hernia, 13, therefore, when about to force the peritonaeum, 6 *a*, near the closed ring, 6 *b*, takes a distinct sac or investment from this membrane. This hernial sac, 6 *e*, will vary as to its position in regard to the tunica vaginalis, 6 **d**, according to the place whereat it dilates the peritonaeum at the ring. The peculiarity of this hernia, as distinguished from the congenital form, is owing to the scrotum containing two sacs,--the tunica vaginalis and the proper sac of the hernia; whereas, in the congenital variety, the tunica vaginalis itself becomes the hernial sac by a direct reception of the naked intestine. If in infantile hernia a hydrocele should form in the tunica vaginalis, the fluid will also distend the pervious serous spermatic tube, 6 *c*, as far up as the closed internal ring, 6 *b*, and will thus invest and obscure the descending herniary sac, 13. This form of hernia is named infantile (*Hey*), owing to the congenital defect in that process, whereby the serous tube lining the cord is normally obliterated. Such a form of hernia may occur at the adult age for the first time, but it is still the consequence of original default.

Fig. 5.

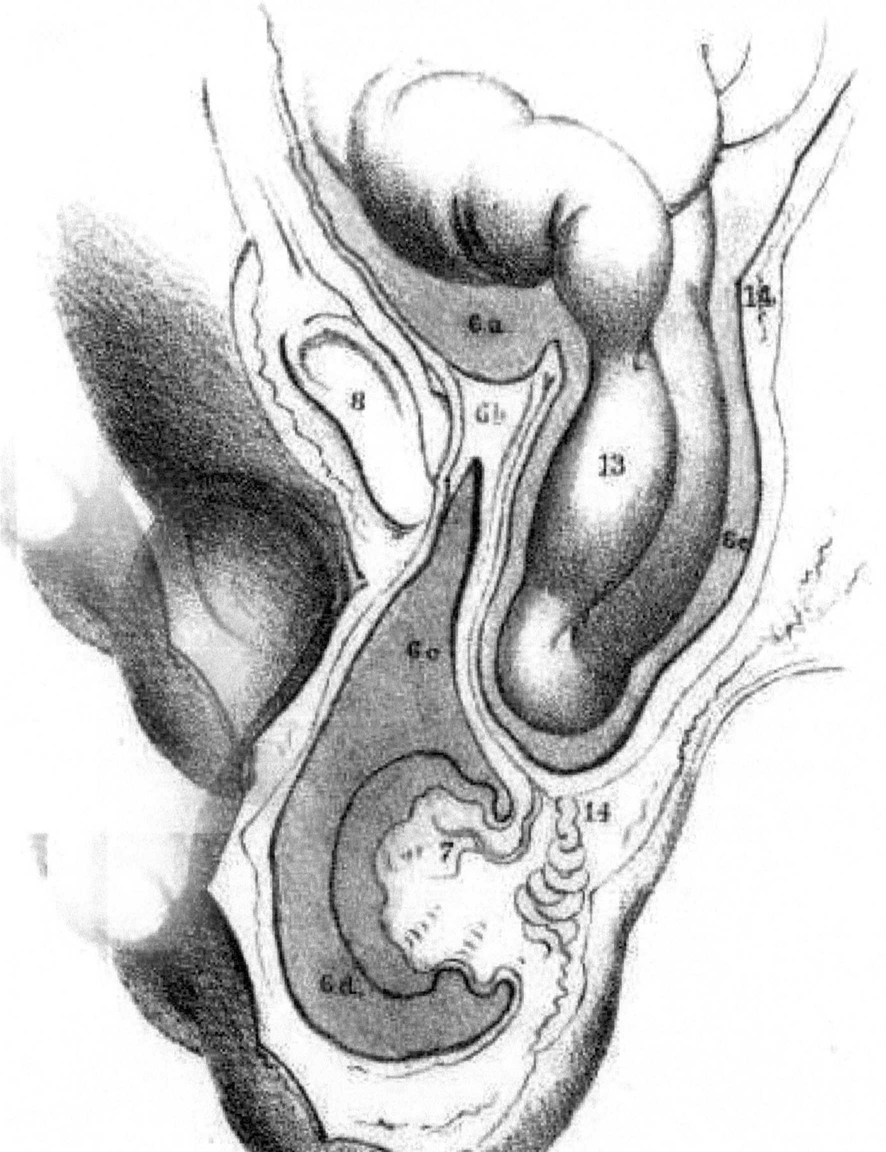

Plate 40--Figure 5.

PLATE 40, Fig. 6.--*Oblique inguinal hernia in the adult.*--This variety of hernia occurs not in consequence of any congenital defect, except inasmuch as the natural weakness of the inguinal wall opposite the internal ring may be attributed to this cause. The serous spermatic tube has been normally obliterated for its whole length between the internal ring and the tunica vaginalis; but the fibrous tube, or spermatic fascia, is open at the internal ring where it joins the transversalis fascia, and remains pervious as far down as the testicle. The intestine, 13, forces and distends the upper end of the closed serous tube; and as this is now wholly obliterated, the herniary sac, 6 *c*, derived anew from the inguinal peritonaeum, enters the fibrous tube, or sheath of the cord, and descends it as far as the tunica vaginalis, 6 *d*, but does not enter this sac, as it is already closed. When we compare this hernia, Fig. 6, Plate 40, with the infantile variety, Fig. 5, Plate 40, we find that they agree in so far as the intestinal sac is distinct from the tunica vaginalis; whereas the difference between them is caused by the fact of the serous cord remaining in part pervious in the infantile hernia; and on comparing Fig. 6, Plate 40, with the congenital variety, Fig. 4, Plate 40, we see that the intestine has acquired a new sac in the former, whereas, in the latter, the intestine has entered the tunica vaginalis. The variable position of the testicle in Figs. 4, 5, & 6, Plate 40, is owing to the variety in the anatomical circumstances under which these herniae have happened.

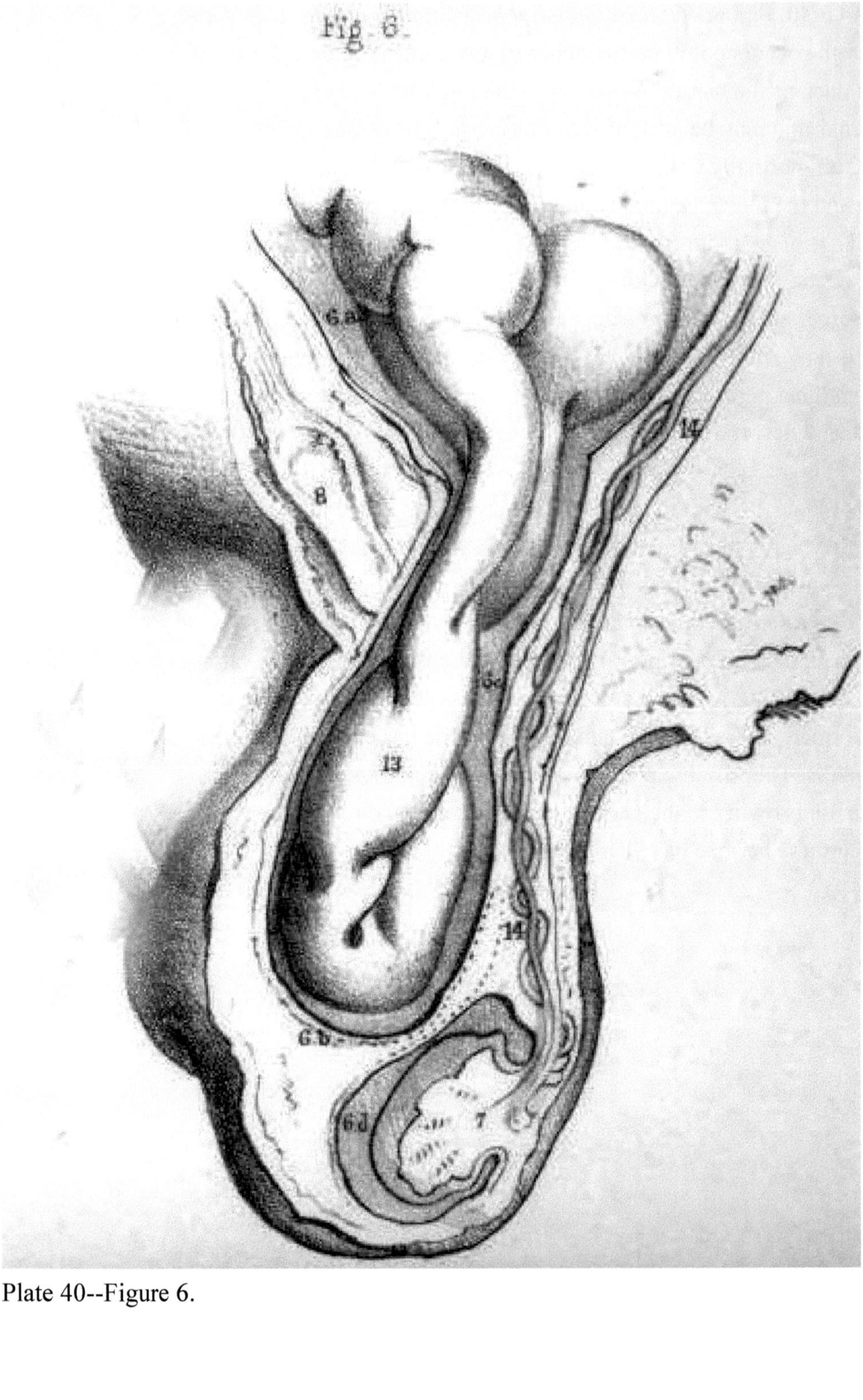

Plate 40--Figure 6.

DEMONSTRATIONS OF THE ORIGIN AND PROGRESS OF INGUINAL HERNIAE IN GENERAL.

Formation of the serous sac. Formation of congenital hernia. Hernia in the canal of Nuck. Formation of infantile hernia. Dilatation of the serous sac. Funnel-shaped investments of the hernia. Descent of the hernia like that of the testicle. Varieties of infantile hernia. Sacculated cord. Oblique internal inguinal hernia--cannot be congenital. Varieties of internal hernia. Direct external hernia. Varieties of the inguinal canal.

PLATE 41, Fig. 1.--When the serous spermatic tube is obliterated for its whole length between the internal ring, 1, and the top of the testicle, 13, a hernia, in order to enter the inguinal canal, 1, 4, must either rupture the peritonaeum at the point 1, or dilate this membrane before it in the form of a sac. [Footnote] If the peritonaeum at the point 1 be ruptured by the intestine, this latter will enter the fibrous spermatic tube, 2, 3, and will pass along this tube devoid of the serous sac. If, on the other hand, the intestine dilates the serous membrane at the point, 1, where it stretches across the internal ring, it will, on entering the fibrous tube, (infundibuliform fascia,) be found invested by a sac of the peritonaeum, which it dilates and pouches before itself. As the epigastric artery, 9, bends in general along the internal border of the ring of the fibrous tube, 2, 2, the neck of the hernial sac which enters the ring at a point external to the artery must be external to it, and remain so despite all further changes in the form, position, and dimensions of the hernia. And as this hernia enters the ring at a point anterior to the spermatic vessels, its neck must be anterior to them. Again, if the bowel be invested by a serous sac, formed of the peritonaeum at the point 1, the neck of such sac must intervene between the protruding bowel and the epigastric and spermatic vessels. But if the intestine enter the ring of the fibrous tube, 2, 2, by having ruptured the peritonaeum at the point 1, then the naked intestine will lie in immediate contact with these vessels.

[Footnote: Mr. Lawrence (op. cit.) remarks, "When we consider the texture of the peritonaeum, and the mode of its connexion to the abdominal parietes, we cannot fancy the possibility of tearing the membrane by any attitude or motion." Cloquet and Scarpa have also expressed themselves to the effect, that the peritonaeum suffers a gradual distention before the protruding bowel.]

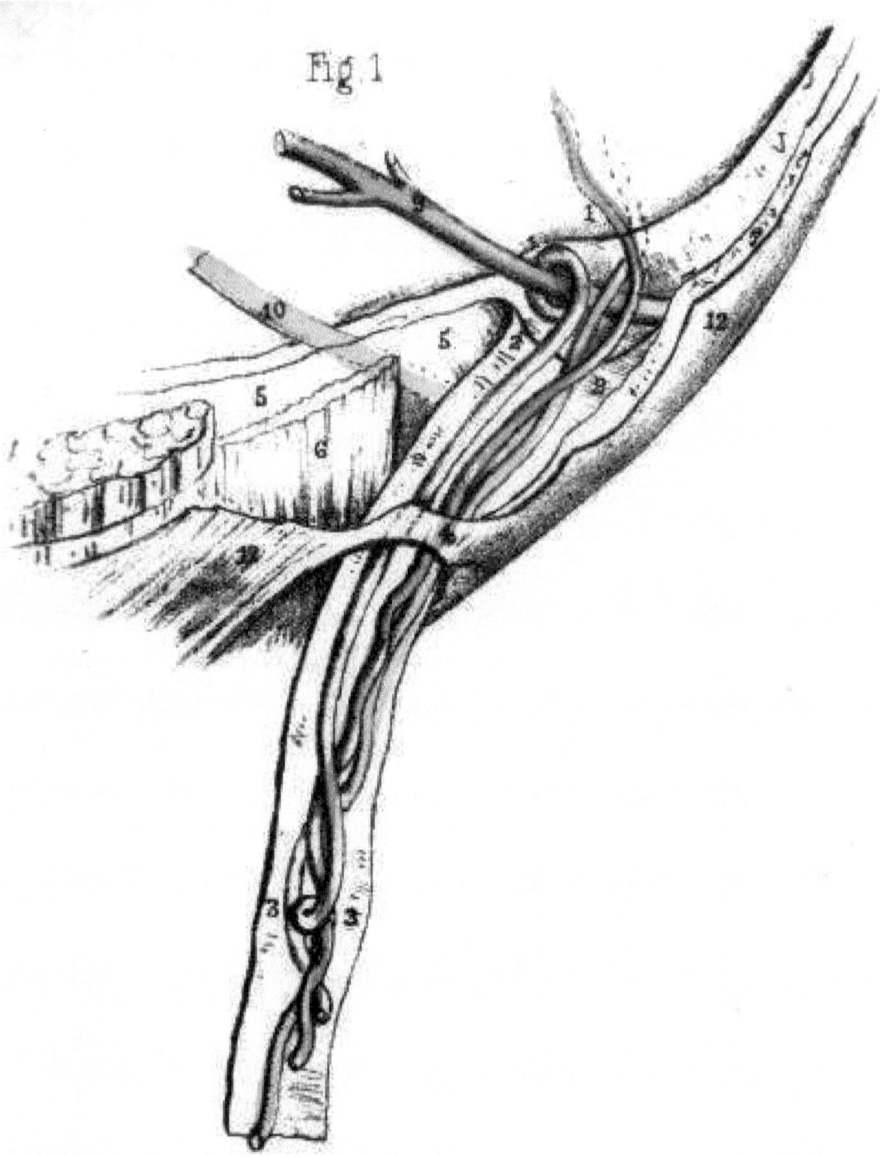

Plate 41--Figure 1

PLATE 41, Fig. 2--When the serous spermatic tube, 11, remains pervious between the internal ring, 1, (where it communicates with the general peritonaeal membrane,) and the top of the testicle, (where it opens into the tunica vaginalis,) the bowel enters this tube directly, without a rupture of the peritonaeum at the point 1. This tube, therefore, becomes one of the investments of the bowel. It is the serous sac, not formed by the protruding bowel, but one already open to receive the bowel. This is the condition necessary to the formation of congenital hernia. This hernia must be one of the external oblique variety, because it enters the open abdominal end of the infantile serous spermatic tube, which is always external to the epigastric artery. Its position in regard to the spermatic vessels is the same as that noticed in Fig, 1, Plate 41. But, as the serous tube through which the congenital hernia descends, still communicates with the tunica vaginalis, so will this form of hernia enter this tunic, and thereby become different to all other herniae, forasmuch as it will lie in immediate contact with the testicle. [Footnote]

[Footnote: A hernia may be truly congenital, and yet the intestine may not enter the tunica vaginalis. Thus, if the serous spermatic tube close only at the top of the testicle, the bowel which traverses the open internal inguinal ring and pervious tube will not enter the tunica vaginalis.]

Fig. 2.

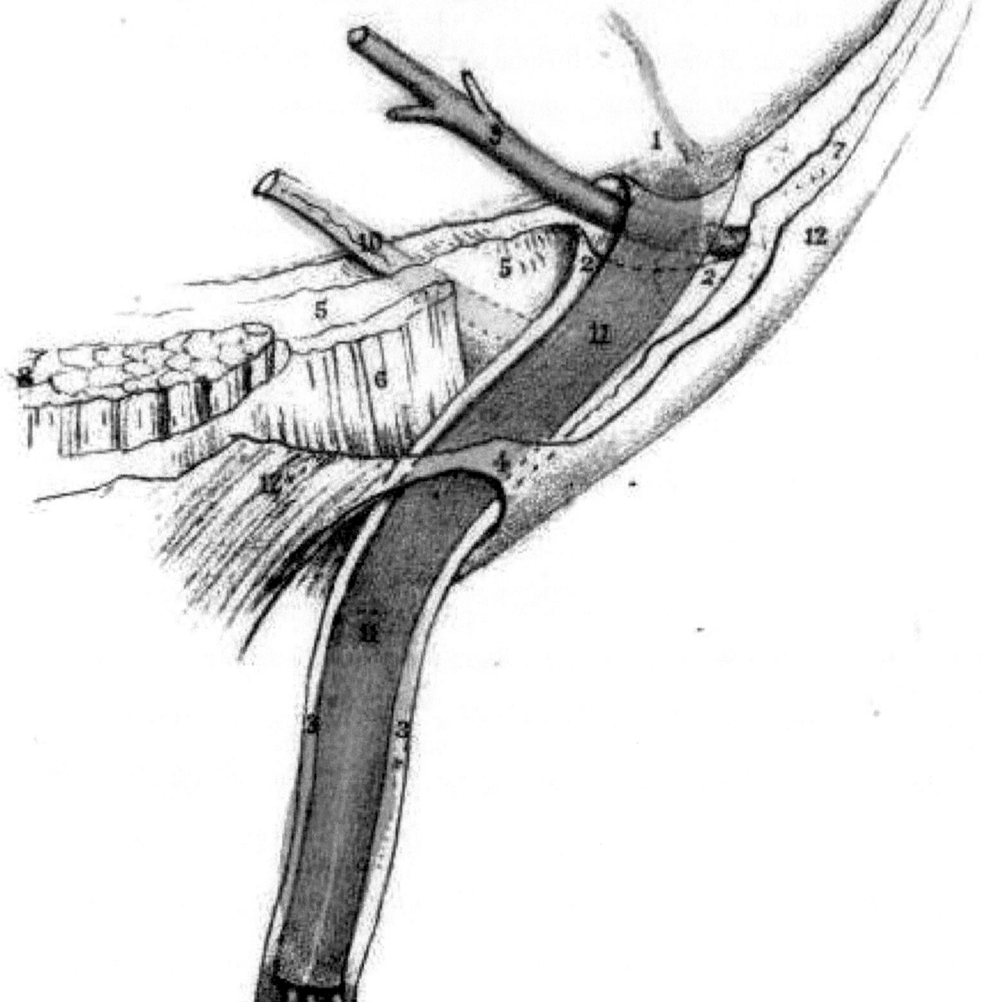

Plate 41--Figure 2

PLATE 41, Fig. 3.--The infantile serous spermatic tube, 11, sometimes remains pervious in the neighbourhood of the internal ring, 1, and a narrow tapering process of the tube (the canal of Nuck) descends within the fibrous tube, 2, 3, and lies in front of the spermatic vessels and epigastric artery. Before this tube reaches the testicle, it degenerates into a mere filament, and thus the tunica vaginalis has become separated from it as a distinct sac. When the bowel enters the open abdominal end of the serous tube, this latter becomes the hernial sac. It is not possible to distinguish by any special character a hernia of this nature, when already formed, from one which occurs in the condition of parts proper to Fig. 1, Plate 41, or that which is described in the note to Fig. 2, Plate 41; for when the intestine dilates the tube, 11, into the form of a sac, this latter assumes the exact shape of the sac, as noticed in Fig. 1, Plate 41. The hernia in question cannot enter the tunica vaginalis. Its position in regard to the epigastric and spermatic vessels is the same as that mentioned above.

Fig. 3.

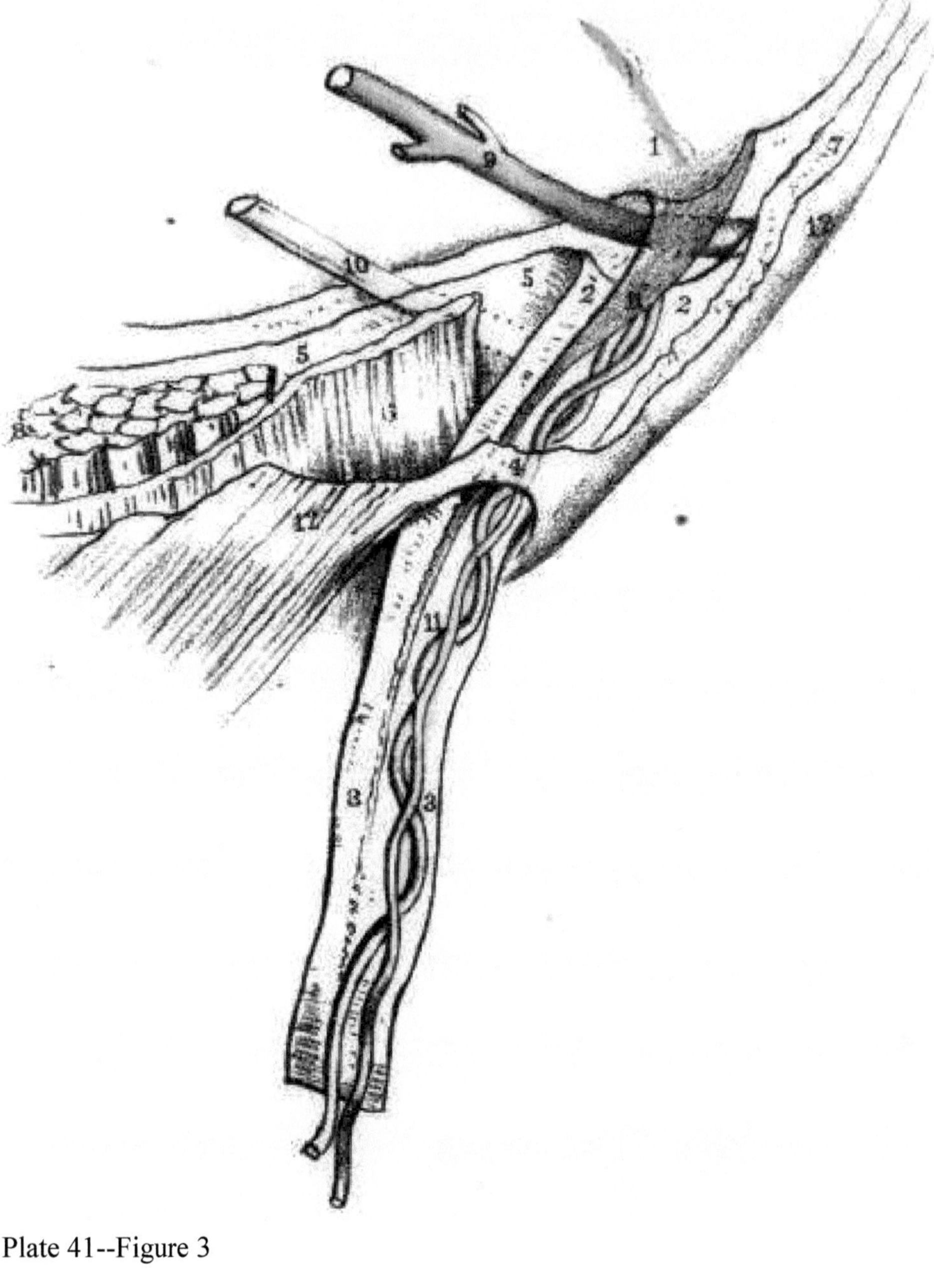

Plate 41--Figure 3

PLATE 41, Fig. 4.--If the serous spermatic tube, 11, be obliterated or closed at the internal ring, 1, thus cutting off communication with the general peritonaeal membrane; and if, at the same time, it remain pervious from this point above to the tunica vaginalis below, then the herniary bowel, when about to protrude at the point 1, must force and dilate the peritonaeum, in order to form its sac anew, as stated of Fig. 1, Plate 41. Such a hernia does not enter either the serous tube or the tunica vaginalis; but progresses from the point 1, in a distinct sac. In this case, there will be found two sacs--one enclosing the bowel; and another, consisting of the serous spermatic tube, still continuous with the tunica vaginalis. This original state of the parts may, however, suffer modification in two modes: 1st, if the bowel rupture the peritonaeum at the point 1, it will enter the serous tube 11, and descend through this into the cavity of the tunica vaginalis, as in the congenital variety. 2nd, if the bowel rupture the peritonaeum near the point 1, and does not enter the serous tube 11, nor the tunica vaginalis, then the bowel will be found devoid of a proper serous sac, while the serous tube and tunica vaginalis still exist in communication. In either case, the hernia will hold the same relative position in regard to the epigastric artery and spermatic vessels, as stated of Fig. 1, Plate 41.

Fig. 4.

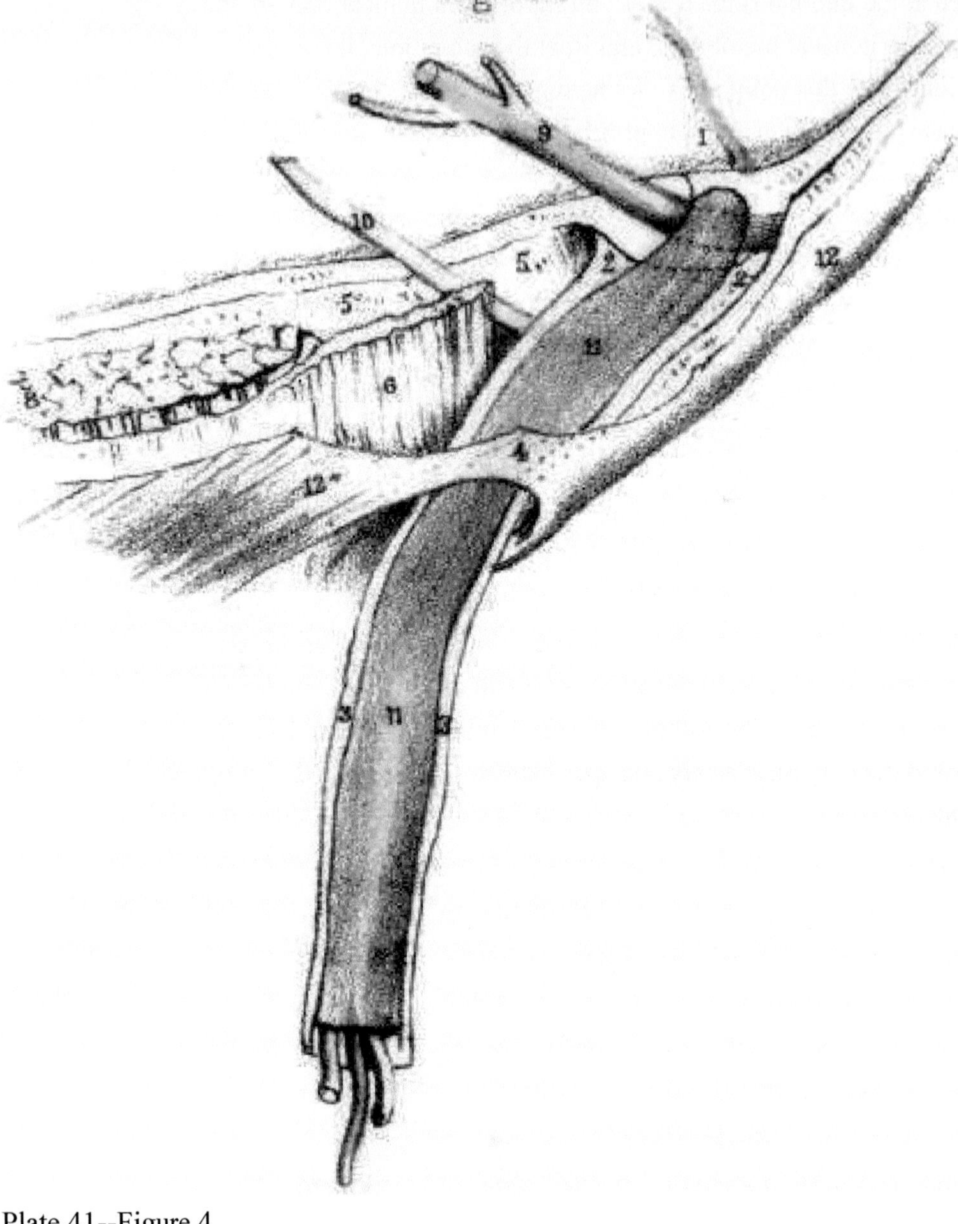

Plate 41--Figure 4

PLATE 41, Fig. 5.--Sudden rupture of the peritonaeum at the closed internal serous ring, 1, though certainly not impossible, may yet be stated as the exception to the rule in the formation of an external inguinal hernia. The aphorism, "natura non facit saltus," is here applicable. When the peritonaeum suffers dilatation at the internal ring, 1, it advances *gradatim* and *pari passu* with the progress of the protruding bowel, and assumes the form, character, position, and dimensions of the inverted curved phases, marked 11, 11, till, from having at first been a very shallow pouch, lying external to the epigastric artery, 9, it advances through the inguinal canal to the external ring, 4, and ultimately traverses this aperture, taking the course of the fibrous tube, 3, down to the testicle in the scrotum.

Fig. 5

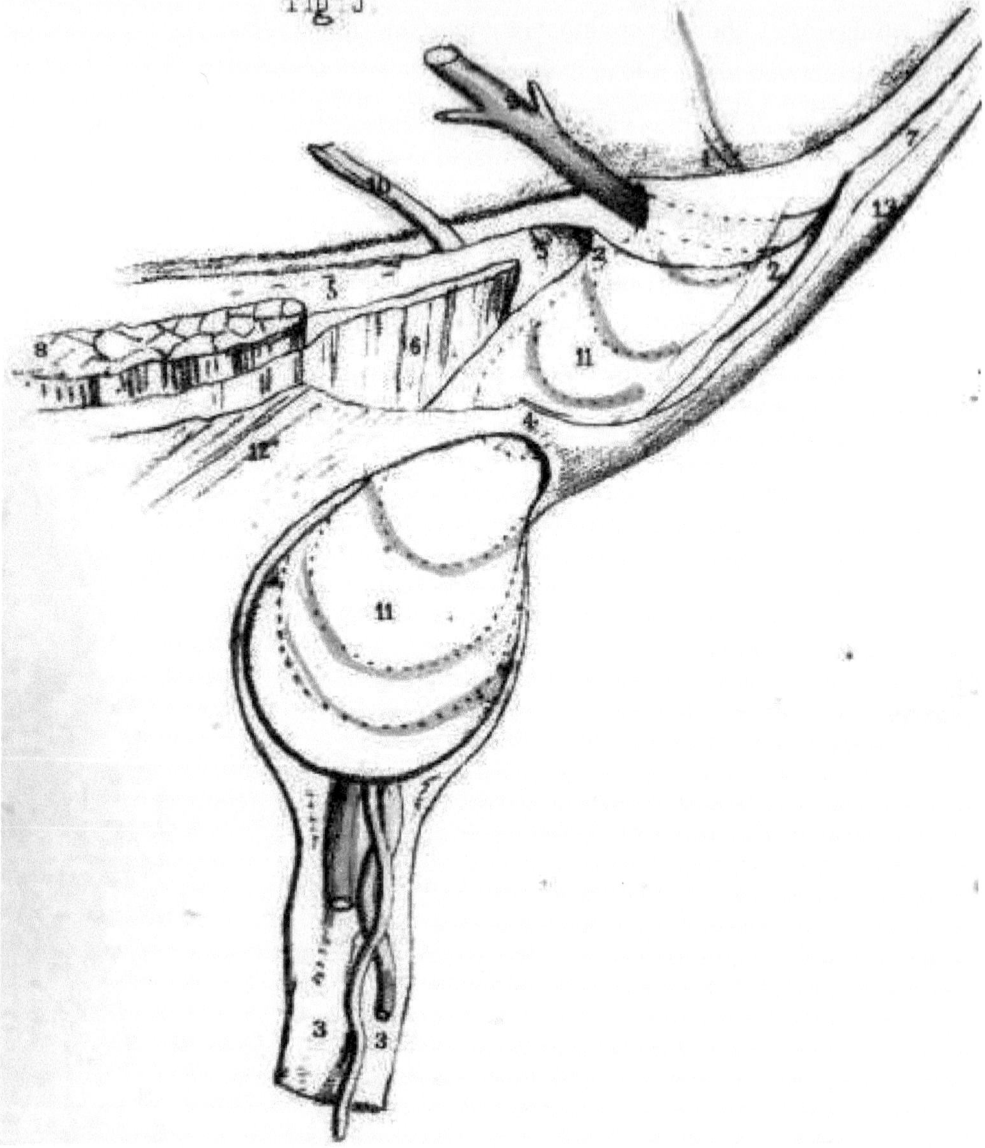

Plate 41--Figure 5

PLATE 41, Fig. 6.--When the bowel dilates the peritonaeum opposite the internal ring, and carries a production of this membrane before it as its sac, then the hernia will occupy the inguinal canal, and become invested by all those structures which form the canal. These structures are severally infundibuliform processes, so fashioned by the original descent of the testicle; and, therefore, as the bowel follows the track of the testicle, it becomes, of course, invested by the selfsame parts in the selfsame manner. Thus, as the infundibuliform fascia, 2, 3, contains the hernia and spermatic vessels, so does the cremaster muscle, extending from the lower margins of the internal oblique and transversalis, invest them also in an infundibuliform manner. [Footnote]

[Footnote: Much difference of opinion prevails as to the true relation which the cord (and consequently the oblique hernia) bears to the lower margins of the oblique and transverse muscles, and their cremasteric prolongation. Mr. Guthrie (Inguinal and Femoral Hernia) has shown that the fibres of the transversalis, as well as those of the internal oblique, are penetrated by the cord. Albinus, Haller, Cloquet, Camper, and Scarpa, record opinions from which it may be gathered that this disposition of the parts is (with some exceptions) general. Sir Astley Cooper describes the lower edge of the transversalis as curved all round the internal ring and cord. From my own observations, coupled with these, I am inclined to the belief that, instead of viewing these facts as isolated and meaningless particulars, we should now fuse them into the one idea expressed by the philosophic Carus, and adopted by Cloquet, that the cremaster is a production of the abdominal muscles, formed mechanically by the testicle, which in its descent dilates, penetrates, and elongates their fibres.]

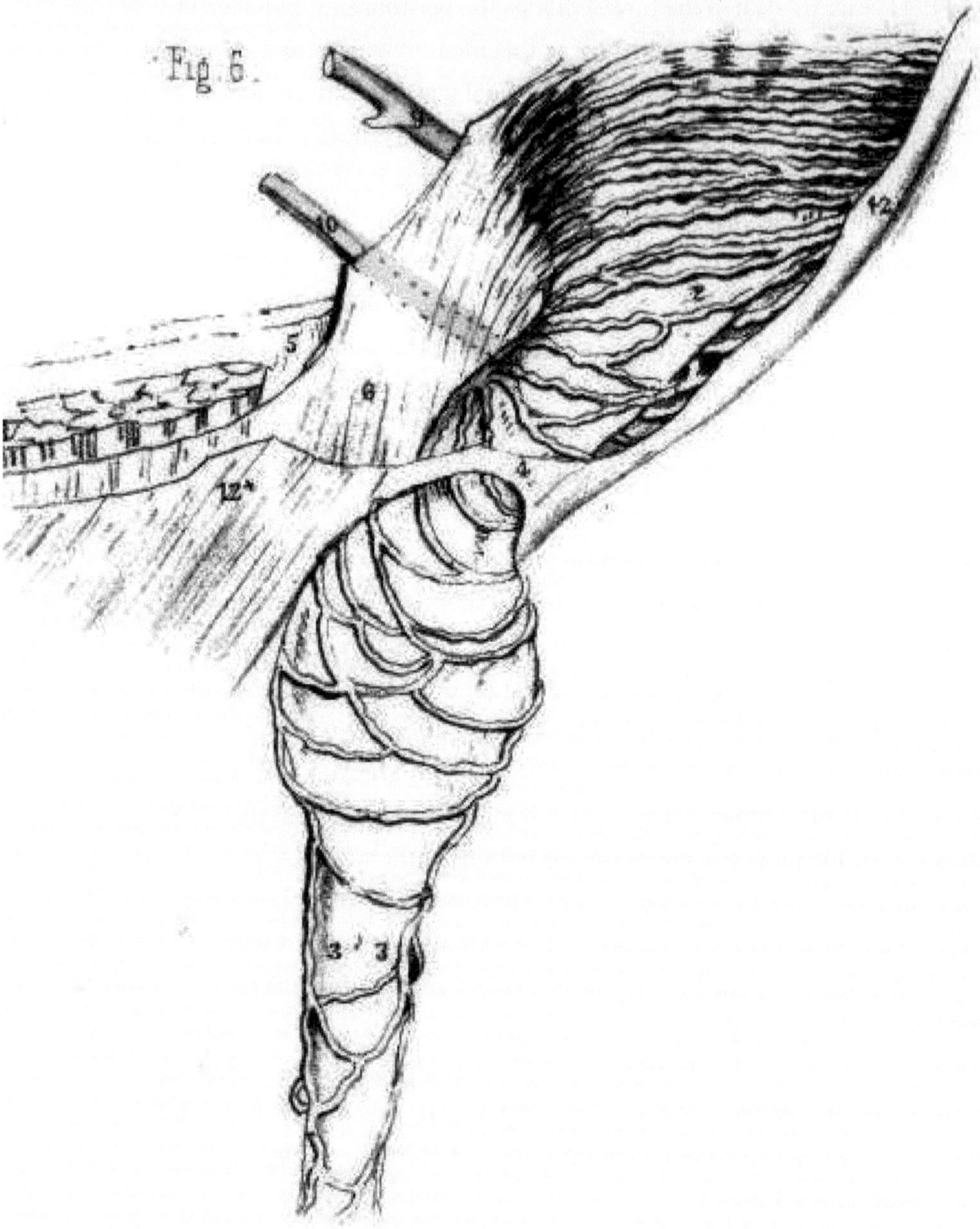

Plate 41--Figure 6

PLATE 41. Fig. 7.--When an external inguinal hernia, 11, dilates and protrudes the peritonaeum from the closed internal ring, 1, and descends the inguinal canal and fibrous tube, 3, 3, it imitates, in most respects, the original descent of the testicle. The difference between both descents attaches alone to the mode in which they become covered by the serous membrane; for the testicle passes through the internal ring *behind* the inguinal peritonaeum, at the same time that it takes a duplicature of this membrane; whereas the bowel encounters this part of the peritonaeum *from within*, and in this mode becomes invested by it on all sides. This figure also represents the form and relative position of a hernia, as occurring in Figs. 1 and 3, 5, and 6, Plate 41.

Fig. 7.

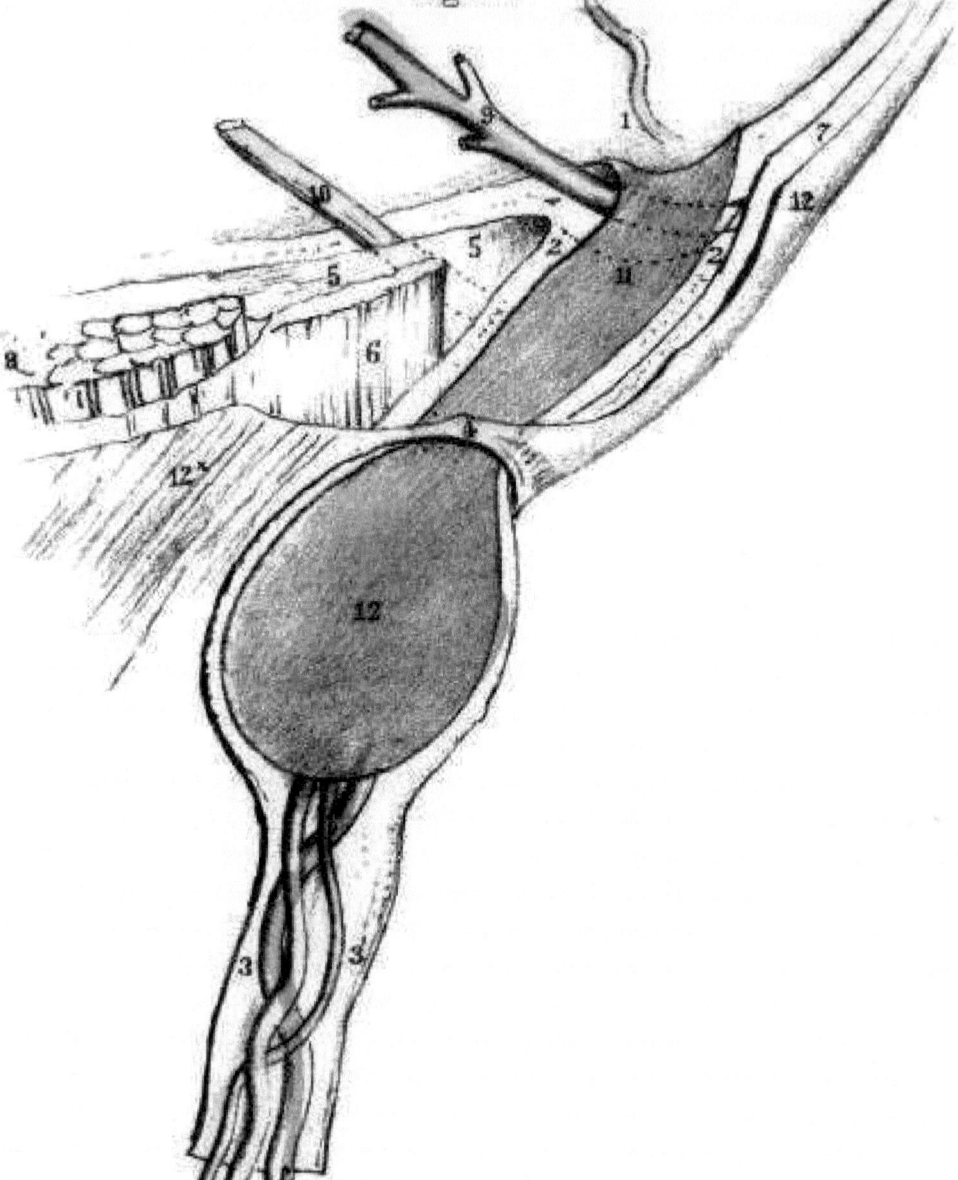

Plate 41--Figure 7

PLATE 41, Fig. 8.--When the serous spermatic tube only closes at the internal ring, as seen at 1, Fig. 4, Plate 41, if a hernia afterwards pouch the peritonaeum at this part, and enter the inguinal canal, we shall then have the form of hernia, Fig. 8, Plate 41, termed infantile. Two serous sacs will be here found, one within the cord, 13, and communicating with the tunica vaginalis, the other, 11, containing the bowel, and being received by inversion into the upper extremity of the first. Thus the infantile serous canal, 13, receives the hernial sac, 11. The inguinal canal and cord may become multicapsular, as in Fig. 8, from various causes, each capsule being a distinct serous membrane. First, independent of hernial formation, the original serous tube may become interruptedly obliterated, as in Plate 40, Fig. 2. Secondly, these sacs may persist to adult age, and have a hernial sac added to their number, whatever this may be. Thirdly, the original serous tube, 13, Fig. 8, may persist, and after having received the hernial sac, 11, the bowel may have been reduced, leaving its sac behind it in the inguinal canal; the neck of this sac may have been obliterated by the pressure of a truss, a second hernia may protrude at the point 1, and this may be received into the first hernial sac in the same manner as the first was received into the original serous infantile tube. The possibility of these occurrences is self-evident, even if they were never as yet experienced. [Footnote]

[Footnote: According to Mr. Lawrence and M. Cloquet, most of the serous cysts found around hernial tumours are ancient sacs obliterated at the neck, and adhering to the new swelling (opera cit.)]

Fig 8

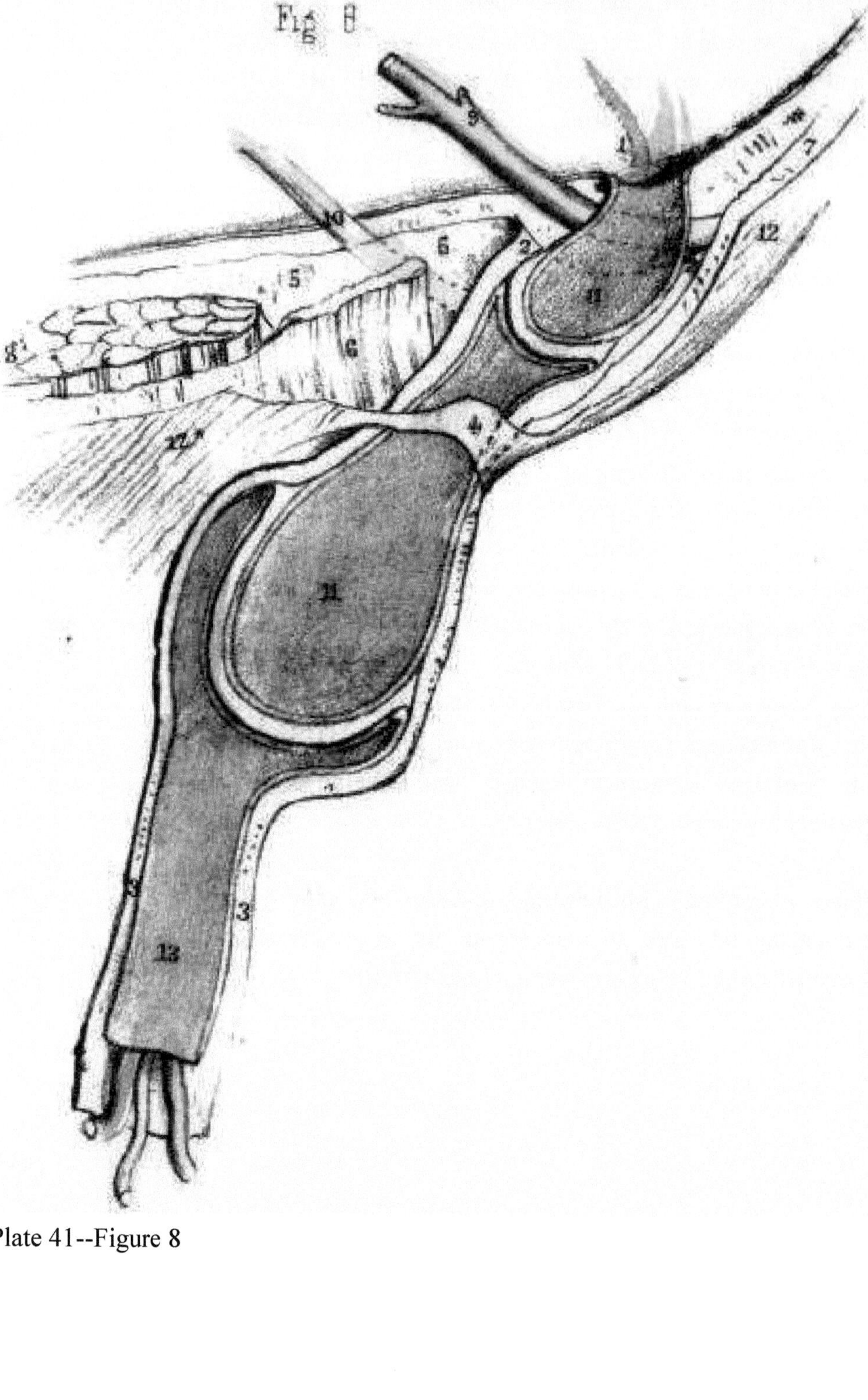

Plate 41--Figure 8

PLATE 42, Fig. 1.--The epigastric artery, 9, being covered by the fascia transversalis, can lend no support to the internal ring, 2, 2, nor to the tube prolonged from it. The herniary bowel may, therefore, dilate the peritonaeum immediately on the inner side of the artery, and enter the inguinal canal. In this way the hernia, 11, although situated internal to the epigastric artery, assumes an oblique course through the canal, and thus closely simulates the external variety of inguinal hernia, Fig. 7, Plate 41. If the hernia enter the canal, as represented in Fig. 1, Plate 42, it becomes invested by the same structures, and assumes the same position in respect to the spermatic vessels, as the external hernia.

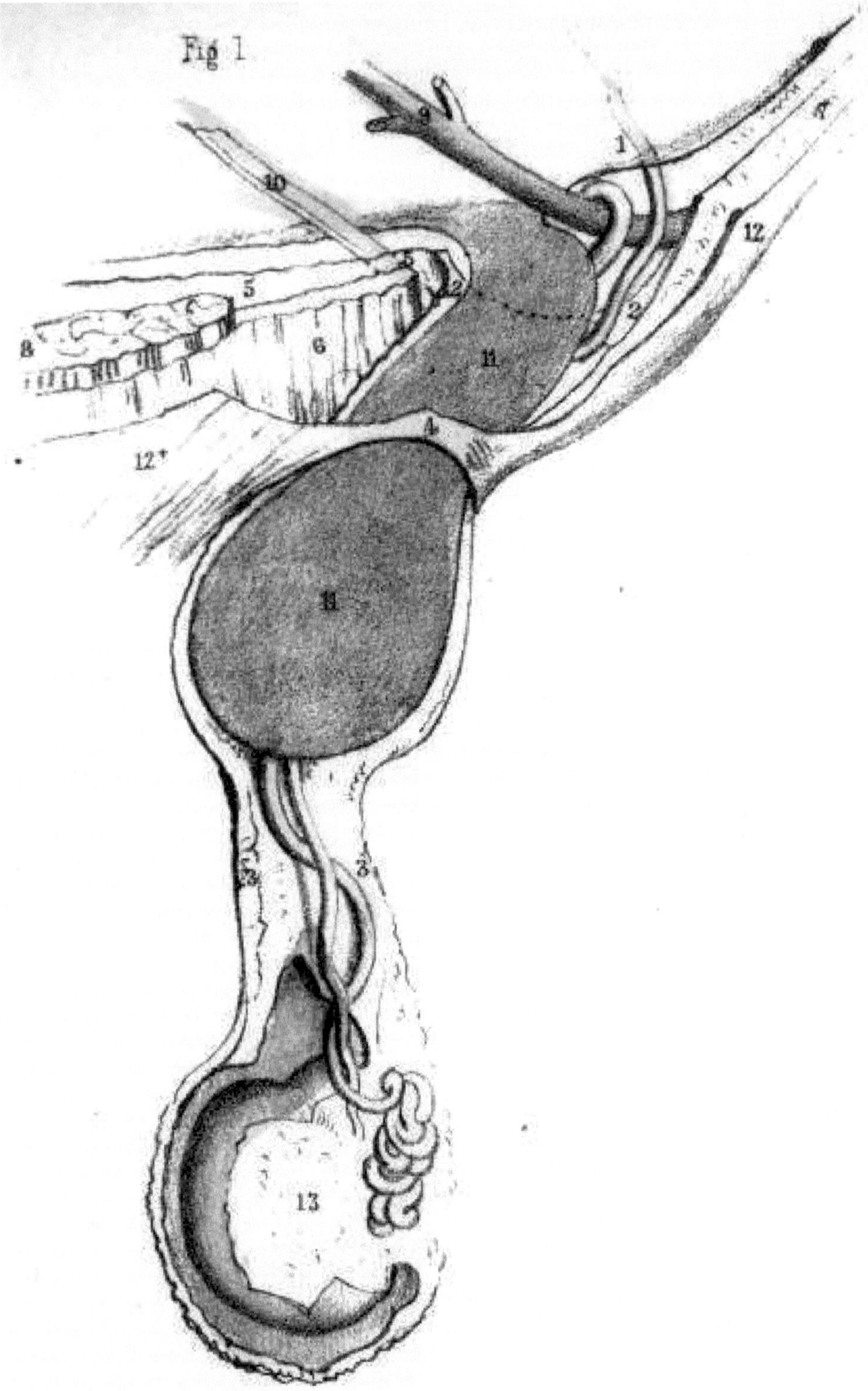

Plate 42--Figure 1

PLATE 42, Fig. 2.--The hernial sac, 11, which entered the ring of the fibrous tube, 2, 2, at a point immediately internal to the epigastric artery, 9, may, from having been at first oblique, as in Fig. 1, Plate 42, assume a direct position. In this case, the ring of the fibrous tube, 2, 2, will be much widened; but the artery and spermatic vessels will remain in their normal position, being in no wise affected by the gravitating hernia. If the conjoined tendon, 6, be so weak as not to resist the gravitating force of the hernia, the tendon will become bent upon itself. If the umbilical cord, 10, be side by side with the epigastric artery at the time that the hernia enters the mouth of the fibrous tube, then, of course, the cord will be found external. If the cord lie towards the pubes, apart from the vessel, the hernia may enter the fibrous tube between the cord, 10, and artery, 9. [Footnote:] It is impossible for any internal hernia to assume the congenital form, because the neck of the original serous spermatic tube, 11, Fig. 2, Plate 41, being external to the epigastric artery, 9, cannot be entered by the hernia, which originates internally to this vessel.

[Footnote: M. Cloquet states that the umbilical cord is always found on the inner side of the external hernia. Its position varies in respect to the internal hernia, (op. cit. prop. 52.)]

Fig. 2

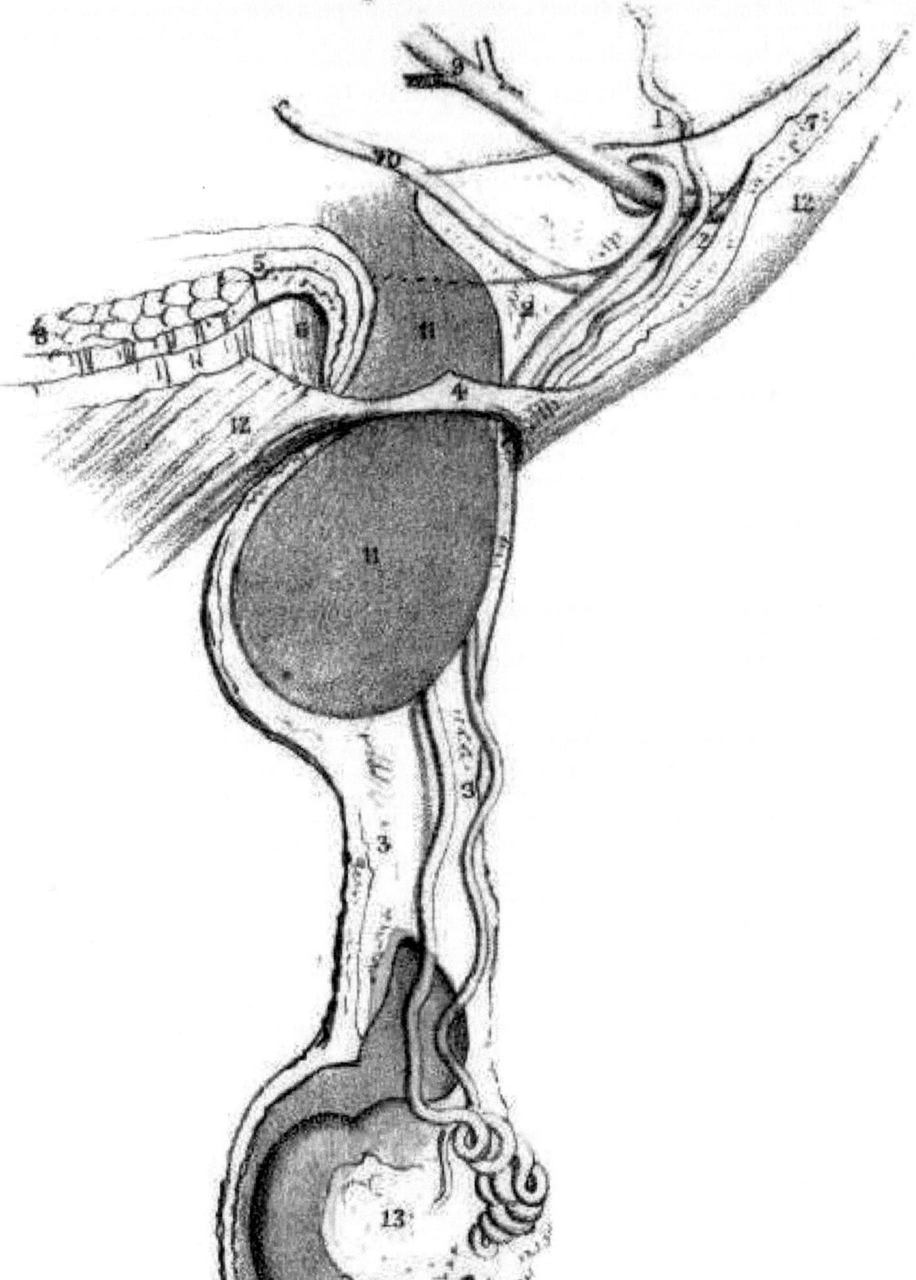

Plate 42--Figure 2

PLATE 42, Fig. 3.--Every internal hernia, which does not rupture the peritonaeum, carries forward a sac produced anew from this membrane, whether the hernia enter the inguinal canal or not. But this is not the case with respect to the fibrous membrane which forms the fascia propria. If the hernia enter the inguinal wall immediately on the inner side of the epigastric artery, Fig. 1, Plate 42, it passes direct into the ring of the fibrous tube, 2, 2, already prepared to receive it. But when the hernia, 11, Fig. 3, Plate 42, cleaves the conjoined tendon, 6, 6, then the artery, 9, and the tube, 2, 2, remain in their usual position, while the bowel carries forward a new investment from the transversalis fascia, 5, 5. That part of the conjoined tendon which stands external to the hernia keeps the tube, 2, 2, in its proper place, and separates it from the fold of the fascia which invests the hernial sac. This is the only form in which an internal hernia can be said to be absolutely distinct from the inguinal canal and spermatic vessels. This hernia, when passing the external ring, 4, has the spermatic cord on its outer side.

Fig. 3

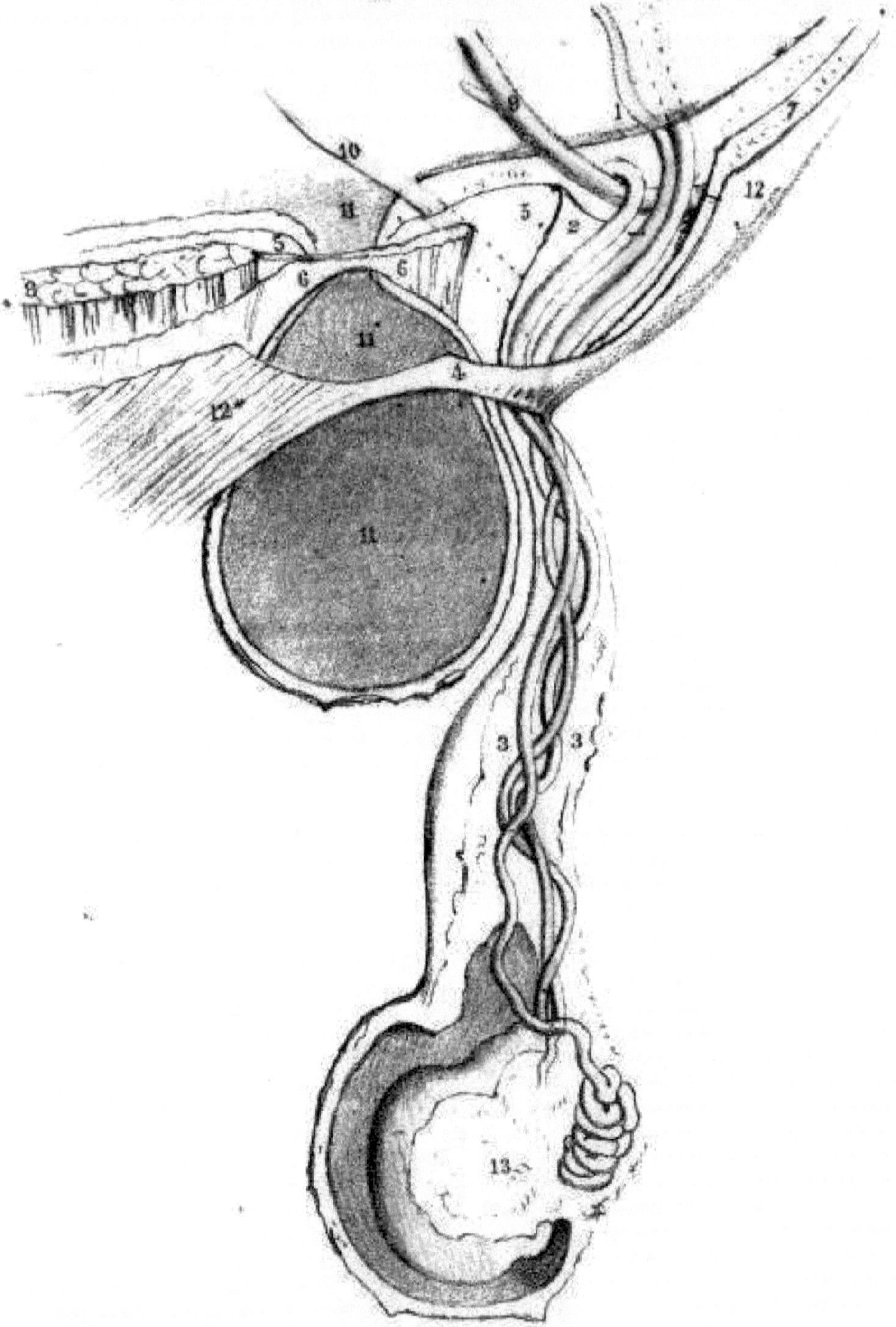

Plate 42--Figure 3

PLATE 42, Fig. 4.--The external hernia, from having been originally oblique, may assume the position of a hernia originally internal and direct. The change of place exhibited by this form of hernia does not imply a change either in its original investments or in its position with respect to the epigastric artery and spermatic vessels. The change is merely caused by the weight and gravitation of the hernial mass, which bends the epigastric artery, 9*, from its first position on the inner margin of the internal ring, 1, till it assumes the place 9. In consequence of this, the internal ring of the fascia transversalis, 2, 2, is considerably widened, as it is also in Fig. 2, Plate 42. It is the inner margin of the fibrous ring which has suffered the pressure; and thus the hernia now projects directly from behind forwards, through, 4, the external ring. The conjoined tendon, 6, when weak, becomes bent upon itself. The change of place performed by the gravitating hernia may disturb the order and relative position of the spermatic vessels; but these, as well as the hernia, still occupy the inguinal canal, and are invested by the spermatic fascia, 3, 3. When an internal hernia, Fig. 1, Plate 42, enters the inguinal canal, it also may descend the cord as far as the testicle, and assume in respect to this gland the same position as the external hernia. [Footnote]

[Footnote: As the external hernia, Fig. 4, Plate 42, may displace the epigastric artery inwards, so may the internal hernia, Fig. 1, Plate 42, displace the artery outwards. Mr. Lawrence, Sir Astley Cooper, Scarpa, Hesselbach, and Langenbeck, state, however, that the internal hernia does not disturb the artery from its usual position three-fourths of an inch from the external ring.]

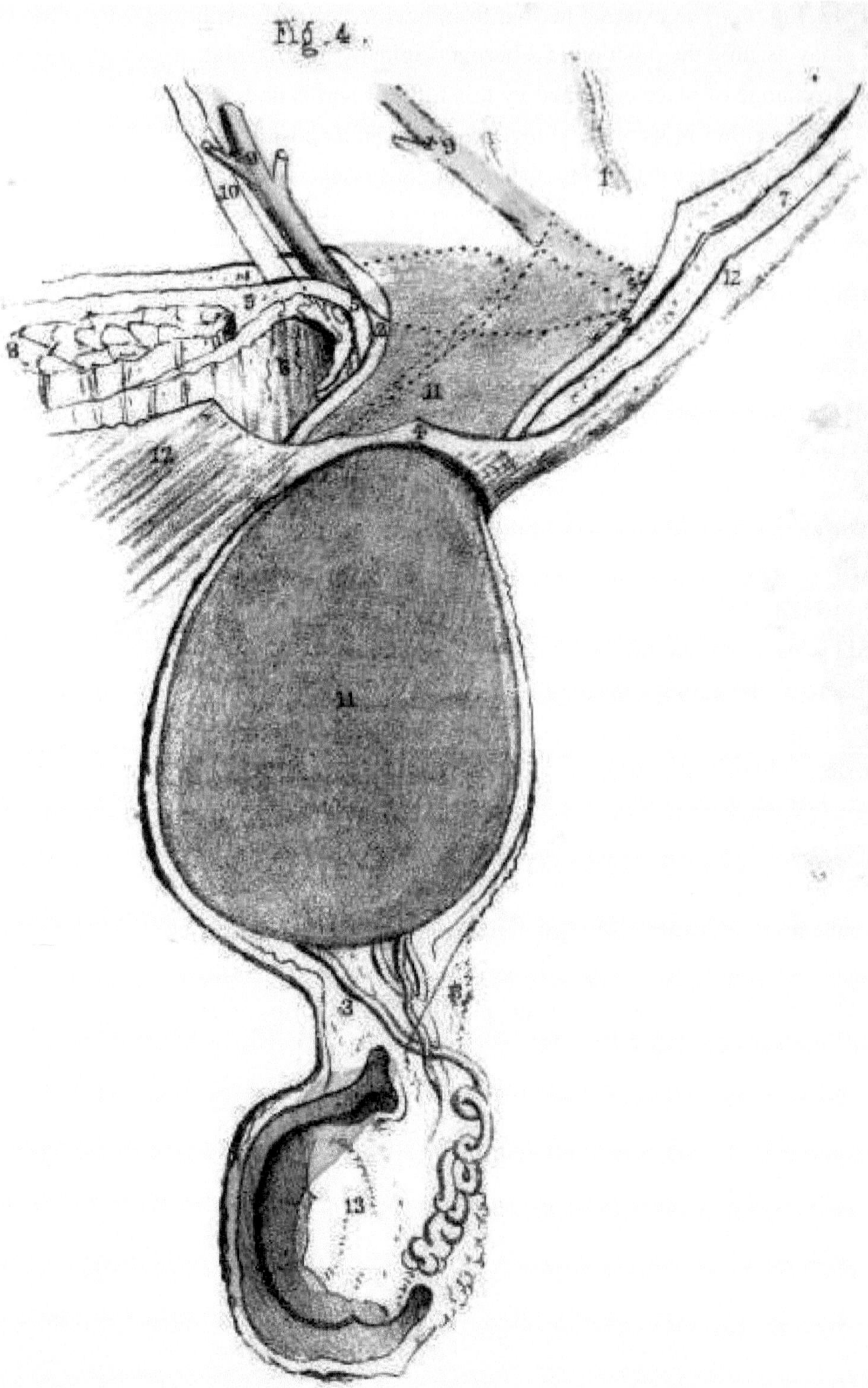

Plate 42--Figure 4

PLATE 42, Figs. 5, 6, 7.--The form and position of the inguinal canal varies according to the sex and age of the individual. In early life, Fig. 6, the internal ring is situated nearly opposite to the external ring, 4. As the pelvis widens gradually in the advance to adult age, Fig. 5, the canal becomes oblique as to position. This obliquity is caused by a change of place, performed rather by the internal than the external ring. [Footnote] The greater width of the female pelvis than of the male, renders the canal more oblique in the former; and this, combined with the circumstance that the female inguinal canal, Fig. 7, merely transmits the round ligament, 14, accounts anatomically for the fact, that this sex is less liable to the occurrence of rupture in this situation.

[Footnote: M. Velpeau (Nouveaux Elemens de med. Operat.) states the length of the inguinal canal in a well-formed adult, measured from the internal to the external ring, to be 1-1/2 or 2 inches, and 3 inches including the rings; but that in some individuals the rings are placed nearly opposite; whilst in young subjects the two rings nearly always correspond. When, in company with these facts, we recollect how much the parts are liable to be disturbed in ruptures, it must be evident that their relative position cannot be exactly ascertained by measurement, from any given point whatever. The judgment alone must fix the general average.]

Fig 5

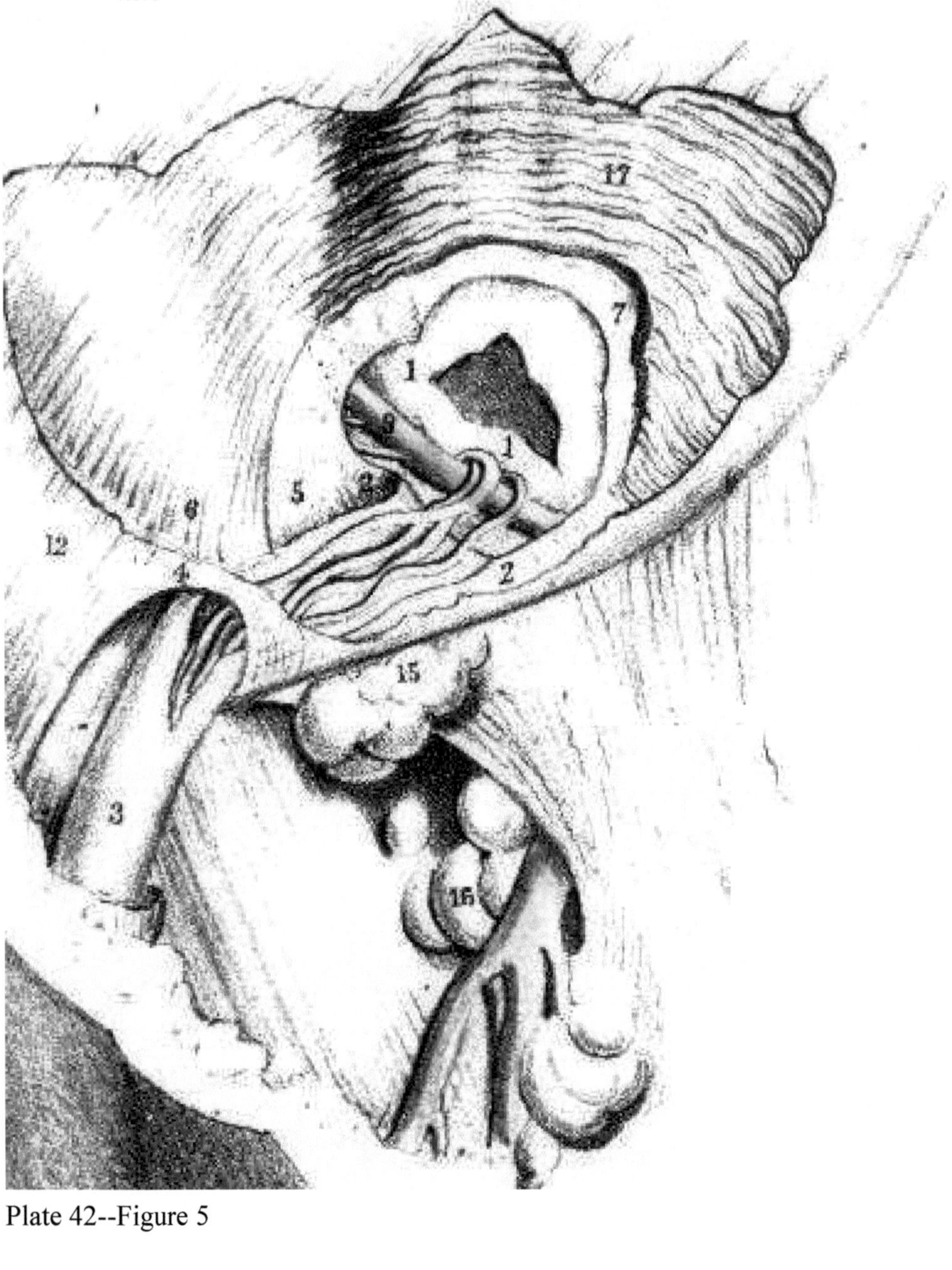

Plate 42--Figure 5

Fig. 6

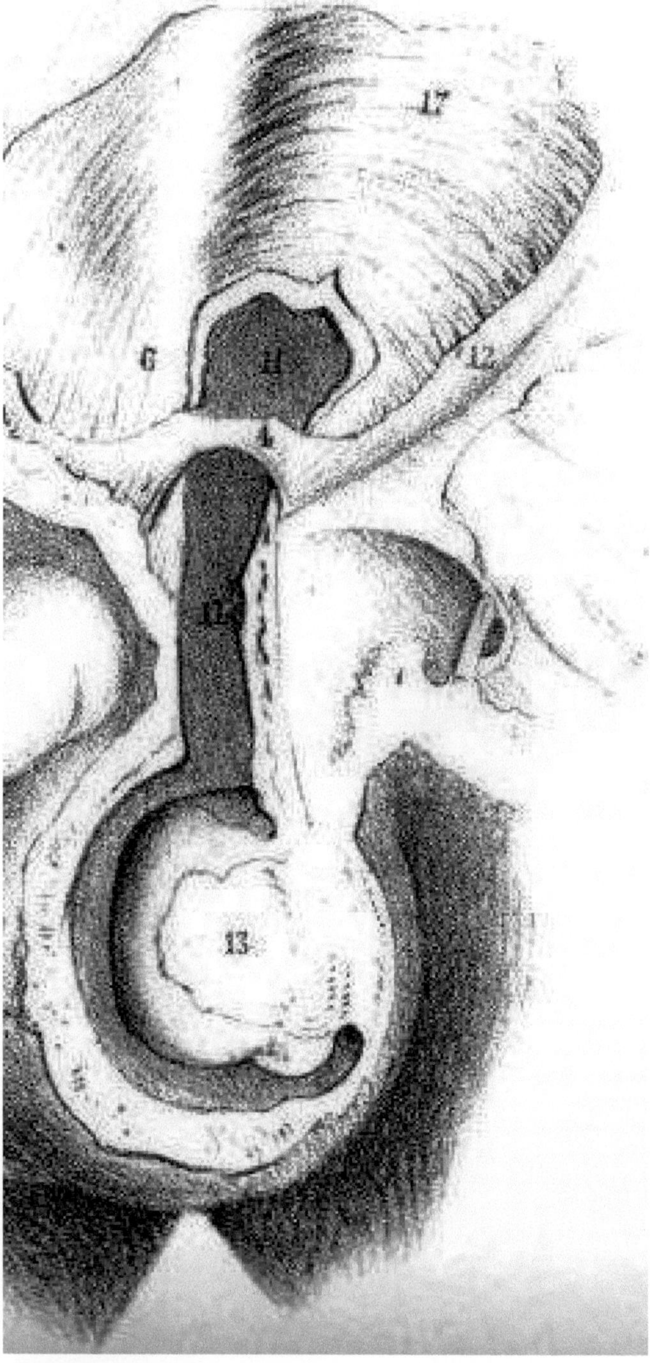

Plate 42--Figure 6

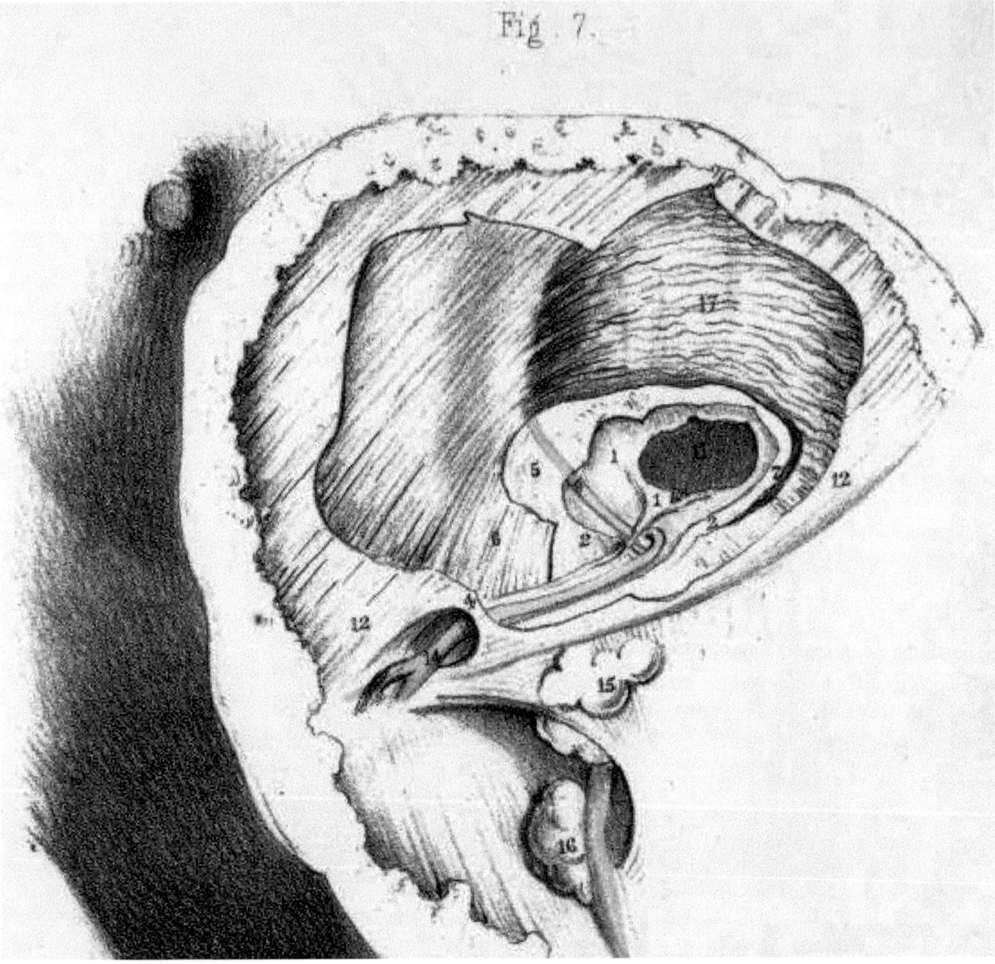

Plate 42--Figure 7

THE DISSECTION OF FEMORAL HERNIA AND THE SEAT OF STRICTURE.

Compared with the inguinal variety. Position and relations. Sheath of the femoral vessels and of the hernia. Crural ring and canal. Formation of the sac. Saphenous opening. Relations of the hernia. Varieties of the obturator and epigastric arteries. Course of the hernia. Investments. Causes and situations of the stricture.

Whilst all forms of inguinal herniae escape from the abdomen at places situated immediately above Poupart's ligament, the femoral hernia, G, Fig. 1, Plate 43, is found to pass from the abdomen immediately below this structure, A I, and between it and the horizontal branch of the pubic bone. The inguinal canal and external abdominal ring are parts concerned in the passage of inguinal herniae, whether oblique or direct, external or internal; whilst the femoral canal and saphenous opening are the parts through which the femoral hernia passes. Both these orders of parts, and of the herniae connected with them respectively, are, however, in reality situated so closely to each other in the inguino-femoral region, that, in order to understand either, we should, examine both at the same time comparatively.

The structure which is named Poupart's ligament in connexion with inguinal herniae, is named the femoral or crural arch (Gimbernat) in relation to femoral hernia. The simple line, therefore, described by this ligament explains the narrow interval which separates both varieties of the complaint. So small is the line of separation described between these herniae by the ligament, that this (so to express the idea) stands in the character of an arch, which, at the same time, supports an aqueduct (the inguinal canal) and spans a road (the femoral sheath.) The femoral arch, A I, Fig. 1, Plate 43, extends between the anterior superior iliac spinous process and the pubic spine. It connects the aponeurosis of the external oblique muscle, D *d*, Fig. 2, Plate 44, with F, the fascia lata. Immediately above and below its pubic extremity appear the external ring and the saphenous opening. On cutting through the falciform process, F, Fig. 1, Plate 44, we find Gimbernat's ligament, R, a structure well known in connexion with femoral hernia. Gimbernat's ligament consists of tendinous fibres which connect the inner end of the femoral arch with the pectineal ridge of the os pubis. The shape of the ligament is acutely triangular, corresponding to the form of the space which it occupies. Its apex is internal, and close to the pubic spine; its base is external, sharp and concave, and in apposition with the sheath of the femoral vessels. It measures an inch, more or less, in width, and it is broader in the male than in the female--a fact which is said to account for the greater frequency of femoral hernia in the latter sex than in the former, (Monro.) Its strength and density also vary in different individuals. It is covered anteriorly by, P, Fig. 1, Plate 44, the upper cornu of the falciform process; and behind, it is in connexion with, *k*, the conjoined tendon. This tendon is inserted with the ligament into the pectineal ridge. The falciform process also blends with the ligament; and thus it is that the femoral hernia, when constricted by either of these three structures, may well be supposed to suffer pressure from the three together.

A second or deep femoral arch is occasionally met with. This structure consists of tendinous fibres, lying deeper than, but parallel with, those of the superficial arch. The deep arch spans the femoral sheath more closely than the superficial arch, and occupies the interval left between the latter and the sheath of the vessels. When the deep arch exists, its inner end blends with the conjoined tendon and Gimbernat's ligament, and with these may also constrict the femoral hernia.

The sheath, *e f,* of the femoral vessels, E F, Fig. 1, Plate 43, passes from beneath the middle of the femoral arch. In this situation, the iliac part of the fascia lata, F G, Fig. 2, Plate 44, covers the sheath. Its inner side is bounded by Gimbernat's ligament, R, Fig. 1, Plate 44, and F, the falciform edge of the saphenous opening. On its outer side are situated the anterior crural nerve, and the femoral parts of the psoas and iliacus muscles. Of the three compartments into which the sheath is divided by two septa in its interior, the external one, E, Fig. 1, Plate 43, is occupied by the femoral artery; the middle one, F, by the femoral vein; whilst the inner one, G, gives passage to the femoral lymphatic vessels; and occasionally, also, a lymphatic body is found in it. The inner compartment, G, is the femoral canal, and through it the femoral hernia descends from the abdomen to the upper and forepart of the thigh. As the canal is the innermost of the three spaces inclosed by the sheath, it is that which lies in the immediate neighbourhood of the saphenous opening, Gimbernat's ligament, and the conjoined tendon, and between these structures and the femoral vein.

The sheath of the femoral vessels, like that of the spermatic cord, is infundibuliform. Both are broader at their abdominal ends than elsewhere. The femoral sheath being broader above than below, whilst the vessels are of a uniform diameter, presents, as it were, a surplus space to receive a hernia into its upper end. This space is the femoral or crural canal. Its abdominal entrance is the femoral or crural ring.

The femoral ring, H, Fig. 2, Plate 43, is, in the natural state of the parts, closed over by the peritonaeum, in the same manner as this membrane shuts the internal inguinal ring. There is, however, corresponding to each ring, a depression in the peritonaeal covering; and here it is that the bowel first forces the membrane and forms of this part its sac.

On removing the peritonaeum from the inguinal wall on the inner side of the iliac vessels, K L, we find the horizontal branch of the os pubis, and the parts connected with it above and below, to be still covered by what is called the subserous tissue. The femoral ring is not as yet discernible on the inner side of the iliac vein, K; for the subserous tissue being stretched across this aperture masks it. The portion of the tissue which closes the ring is named the crural septum, (Cloquet.) When we remove this part, we open the femoral ring leading to the corresponding canal. The ring is the point of union between the fibrous membrane of the canal and the general fibrous membrane which lines the abdominal walls external to the peritonaeum. This account of the continuity between the canal and abdominal fibrous membrane equally applies to the connexion existing between the general sheath of the vessels and the abdominal membrane. The difference exists in the fact, that the two outer compartments of the sheath are occupied by the vessels, whilst the inner one is vacant. The neck or inlet of the hernial sac, H, Fig. 2, Plate 43, exactly represents the natural form of the crural ring, as formed in the fibrous membrane external to, or (as seen in this view) beneath the peritonaeum.

The femoral ring, H, is girt round on all sides by a dense fibrous circle, the upper arc being formed by the two femoral arches; the outer arc is represented by the septum of the femoral sheath, which separates the femoral vein from the canal; the inner arc is formed by the united dense fibrous bands of the conjoined tendon and Gimbernat's ligament; and the inferior arc is formed by the pelvic fascia where this passes over the pubic bone to unite with the under part of the femoral canal and sheath. The ring thus bound by dense resisting fibrous structure, is rendered sharp on its pubic and upper sides by the salient edges of the conjoined tendon and Gimbernat's ligament, &c. From the femoral ring the canal extends down the thigh for an inch and a-half or two inches in a tapering form, supported by the pectineus muscle, and covered by the iliac part of the fascia lata. It lies side by side with the saphenous opening, but does not communicate with this place. On a level with the lower cornu of the saphenous opening, the walls of the canal become closely applied to the femoral vessels, and here it may be said to terminate.

The bloodvessels which pass in the neighbourhood of the femoral canal are, 1st. the femoral vein, F, Fig. 1, Plate 43, which enclosed in its proper sheath lies parallel with and close to the outer side of the passage. 2nd, Within the inguinal canal above are the spermatic vessels, resting on the upper surface of the femoral arch, which alone separates them from the upper part or entrance of the femoral canal. 3rd, The epigastric artery, F, Fig. 2, Plate 43, which passes close to the outer and upper border of, H, the femoral ring. This vessel occasionally gives off the obturator artery, which, when thus derived, will be found to pass towards the obturator foramen, in close connexion with the ring; that is, either descending by its outer border, G*, between this point and the iliac vein, K; or arching the ring, G, so as to pass down close to its inner or pubic border. In some instances, the vessel crosses the ring; a vein generally accompanies the artery. These peculiarities in the origin and course of the obturator artery, especially that of passing on the pubic side of the ring, behind Gimbernat's ligament and the conjoined tendon, E H, are fortunately very rare.

As the course to be taken by the bowel, when a femoral hernia is being formed, is through the crural ring and canal, the structures which have just now been enumerated as bounding this passage, will, of course, hold the like relation to the hernia. The manner in which a femoral hernia is formed, and the way in which it becomes invested in its descent, may be briefly stated thus: The bowel first dilates the peritonaeum opposite the femoral ring, H, Fig. 2, Plate 43, and pushes this membrane before it into the canal. This covering is the hernial sac. The crural septum has, at the same time, entered the canal as a second investment of the bowel. The hernia is now enclosed by the sheath, G, Fig. 1, Plate 43, of the canal itself. [Footnote 1] Its further progress through the saphenous opening, B F, Fig. 1, Plate 44, must be made either by rupturing the weak inner wall of the canal, or by dilating this part; in one or other of these modes, the herniary sac emerges from the canal through the saphenous opening. In general, it dilates the side of the canal, and this becomes the fascia propria, B G. If it have ruptured the canal, the hernial sac appears devoid of this covering. In either case, the hernia, increasing in size, turns up over the margin of F, the falciform process, [Footnote 2] and ultimately rests upon the iliac fascia lata, below the pubic third of Poupart's ligament. Sometimes the hernia rests upon this ligament, and simulates, to all outward appearance, an oblique inguinal hernia. In this course, the femoral hernia will have its three parts--neck, body, and fundus--forming nearly right angles with each other: its neck [Footnote 3] descends the crural canal, its body is directed to the pubis through the saphenous opening, and its fundus is turned upwards to the femoral arch.

[Footnote 1: The sheath of the canal, together with the crural septum, constitutes the "fascia propria" of the hernia (Sir Astley Cooper). Mr. Lawrence denies the existence of the crural septum.]

[Footnote 2: The "upper cornu of the saphenous opening," the "falciform process" (Burns), and the "femoral ligament" (Hey), are names applied to the same part. With what difficulty and perplexity does this impenetrable fog of surgical nomenclature beset the progress of the learner!]

[Footnote 3: The neck of the sac at the femoral ring lies very deep, in the undissected state of the parts (Lawrence).]

The crural hernia is much more liable to suffer constriction than the inguinal hernia. The peculiar sinuous course which the former takes from its point of origin, at the crural ring, to its place on Poupart's ligament, and the unyielding fibrous structures which form the canal through which it passes, fully account for the more frequent occurrence of this casualty. The neck of the sac may, indeed, be supposed always to suffer more or less constriction at the crural ring. The part which occupies the canal is also very much compressed; and again, where the hernia turns over the falciform process, this structure likewise must cause considerable compression on the bowel in the sac. [Footnote] This hernia suffers stricture of the passive kind always; for the dense fibrous bands in its neighbourhood compress it rather by withstanding the force of the herniary mass than by reacting upon it. There are no muscular fibres crossing the course of this hernia; neither are the parts which constrict it likely to change their original position, however long it may exist. In the inguinal hernia, the weight of the mass may in process of time widen the canal by gravitating; but the crural hernia, resting on the pubic bone, cannot be supposed to dilate the crural ring, however greatly the protrusion may increase in size and weight.

[Footnote: Sir A. Cooper (Crural Hernia) is of opinion that the stricture is generally in the neck of the sheath. Mr. Lawrence remarks, "My own observations of the subject have led me to refer the cause of stricture to the thin posterior border (Gimbernat's ligament) of the crural arch, at the part where it is connected to the falciform process." (Op. cit.) This statement agrees also with the experience of Hey, (Practical Obs.)]

DESCRIPTION OF THE FIGURES OF PLATES 43 & 44.

PLATE 43.

FIGURE 1.

A. Anterior superior iliac spine.

B. Iliacus muscle, cut.

C. Anterior crural nerve, cut.

D. Psoas muscle, cut.

E. Femoral artery enclosed in *e*, its compartment of the femoral sheath.

F. Femoral vein in its compartment, *f*, of the femoral sheath.

G. The fascia propria of the hernia; *g*, the contained sac.

H. Gimbernat's ligament.

I. Round ligament of the uterus.

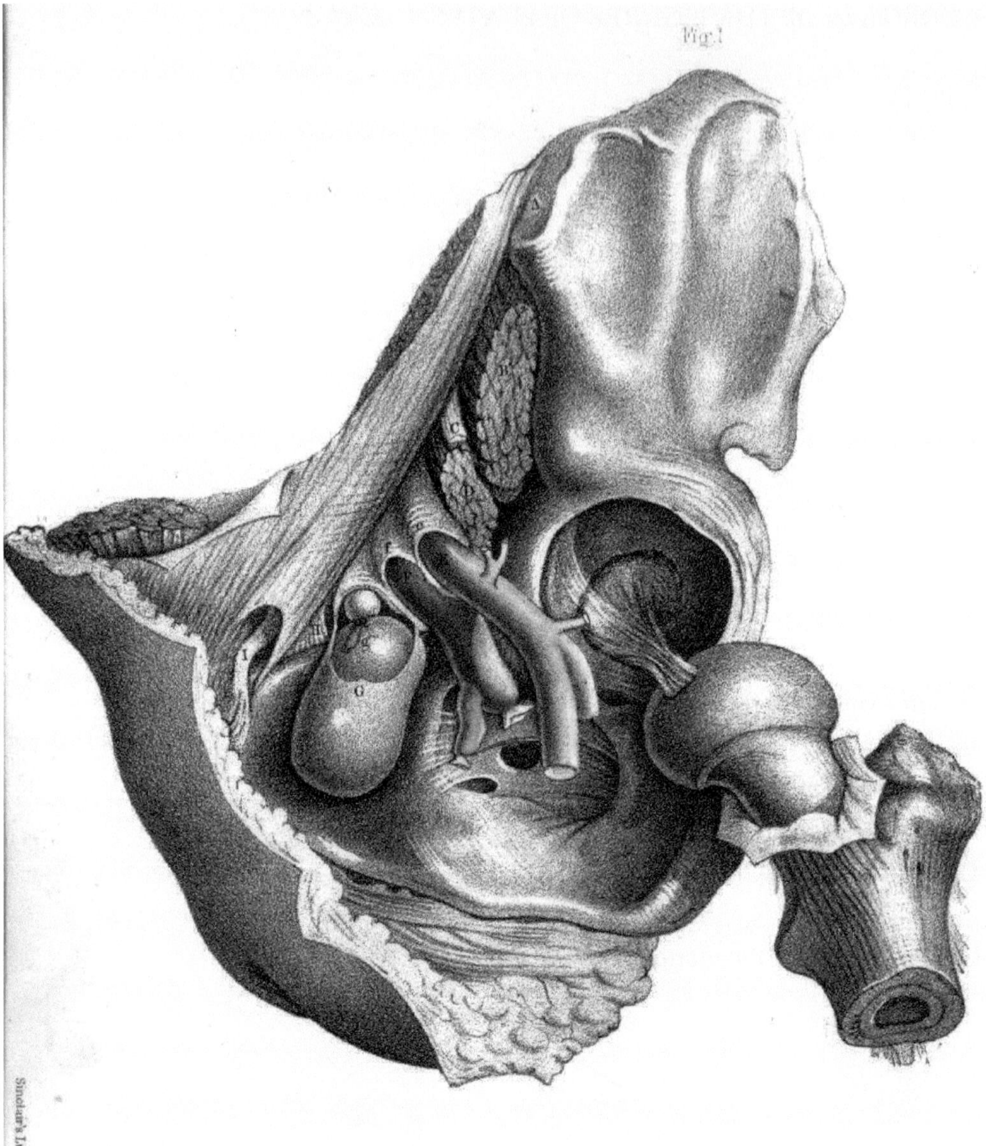

PLATE 43.--FIGURE 1.

FIGURE 2.

A. Anterior superior iliac spine.

B. Symphysis pubis.

C. Rectus abdominis muscle.

D. Peritonaeum.

E. Conjoined tendon.

F. Epigastric artery.

G* G. Positions of the obturator artery when given off from the epigastric.

H. Neck of the sac of the crural hernia.

I. Round ligament of the uterus.

K. External iliac vein.

L. External iliac artery.

M. Tendon of the psoas parvus muscle, resting on the psoas magnus.

N. Iliacus muscle.

O. Transversalis fascia.

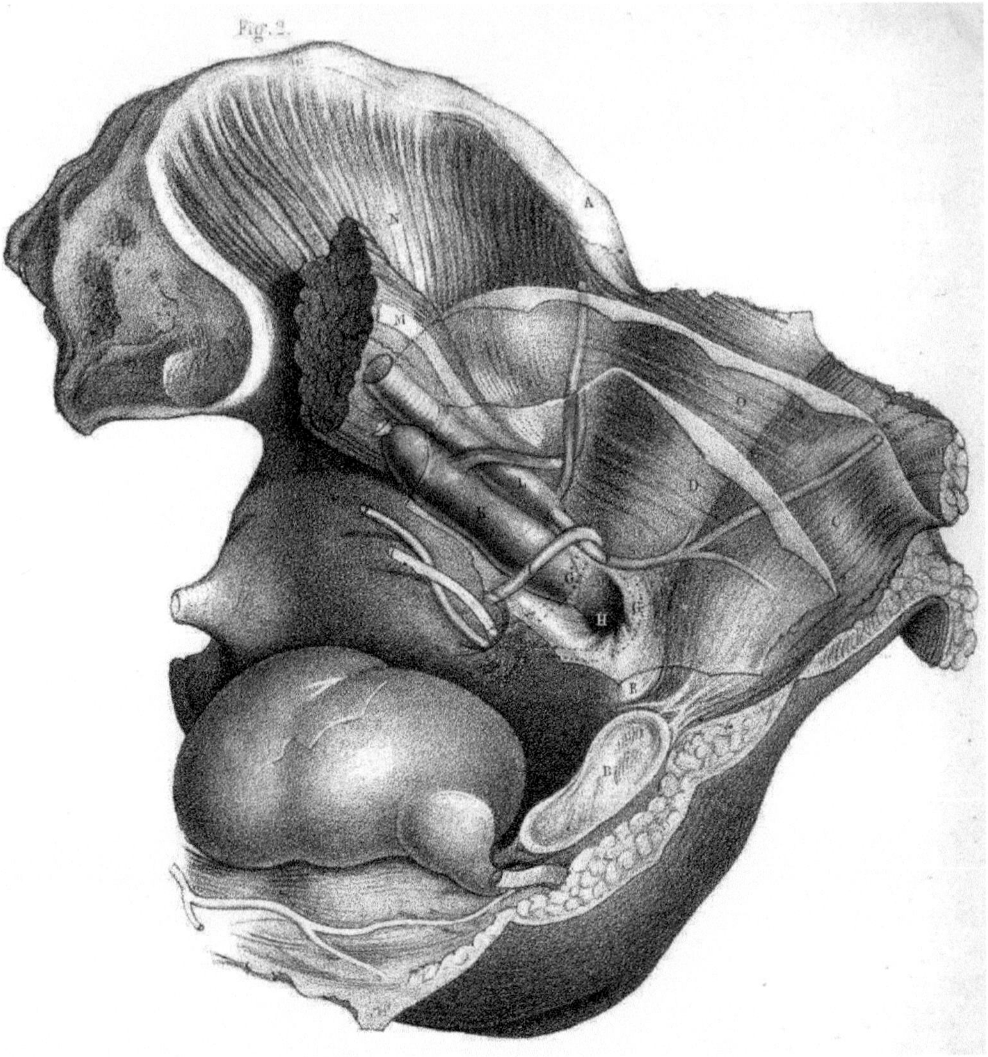

PLATE 43.--FIGURE 2.

PLATE 44.

FIGURE 1.

A. Anterior superior iliac spine.

B. The crural hernia.

C. Round ligament of the uterus.

D. External oblique muscle; *d*, Fig. 2, its aponeurosis.

E. Saphaena vein.

F. Falciform process of the saphenous opening.

G. Femoral artery in its sheath.

H. Femoral vein in its sheath.

I. Sartorius muscle.

K. Internal oblique muscle; *k*, conjoined tendon.

L L. Transversalis fascia.

M. Epigastric artery.

N. Peritonaeum.

O. Anterior crural nerve.

P. The hernia within the crural canal.

Q Q. Femoral sheath.

R. Gimbernat's ligament.

FIGURE 2.

The other letters refer to the same parts as seen in Fig. 1.

G. Glands in the neighbourhood of Poupart's ligament.

H. Glands in the neighbourhood of the saphenous opening.

I. The sartorius muscle seen through its fascia.

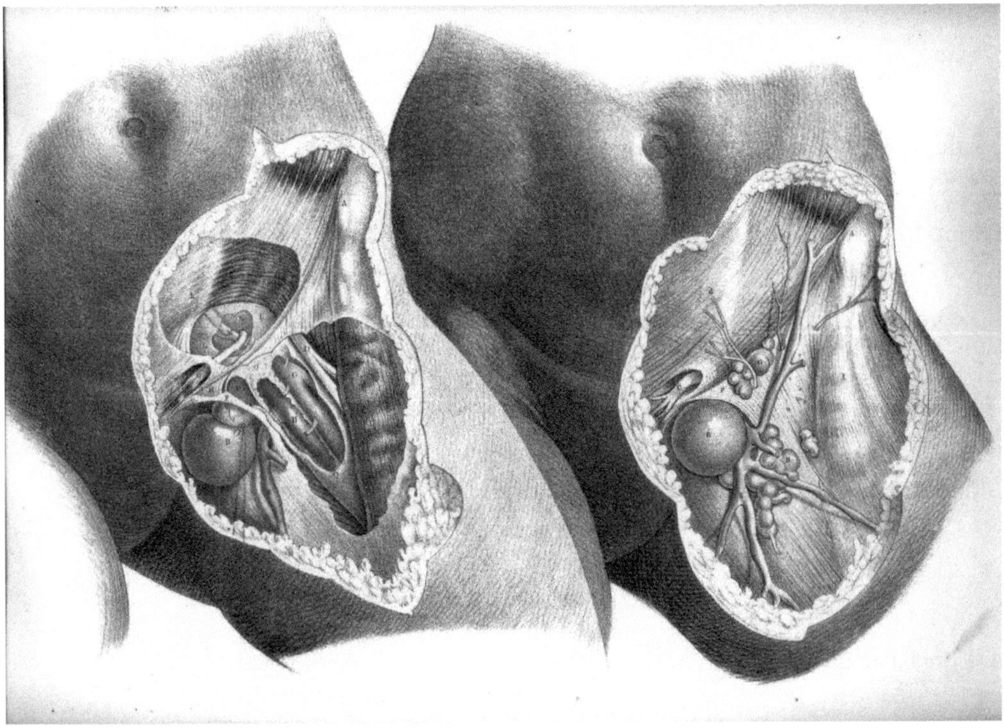

PLATE 44.--FIGURE 1, 2.

DEMONSTRATIONS OF THE ORIGIN AND PROGRESS OF FEMORAL HERNIA; ITS DIAGNOSIS, THE TAXIS, AND THE OPERATION.

Its course compared with that of the inguinal hernia. Its investments and relations. Its diagnosis from inguinal hernia, &c. Its varieties. Mode of performing the taxis according to the course of the hernia. The operation for the strangulated condition. Proper lines in which incisions should be made. Necessity for and mode of opening the sac.

PLATE 45, Fig. 1.--The point, 3, from which an external inguinal hernia first progresses, and the part, 5, within which the femoral hernia begins to be formed, are very close to each other. The inguinal hernia, 3, arising above, 5, the crural arch, descends the canal, 3, 3, under cover of the aponeurosis of the external oblique muscle, obliquely downwards and inwards till it gains the external abdominal ring formed in the aponeurosis, and thence descends to the scrotum. The femoral hernia, commencing on a level with, 5, the femoral arch, descends the femoral canal, under cover of the fascia lata, and appears on the upper and forepart of the thigh at the saphenous opening, 6, 7, formed in the fascia lata; and thence, instead of descending to the scrotum, like the inguinal hernia, turns, on the contrary, up over the falciform process, 6, till its fundus rests near, 5, the very place beneath which it originated. Such are the peculiarities in the courses of these two hernial; and they are readily accounted for by the anatomical relations of the parts concerned.

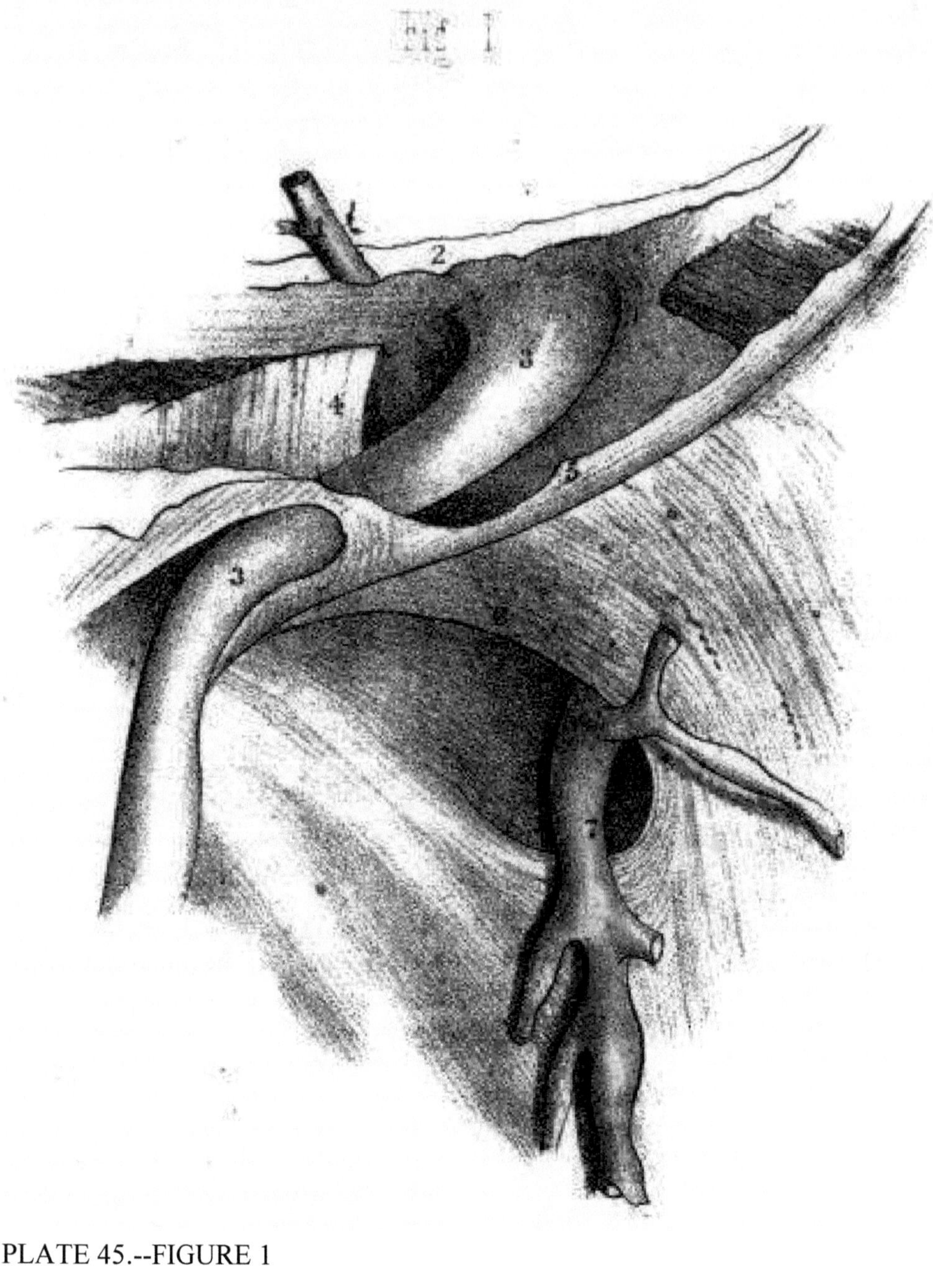

PLATE 45.--FIGURE 1

PLATE 45, Fig. 2.--There exists a very evident analogy between the canals through which both herniae pass. The infundibuliform fascia, 3, 3, of the spermatic vessels is like the infundibuliform sheath, 9, 9, of the femoral vessels. Both sheaths are productions of the general fibrous membrane of the abdomen. They originate from nearly the same locality. The ring of the femoral canal, 12, is situated immediately below, but to the inner side of the internal inguinal ring, 3. The epigastric artery, 1, marks the width of the interval which separates the two rings. Poupart's ligament, 5, being the line of union between the oblique aponeurosis of the abdominal muscle and the fascia lata, merely overarches the femoral sheath, and does not separate it absolutely from the spermatic sheath.

Fig. 2

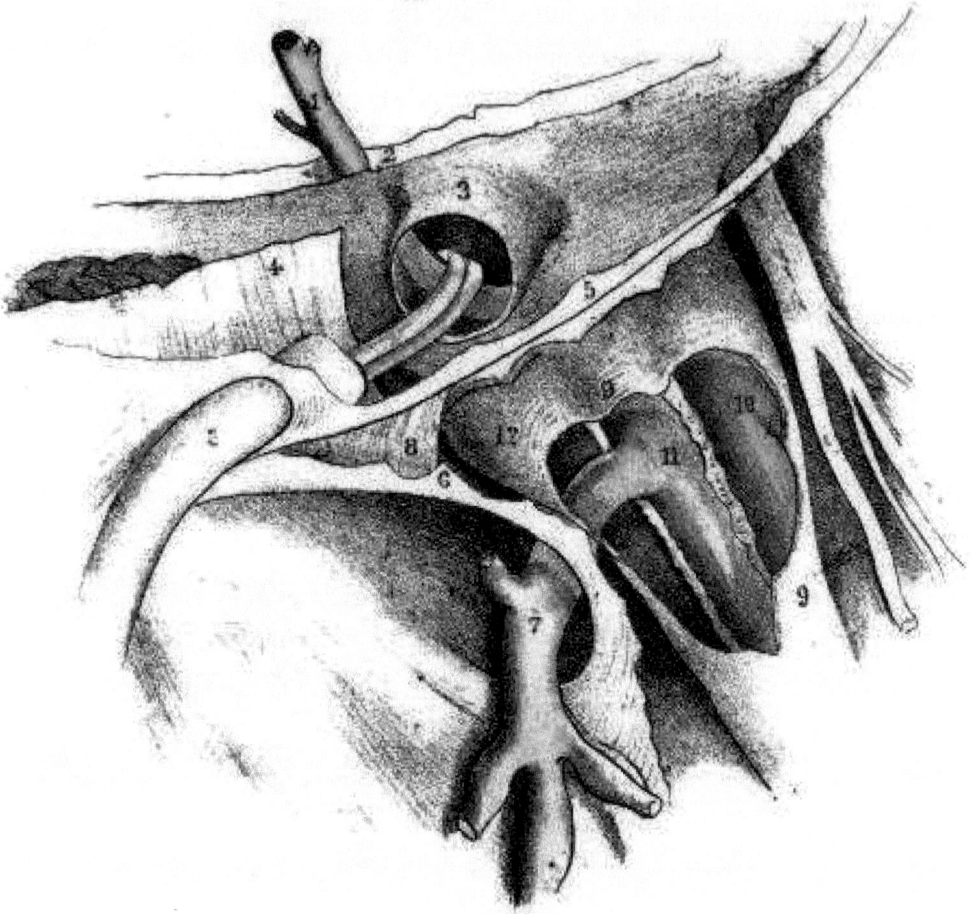

PLATE 45.--FIGURE 2

PLATE 45, Fig. 3.--The peritonaeum, 2, 3, closes the femoral canal, 12, at the femoral ring, in the same way as this membrane closes the inguinal canal at the internal inguinal ring, 3, Fig. 2, Plate 45. The epigastric artery always holds an intermediate position between both rings. The spermatic vessels in the inguinal tube, 3, 3, Fig. 2, Plate 45, are represented by the round ligament in the female inguinal canal, Fig. 3, Plate 45. When the bowel is about to protrude at either of the rings, it first dilates the peritonaeum, which covers these openings.

Fig. 3

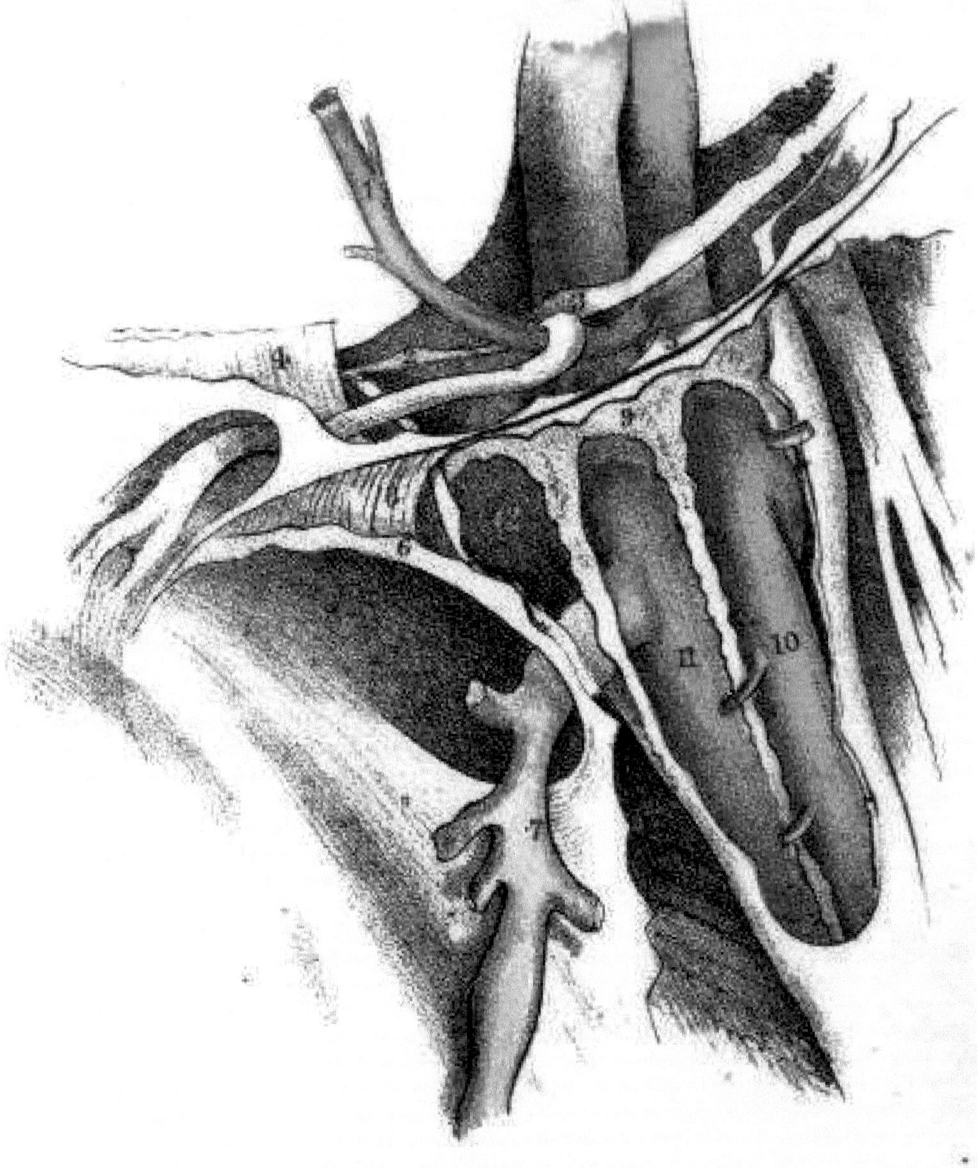

PLATE 45.--FIGURE 3

PLATE 45, Fig. 4.--The place of election for the formation of any hernia is that which is structurally the weakest. As the space which the femoral arch spans external to the vessels is fully occupied by the psoas and iliacus muscles, and, moreover, as the abdominal fibrous membrane and its prolongation, the femoral sheath, closely embrace the vessels on their outer anterior and posterior sides, whilst on their inner side the membrane and sheath are removed at a considerable interval from the vessels, it is through this interval (the canal) that the hernia may more readily pass. The peritonaeum, 2, and crural septum, 13, form at this place the only barrier against the protrusion of the bowel into the canal.

Fig. 4.

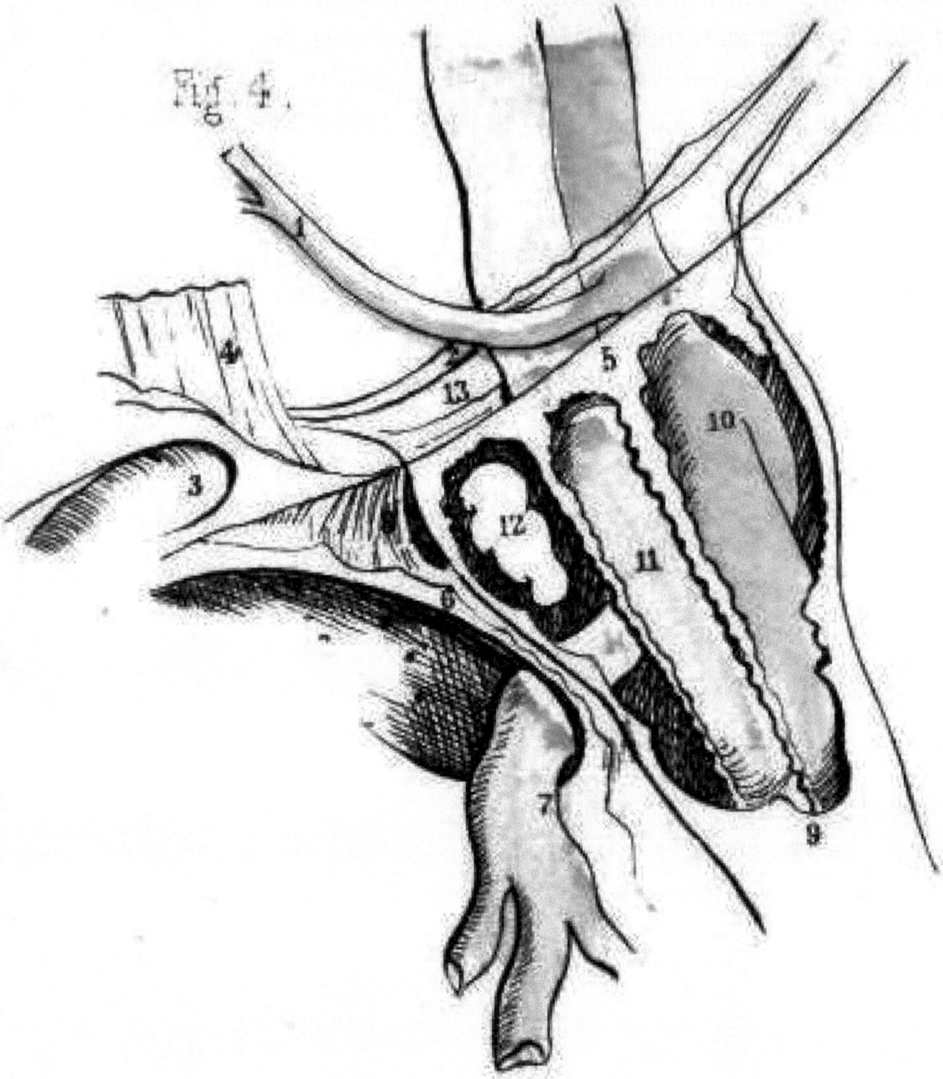

PLATE 45.--FIGURE 4

PLATE 45, Fig. 5.--The hernia cannot freely enter the compartment, 10, occupied by the artery, neither can it enter the place 11, occupied as it is by the vein. It cannot readily pass through the inguinal wall at a point internal to, 9, the crural sheath, for here it is opposed by, 4, the conjoined tendon, and by, 8, Gimbernat's ligament. Neither will the hernia force a way at a point external to the femoral vessels in preference to that of the crural canal, which is already prepared to admit it. [Footnote] The bowel, therefore, enters the femoral canal, 9, and herein it lies covered by its peritonaeal sac, derived from that part of the membrane which once masked the crural ring. The septum crurale itself, having been dilated before the sac, of course invests it also. The femoral canal forms now the third covering of the bowel. If in this stage of the hernia it should suffer constriction, Gimbernat's ligament, 8, is the cause of it. An incipient femoral hernia of the size of 2, 12, cannot, in the undissected state of the parts, be detected by manual operation; for, being bound down by the dense fibrous structures which gird the canal, it forms no apparent tumour in the groin.

[Footnote: The mode in which the femoral sheath, continued from the abdominal membrane, becomes simply applied to the sides of the vessels, renders it of course not impossible for a hernia to protrude into the sheath at any point of its abdominal entrance. Mr. Stanley and M. Cloquet have observed a femoral hernia external to the vessels. Hesselbach has also met with this variety. A hernia of this nature has come under my own observation. Cloquet has seen the hernia descend the sheath *once* in front of the vessels, and *once* behind them. These varieties, however, must be very rare. The external form has never been met with by Hey, Cooper, or Scarpa; whilst no less than six instances of it have come under the notice of Mr. Macilwain, (on Hernia, p. 293.)]

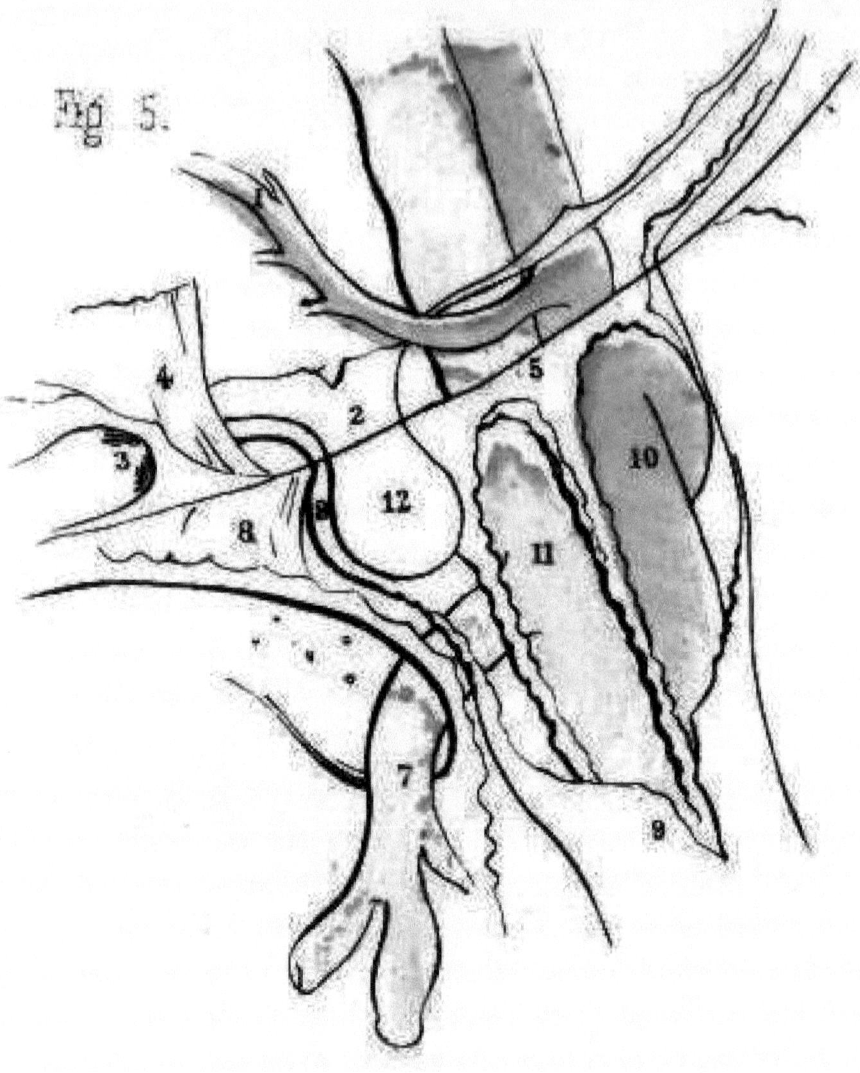

PLATE 45.--FIGURE 5

PLATE 45, Fig. 6.--The hernia, 2, 12, increasing gradually in size, becomes tightly impacted in the crural canal, and being unable to dilate this tube uniformly to a size corresponding with its own volume, it at length bends towards the saphenous opening, 6, 7, this being the more easy point of egress. Still, the neck of the sac, 2, remains constricted at the ring, whilst the part which occupies the canal is also very much narrowed. The fundus of the sac, 9*, 12, alone expands, as being free of the canal; and covering this part of the hernia may be seen the fascia propria, 9*. This fascia is a production of the inner wall of the canal; and if we trace its sides, we shall find its lower part to be continuous with the femoral sheath, whilst its upper part is still continuous with the fascia transversalis. When the hernia ruptures the saphenous side of the canal, the fascia propria is, of course, absent.

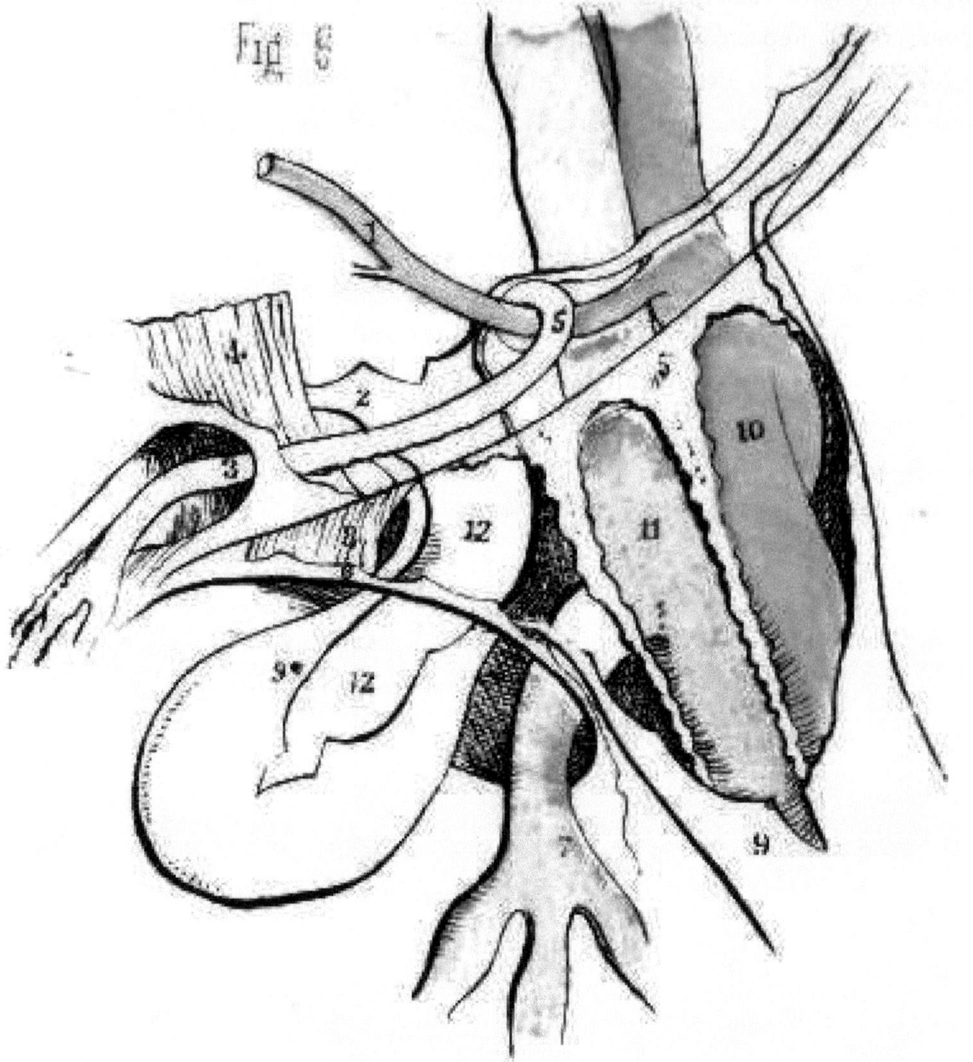

PLATE 45.--FIGURE 6

PLATE 46, Fig. 1.--The anatomical circumstances which serve for the diagnosis of a femoral from an inguinal hernia may be best explained by viewing in contrast the respective positions assumed by both complaints. The direct hernia, 13, traverses the inguinal wall from behind, at a situation corresponding with the external ring; and from this latter point it descends the scrotum. An oblique external inguinal hernia enters the internal ring, 3, which exists further apart from the general median line, and, in order to gain the external ring, has to take an oblique course from without inwards through the inguinal canal. A femoral hernia enters the crural ring, 2, immediately below, but on the inner side of, the internal inguinal ring, and descends the femoral canal, 12, vertically to where it emerges through, 6, 7, the saphenous opening. The direct inguinal hernia, 13, owing to its form and position, can scarcely ever be mistaken for a femoral hernia. But in consequence of the close relationship between the internal inguinal ring, 3, and the femoral ring, 2, through which their respective herniae pass, some difficulty in distinguishing between these complaints may occur. An incipient femoral hernia, occupying the crural canal between the points, 2, 12, presents no apparent tumour in the undissected state of the parts; and a bubonocele, or incipient inguinal hernia, occupying the inguinal canal, 3, 3, where it is braced down by the external oblique aponeurosis, will thereby be also obscured in some degree. But, in most instances, the bubonocele distends the inguinal canal somewhat; and the impulse which on coughing is felt at a place above the femoral arch, will serve to indicate, by negative evidence, that it is not a femoral hernia.

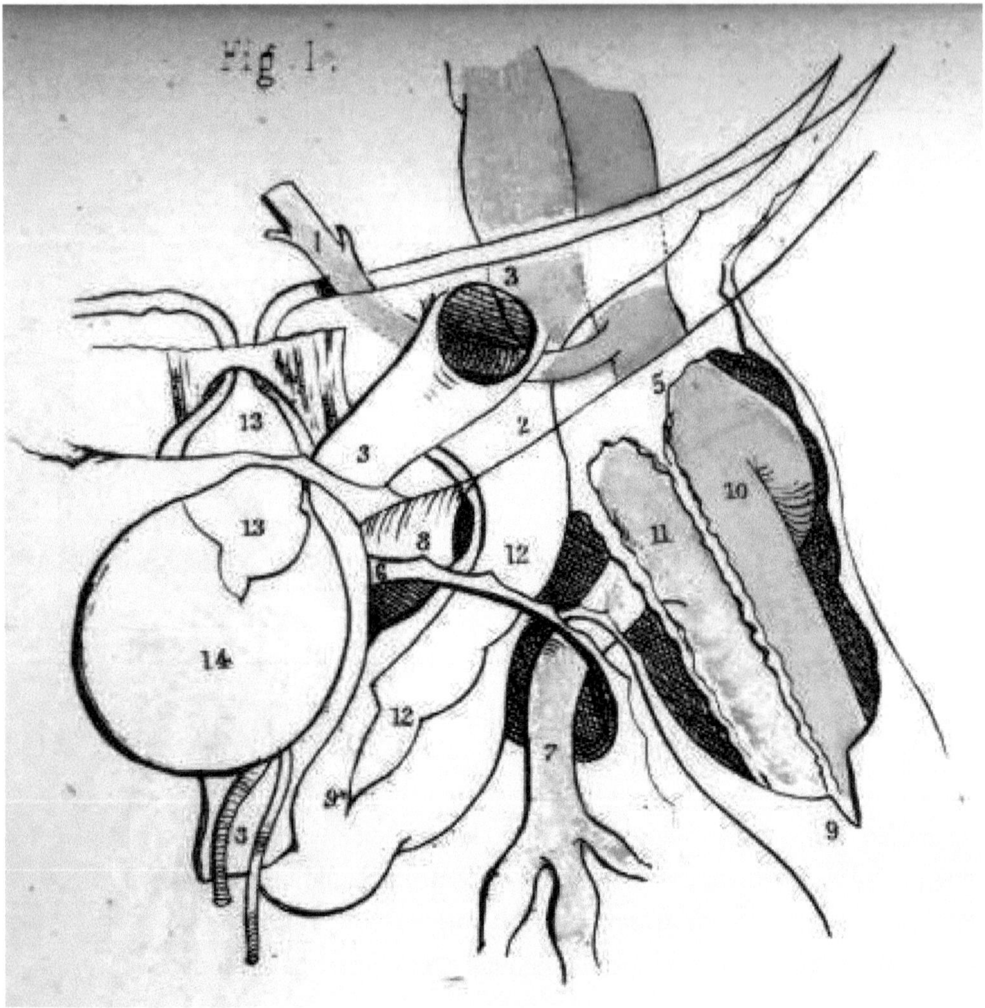

PLATE 46.--FIGURE 1

PLATE 46, Fig. 2.--When the inguinal and femoral herniae are fully produced, they best explain their distinctive nature. The inguinal hernia, 13, descends the scrotum, whilst the femoral hernia, 9*, turns over the falciform process, 6, and rests upon the fascia lata and femoral arch. Though in this position the fundus of a femoral hernia lies in the neighbourhood of the inguinal canal, 3, yet the swelling can scarcely be mistaken for an inguinal rupture, since, in addition to its being superficial to the aponeurosis which covers the inguinal canal, and also to the femoral arch, it may be withdrawn readily from this place, and its body, 12, traced to where it sinks into the saphenous opening, 6, 7, on the upper part of the thigh. An inguinal hernia manifests its proper character more and more plainly as it advances from its point of origin to its termination in the scrotum. A femoral hernia, on the contrary, masks its proper nature, as well at its point of origin as at its termination. But when a femoral hernia stands midway between these two, points--viz. in the saphenous opening, 6, 7, it best exhibits its special character; for here it exists below the femoral arch, and considerably apart from the external abdominal ring.

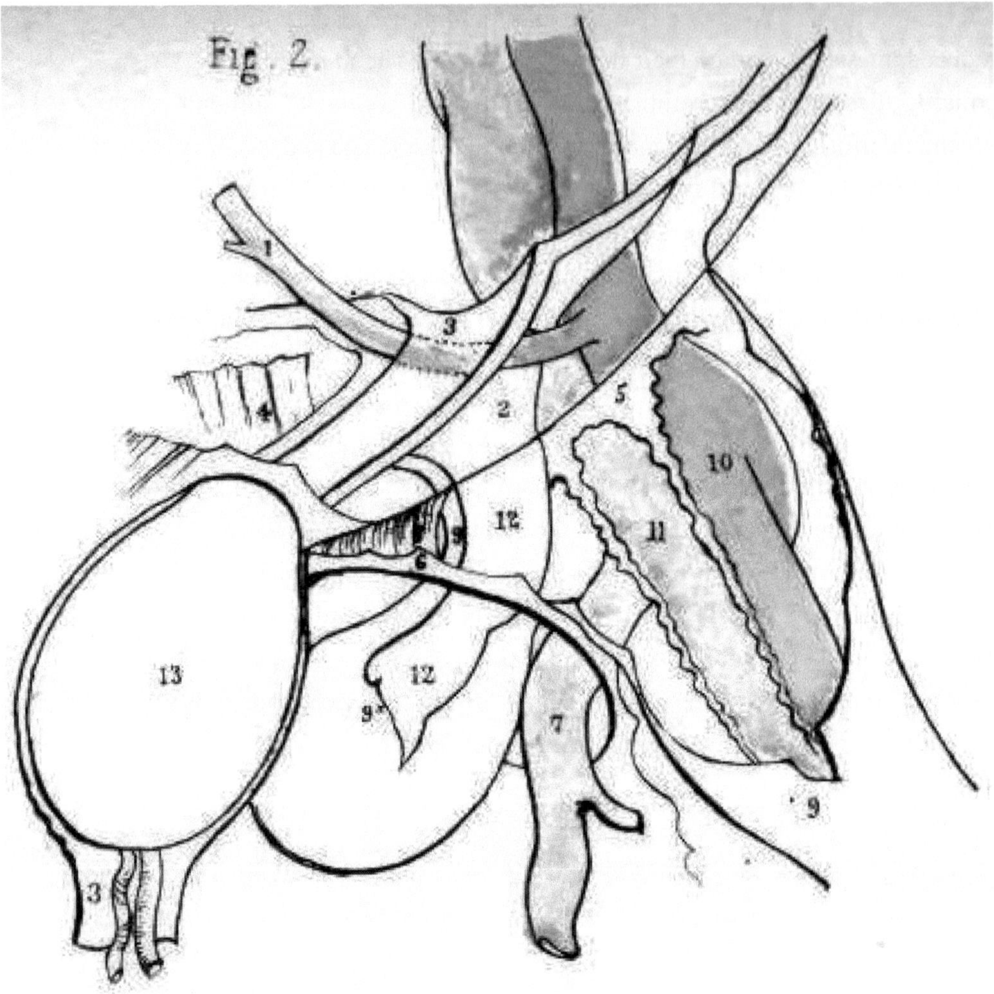

PLATE 46.--FIGURE 2

PLATE 46, Fig. 3.--The neck of the sac of a femoral hernia, 2, lies always close to, 3, the epigastric artery. When the obturator artery is derived from the epigastric, if the former pass internal to the neck behind, 8, Gimbernat's ligament, it can scarcely escape being wounded when this structure is being severed by the operator's knife. If, on the other hand, the obturator artery descend external to the neck of the sac, the vessel will be comparatively remote from danger while the ligament is being divided. In addition to the fact that the cause of stricture is always on the pubic side, 8, of the neck of the sac, 12, thereby requiring the incision to correspond with this situation only, other circumstances, such as the constant presence of the femoral vein, 11, and the epigastric artery, 1, determine the avoidance of ever incising the canal on its outer or upper side. And if the obturator artery, [Footnote] by rare occurrence, happen to loop round the inner side of the neck of the sac, supposing this to be the seat of stricture, what amount of anatomical knowledge, at the call of the most dexterous operator, can render the vessel safe from danger?

[Footnote: M. Velpeau (Medecine Operatoire), in reference to the relative frequency of cases in which the obturator artery is derived from the epigastric, remarks, "L'examen que j'ai pu en faire sur plusieurs milliers de cadavres, ne me permet pas de dire qu'elle se rencontre un fois sur trois, ni sur cinq, ni meme sur dix, mais bien seulement sur quinze a vingt." Monro (Obs. on Crural Hernia) states this condition of the obturator artery to be as 1 in 20-30. Mr. Quain (Anatomy of the Arteries) gives, as the result of his observations, the proportion to be as 1 in 3-1/2, and in this estimate he agrees to a great extent with the observations of Cloquet and Hesselbach. Numerical tables have also been drawn up to show the relative frequency in which the obturator descends on the outer and inner borders of the crural ring and neck of the sac. Sir A. Cooper never met with an example where the vessel passed on the inner side of the sac, and from this alone it may be inferred that such a position of the vessel is very rare. It is generally admitted that the obturator artery, when derived from the epigastric, passes down much more frequently between the iliac vein and outer border of the ring. The researches of anatomists (Monro and others) in reference to this point have given rise to the question,

"What determines the position of the obturator artery with respect to the femoral ring?" It appears to me to be one of those questions which do not admit of a precise answer by any mode of mathematical computation; and even if it did, where then is the practical inference?]

The taxis, in a case of crural hernia, should be conducted in accordance with anatomical principles. The fascia lata, Poupart's ligament, and the abdominal aponeurosis, are to be relaxed by bending the thigh inwards to the hypogastrium. By this measure, the falciform process, 6, is also relaxed; but I doubt whether the situation occupied by Gimbernat's ligament allows this part to be influenced by any position of the limb or abdomen. The hernia is then to be drawn from its place above Poupart's ligament, (if it have advanced so far,) and when brought opposite the saphenous opening, gentle pressure made outwards, upwards, and backwards, so as to slip it beneath the margin of the falciform process, will best serve for its reduction. When this cannot be effected by the taxis, and the stricture still remains, the cutting operation is required.

The precise seat of the stricture cannot be known except during the operation. But it is to be presumed that the sac and contained intestine suffer constriction throughout the whole length of the canal. [Footnote] Previously to the commencement of the operation, the urinary bladder should be emptied; for this organ, in its distended state, rises above the level of the pubic bone, and may thus be endangered by the incision through the stricture--especially if Gimbernat's ligament be the structure which causes it.

[Footnote: "The seat of the stricture is not the same in all cases, though, in by far the greater number of instances, the constriction is relieved by the division upwards and inwards of the falciform process of the fascia lata, and the lunated edge of Gimbernat's ligament, where they join with each other. In some instances, it will be the fibres of the deep crescentic (femoral) arch; in others, again, the neck of the sac itself, and produced by a thickening and contraction of the subserous and peritonaeal membranes where they lie within the circumference of the crural ring."--*Morton* (Surgical Anatomy of the Groin p. 148).]

An incision commencing a little way above Poupart's ligament, is to be carried vertically over the hernia, parallel with, but to the inner side of its median line. This incision divides the skin and subcutaneous adipose membrane, which latter varies considerably in quantity in several individuals. One or two small arteries (superficial pubic, &c.) may be divided, and some lymphatic bodies exposed. On cautiously turning aside the incised adipose membrane contained between the two layers of the superficial fascia, the fascia propria, 9, Figs. 4, 5, Plate 46, of the hernia is exposed. This envelope, besides varying in thickness in two or more cases, may be absent altogether. The fascia closely invests the sac, 12; but sometimes a layer of fatty substance interposes between the two coverings, and resembles the omentum so much, that the operator may be led to doubt whether or not the sac has been already opened. The fascia is to be cautiously slit open on a director; and now the sac comes in view. The hernia having been drawn outwards, so as to separate it from the inner wall of the crural canal, a director [Footnote] is next to be passed along the interval thus left, the groove of the instrument being turned to the pubic side. The position of the director is now between the neck of the sac and the inner wall of the canal. The extent to which the director passes up in the canal will vary according to the suspected level of the stricture. A probe-pointed bistoury is now to be slid along the director, and with its edge turned upwards and inwards, according to the seat of stricture, the following mentioned parts are to be divided--viz., the falciform process, 6; the inner wall of the canal, which is continuous with the fascia propria, 9; Gimbernat's ligament, 8; and the conjoined tendon, 4; where this is inserted with the ligament into the pectineal ridge. By this mode of incision, which seems to be all-sufficient for the liberation of the stricture external to the neck of the sac, we avoid Poupart's ligament; and thereby the spermatic cord, 3, and epigastric artery, 1, are not endangered. The crural canal being thus laid open on its inner side, and the constricting fibrous bands being severed, the sac may now be gently manipulated, so as to restore it and its contents to the cavity of the abdomen; but if any impediment to the reduction still remain, the cause, in all probability, arises either from the neck of the sac having become strongly adherent to the crural ring, or from the bowel being bound by bands of false membrane to the sac. In either case, it will be necessary to open the sac, and examine its contents. The neck of the sac is then to be exposed by an incision carried through the integument across the upper end of the first incision, and parallel with Poupart's ligament. The neck is then to be divided on its inner side, and the exposed intestine may now be restored to the abdomen.

[Footnote: The finger is the safest director; for at the same time that it guides the knife it feels the stricture and protects the bowel. As all the structures which are liable to become the seat of stricture--viz., the falciform process, Gimbernat's ligament, and the conjoined tendon, lie in very close apposition, a very short incision made upwards and inwards is all that is required.]

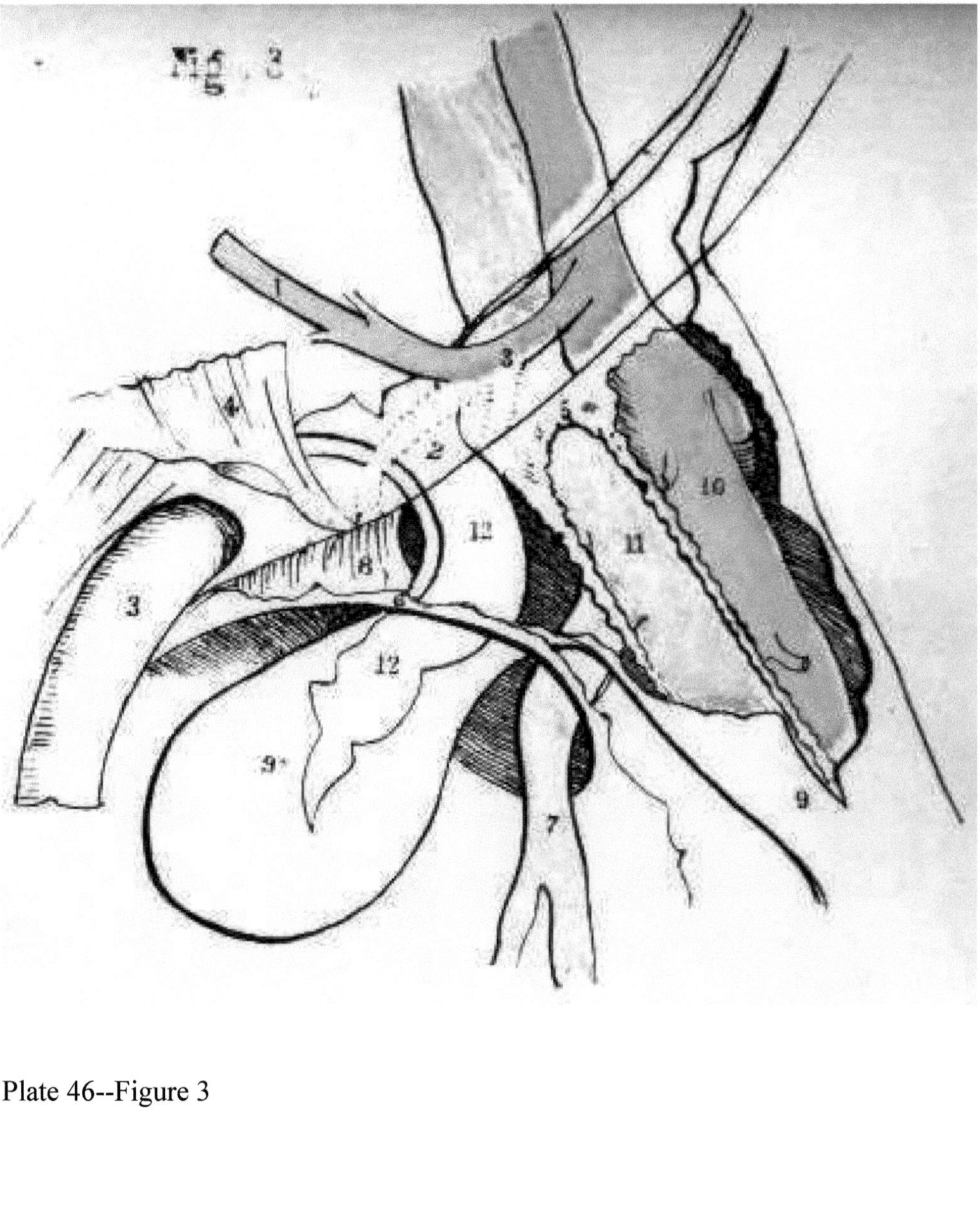

Plate 46--Figure 3

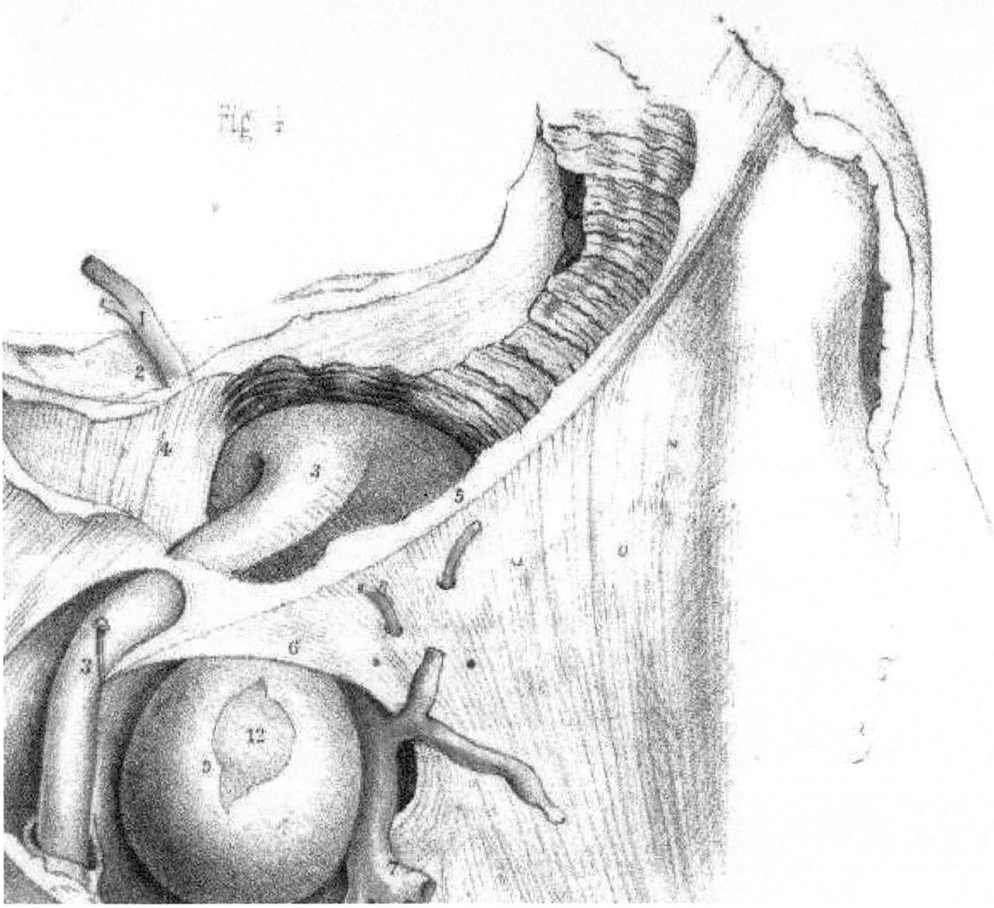

Plate 46--Figure 4

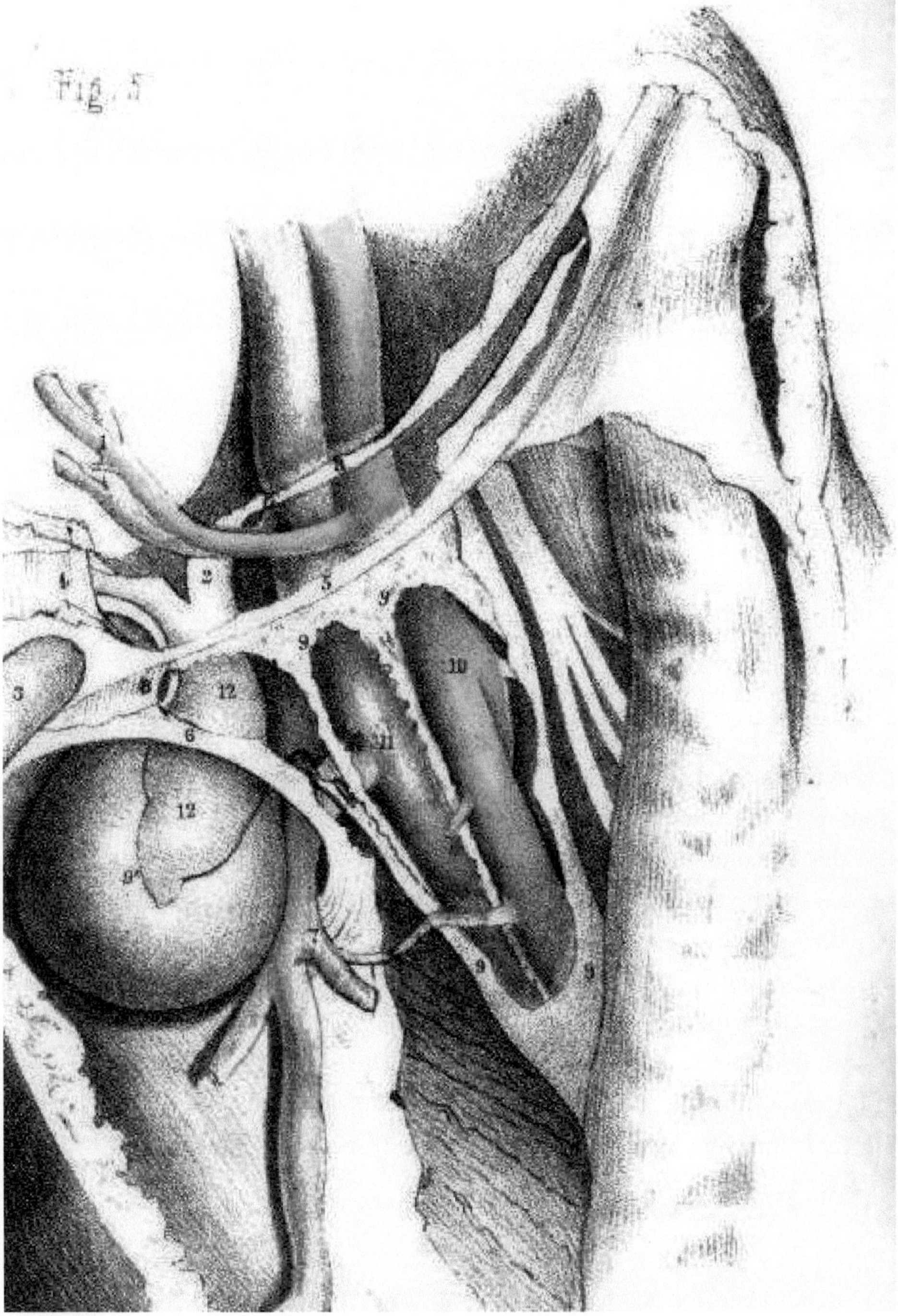

Plate 46--Figure 5

THE SURGICAL DISSECTION OF THE PRINCIPAL BLOODVESSELS AND NERVES OF THE ILIAC AND FEMORAL REGIONS.

The femoral triangle. Eligible place for tying the femoral artery. The operations of Scarpa and Hunter. Remarks on the common femoral artery. Ligature of the external iliac artery according to the seat of aneurism.

Through the groin, as through the axilla, the principal blood vessels and nerves are transmitted to, the corresponding limb. The main artery of the lower limb frequently becomes the subject of a surgical operation. The vessel is usually described as divisible into parts, according to the regions which it traverses. But, as in examining any one of those parts irrespective of the others, many facts of chief surgical importance are thereby obscured and overlooked, I propose to consider the vessel *as a whole*, continuous from the aorta to where it enters the popliteal space. The general course and position of the main artery may be described as follows:--The abdominal aorta, A, bifurcates on the body of the fourth lumbar vertebra. The level of the aortic bifurcation corresponds with the situation of the navel in front, and the crista ilii laterally. The aorta is in this situation borne so far forwards by the lumbar spine as to occupy an almost central position in the cavity of the abdomen. If the abdomen were pierced by two lines, one extending from a little to the left side of the navel, horizontally backwards to the fourth lumbar vertebra, and the other from immediately over the middle of one crista ilii, transversely to a corresponding point in the opposite side, these lines would intersect at the aortic bifurcation. The two arteries, G G,* into which the aorta divides symmetrically at the median line, diverge from one another in their descent towards the two groins. As both vessels correspond in form and relative position, the description of one will serve for the other.

While the thigh is abducted and rotated outwards, if a line be drawn from the navel to a point, D, of the inguinal fold, midway between B, the anterior iliac spine, and C, the symphysis pubis, and continued thence to the inner condyle of the femur, it would indicate the general course of the artery, G I W. In this course, the vessel may be regarded as a main trunk, giving off at intervals large branches for the supply of the pelvic organs, the abdominal parietes, and the thigh. From the point where the vessel leaves the aorta, A, down to the inguinal fold, D, it lies within the abdomen, and here, therefore, all operations affecting the vessel are attended with more difficulty and danger than elsewhere, in its course.

The artery of the lower limb, arising at the bifurcation of the aorta on the fourth lumbar vertebra, descends obliquely outwards to the sacra-iliac junction, and here it gives off its first branch, G, (internal iliac,) to the pelvic organs. The main vessel is named *common iliac*, at the interval between its origin from the aorta and the point where it gives off the internal iliac branch. This interval is very variable as to its length, but it is stated to be usually two inches. The artery, I, continuing to diverge in its first direction from its fellow of the opposite side, descends along the margin of the true pelvis as far as Poupart's ligament, D, where it gives off its next principal branches,--viz., the epigastric and circumflex iliac. At the interval between the internal iliac and epigastric branches, the main artery, I, is named *external iliac*; and the surgical length of this part is also liable to vary, in consequence of the epigastric or circumflex iliac branches arising higher up or lower down than usual. The main vessel, after passing beneath the middle of Poupart's ligament, D, next gives off the profundus branch, N, to supply the thigh. This branch generally arises at a point an inch and half or two inches below the fold of the groin; and between it and the epigastric above, the main artery is named *common femoral.* From the point where the profundus branch arises, down to the popliteal space, the vessel remains as an undivided trunk, being destined to supply the leg and foot. In this course, the artery is accompanied by the vein, H K O, which, according to the region in which it lies, assumes different names, corresponding to those applied to the artery. Both vessels may now be viewed in relation to each other, and to the several structures which lie in connexion with them.

The two vessels above Poupart's ligament lie behind the intestines, and are closely invested by the serous membrane. The origin of the vena cava, F, lies close to the right side of the bifurcation of the aorta, A; and here both vessels are supported by the lumbar spine. Each of the two arteries, G G,* into which the aorta divides, has its accompanying vein, H, on its inner side, but the common iliac part of the right artery is seen to lie upon the upper portions of both the veins, as these joining beneath it form the commencement of the vena cava. The external iliac part, I, of each artery has its vein, K, on its inner side. At the point, G, where the artery gives off its internal iliac branch, the ureter, *g*, crosses it, and thence descends to the bladder. The internal iliac branch subdivides in general so soon after its origin, that it may be regarded as for the most part an unsafe proceeding to place a ligature upon it.

The iliac vessels, A G I, in approaching Poupart's ligament along the border of the true pelvis, are supported by the psoas muscle, and invested and bound to their place by the peritonaeum, and a thin process of the iliac fascia. Some lymphatic glands are here found to lie over the course of the vessels. The spermatic artery and vein, together with the genito-crural nerve, descend along the outer border of the iliac artery. When arrived at Poupart's ligament, the iliac vessels, I K, become complicated by their own branches, and also by the spermatic vessels, as these are about to pass from the abdomen through the internal inguinal ring. While passing beneath the middle of Poupart's ligament, D, the iliac artery, I, having its vein, K, close to its inner side, rests upon the inner border of the psoas muscle, and in this place it may be effectually compressed against the os pubis. The anterior crural nerve, P, which in the iliac region lies concealed by the psoas muscle, and separated by this from the vessels, now comes into view, lying on the outer side of the artery. When the vessels have passed from beneath Poupart's ligament, the serous membrane no longer covers them, but the fibrous membrane is seen to invest them in the form of a sheath, divided into two compartments, one of which (internal) receives the vein, the other the artery. The iliac vessels, in passing to the thigh, assume the name of femoral.

The femoral vessels, O N W, in the upper third of the thigh traverse a triangular space, the base of which is formed by Poupart's ligament, D, whilst the sides and apex are formed by the sartorius, Q, and adductor longus muscles, T, approaching each other. In the undissected state of the part, the structures which bound this space can in general be easily recognised. A central depression extends from the middle of its base, D, to its apex, V, and marks the course of the vessels. Near the middle of Poupart's ligament, the vessels are comparatively superficial, and here the artery may be felt pulsating; but lower down, as they approach the apex of the triangle, the vessels become gradually deeper, till the sartorius muscle inclining from its origin obliquely inwards to the centre of the thigh, w, at length overlaps them. The inner border of the sartorius muscle at the lower part of the upper third of the thigh, W, guides to the position of the artery. Whilst traversing the femoral triangle, the vessels enclosed in their proper sheath are covered by the fascia lata, adipose membrane, and integument. In this place they lie imbedded in loose cellular and adipose tissue. The femoral vein, O, is on the same plane with the artery near Poupart's ligament; but from this place downwards through the thigh, the vein gradually winds from the inner to the back part of the artery; and when both vessels pass under cover of the sartorius, they enter a strong fibrous sheath, V, derived from the tendons of the adductor muscles upon which they lie. The artery approaches the shaft of the femur near its middle; and in this place it may be readily compressed against the bone by the hand. The anterior crural nerve, P, dividing on the outer side of the artery, sends some of its branches coursing over the femoral sheath; and one of these--the long saphenous nerve--enters the sheath and follows the artery as far as the opening in the great adductor tendon. The femoral artery, before it passes through this opening into the popliteal space, gives off its anastomatic branch. The profundus branch, N, springs from the outer side of the femoral artery usually at a distance of from one to two inches (seldom more) below Poupart's ligament, and soon subdivides. [Footnote] The femoral artery in a few instances has been found double.

[Footnote: The ordinary length of each part of the main artery is stated on the authority of Mr. Quain. See "Anatomy of the Arteries," &c.]

The main artery of the lower limb may be exposed and tied in any part of its course from the aorta to the popliteal space. But the situation most eligible for performing such an operation depends of course upon circumstances, both anatomical and pathological. If an aneurism affect the popliteal part of the vessel, or if, from whatever cause arising, it be found expedient to tie the femoral above this part, the place best suited for the operation is that where the artery, W, first passes under cover of the sartorius muscle. [Footnote] For, considering that the vessel gives off no important branch destined to supply any part of the thigh or leg between the profundus branch and those into which it divides below the popliteal space, the arrest to circulation will be the same in amount at whichever part of the vessel between these two points the ligature be applied. But since the vessel in the situation specified can be reached with greater facility here than elsewhere lower down; and since, moreover, a ligature applied to it here will be sufficiently removed from the profundus branch above, and the seat of disease below, to produce the desired result, the choice of the operator is determined accordingly. The steps of the operation performed at the situation W, where the artery is about to pass beneath the sartorius, are these: an incision of sufficient length--from two to three inches--is to be made over the course of the vessel, so as to divide the skin and adipose membrane, and expose the fascia lata, through which the inner edge of the sartorius muscle becomes now readily discernible. A vein (anterior saphena) may be found to cross in this situation, but the saphena vein proper is not met with, as this lies nearer the inner side of the thigh. The fascia having been next divided, the edge of the sartorius is to be turned aside, and now the pulsation of the artery in its sheath will indicate its exact position. The sheath is next to be opened, for an extent sufficient only to carry the point of the ligature-needle safely around the artery, care being taken not to injure the femoral vein, which lies close behind it, and also to exclude any nerve which may lie in contact with the vessel.

[Footnote: This is the situation chosen by Scarpa for arresting by ligature the circulation through the femoral artery in cases of popliteal aneurism. The reasons stated in the text are those which determine the surgeon to perform the operation in this place in preference to that (the lower third of the thigh) where Mr. Hunter first proposed to tie the vessel.]

If an aneurism affect the common femoral portion of the artery, the external iliac part would require to be tied, because, between the seat of the tumour and the epigastric and circumflex ilii branches above, there would not be sufficient space to allow the ligature to rest undisturbed; and even if the aneurism arose from the femoral below the profundus branch in the upper third of the thigh, or if, after amputation of the thigh, a secondary haemorrhage took place from the femoral and the profunda arteries, a ligature would with more safety be applied to the external iliac part than to the common femoral; because of this latter, even when of its clear normal length, presenting so small an interval between the epigastric and profundus branches. In addition to this, it must be noticed, that occasionally the profundus itself, or some one of its branches, (external and internal circumflex, &c.), arises as high up as Poupart's ligament, close to the origin of the epigastric and circumflex iliac. [Footnote]

[Footnote: The main artery (Plate 47) has been exposed in the iliac and femoral regions with the object of showing the relation which its parts bear to each other and to the whole; all the other dissections have been made upon the same plan, the practical tendency of which will be illustrated when considering the subject of arterial anastomosis.]

The external iliac part of the artery, G I, when requiring to be tied, may be reached in the following way: an incision, commencing above the anterior iliac spine, B, is to be carried inwards parallel to, and above, Poupart's ligament, D, as far as the outer margin of the internal abdominal ring. This incision is the one best calculated for avoiding the epigastric artery, and for not disturbing the peritonaeum more than is necessary. The skin and the three abdominal muscles having been successively incised, the fibrous transversalis fascia is next to be carefully divided, so as to expose the peritonaeum. This membrane is then to be gently raised by the fingers, from off the iliacus and psoas muscles as far inwards as the margin of the true pelvis where the artery lies. On raising the peritonaeum the spermatic vessels will be found adhering to it. The iliac artery itself is liable to be displaced by adhering to the serous membrane, when this is being detached from the inner side of the psoas muscle. [Footnote] The artery having been divested of its serous covering as far up as a point midway between I G, the epigastric and internal iliac branches, the ligature is to be passed around it in this place, as being equidistant from these two sources of disturbance. As the vein, K, lies close along the inner side of the artery, the point of the instrument should first be inserted between them, and passed from within outwards, in order to avoid wounding the vein. If an aneurism affect the upper end of the external iliac artery, it is proposed to tie the common iliac; but this is an operation of so serious a nature, that it can in this respect be exceeded only by tying the aorta itself. The common iliac artery is so situated, that it can as easily be reached from the groin upwards as from the side of the abdomen inwards, and in both directions the peritonaeum would have to be disturbed to an equal extent.

[Footnote: The student, in operating upon the dead subject, is often puzzled to find that the iliac artery does not appear in its usual situation, unaware at the time that he has lifted the vessel in connexion with the peritonaeum. I have once seen a very distinguished surgeon, whilst performing this operation on the living body, at fault owing to the same cause.]

DESCRIPTION OF PLATE 47.

A. The aorta at its point of bifurcation.

B. The anterior superior iliac spine.

C. The symphysis pubis.

D. Poupart's ligament, immediately above which are seen the circumflex ilii and epigastric arteries, with the vas deferens and spermatic vessels.

E E*. The right and left iliac muscles covered by the peritonaeum; the external cutaneous nerve is seen through the membrane.

F. The vena cava.

G G*. The common iliac arteries giving off the internal iliac branches on the sacro-iliac symphyses; *g g*, the right and left ureters.

H H*. The right and left common iliac veins.

I I*. The right and left external iliac arteries, each is crossed by the circumflex ilii vein.

K K *. The right and left external iliac veins.

L. The urinary bladder covered by the peritonaeum.

M. The rectum intestinum.

N. The profundus branch of the femoral artery.

O. The femoral vein; *O*, the saphena vein.

P. The anterior crural nerve.

Q. The sartorius muscle, cut.

S. The pectinaeus muscle.

T. The adductor longus muscle.

U. The gracilis muscle.

V. The tendinous sheath given off from the long adductor muscle, crossing the vessels, and becoming adherent to the vastus internus muscle.

W. The femoral artery. The letter is on the part where the vessel becomes first covered by the sartorius muscle.

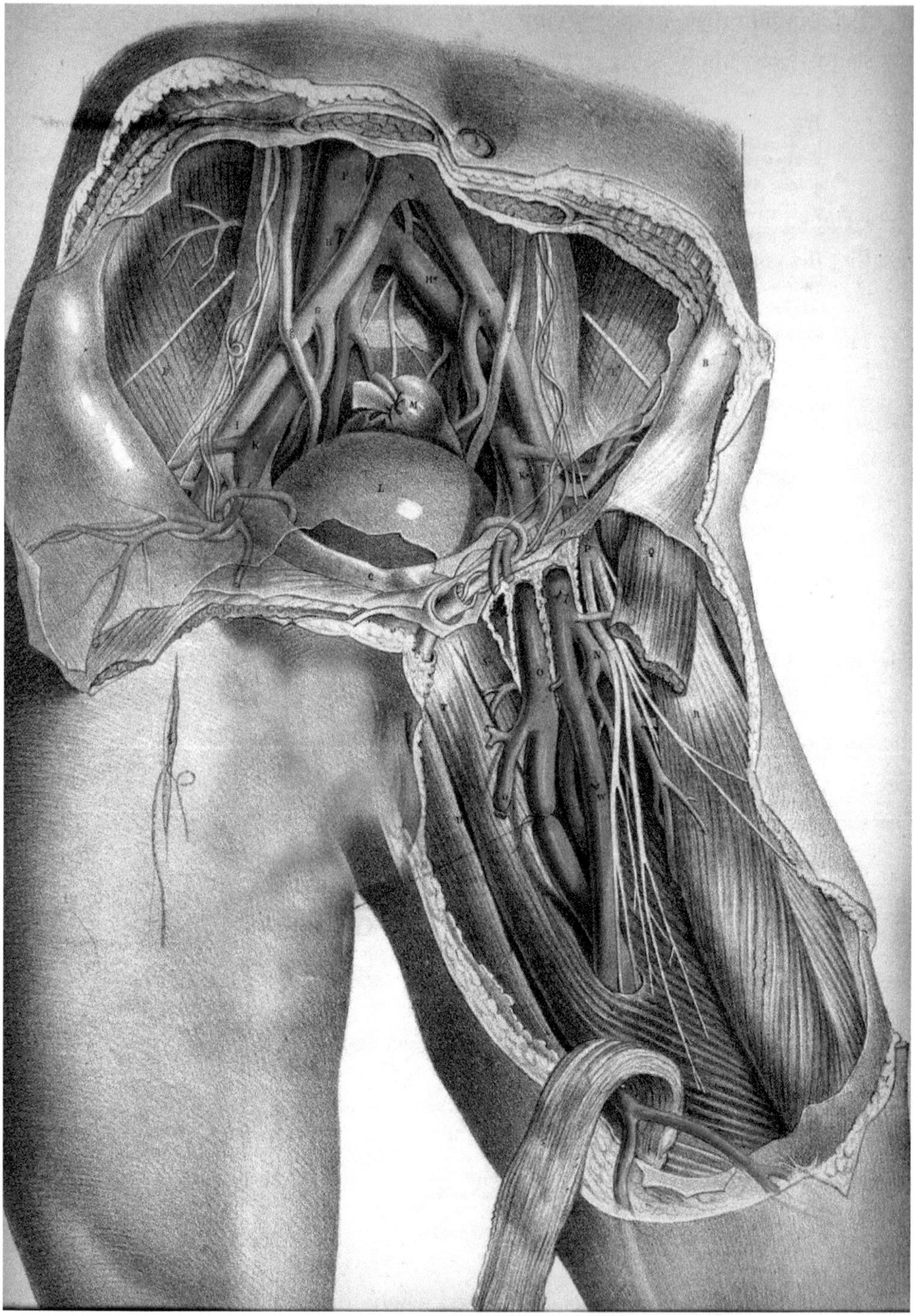

Plate 47.

THE RELATIVE ANATOMY OF THE MALE PELVIC ORGANS.

Physiological remarks on the functions of the abdominal muscles. Effects of spinal injuries on the processes of defecation and micturition. Function of the bladder. Its change of form and position in various states. Relation to the peritonaeum. Neck of the bladder. The prostate. Puncturation of the bladder by the rectum. The pudic artery.

As the abdomen and pelvis form one general cavity, the organs contained in both regions are thereby intimately related. The viscera of the abdomen completely fill this region, and transmit to the pelvic organs all the impressions made upon them by the diaphragm and abdominal walls. The expansion of the lungs, the descent of the diaphragm, and the contraction of the abdominal muscles, cause the abdominal viscera to descend and compress the pelvic organs; and at the same time the muscles occupying the pelvic outlet, becoming relaxed or contracted, allow the perinaeum to be protruded or sustained voluntarily according to the requirements. Thus it is that the force originated in the muscular parietes of the thorax and abdomen is, while opposed by the counterforce of the perinaeal muscles, brought so to bear upon the pelvic organs as to become the principal means whereby the contents of these are evacuated. The abdominal muscles are, during this act, the antagonists of the diaphragm, while the muscles which guard the pelvic outlet become at the time the antagonists of both. As the pelvic organs appear therefore to be little more than passive recipients of their contents, the *voluntary* processes of defecation and micturition may with more correctness be said to be performed rather for them than by them. The relations which they bear to the abdomen and its viscera, and their dependence upon these relations for the due performance of the processes in which they serve, are sufficiently explained by pathological facts. The same system of muscles comprising those of the thorax, abdomen and perinaeum, performs consentaneously the acts of respiration, vomiting, defecation and micturition. When the spinal cord suffers injury above the origin of the phrenic nerve, immediate death supervenes, owing to a cessation of the respiratory act. Considering, however, the effect of such an injury upon the pelvic organs alone, these may be regarded as being absolutely excluded from the pale of voluntary influence in consequence of the paralysis of the diaphragm, the abdominal and perinaeal muscles. The expulsory power over the bladder and rectum being due to the opposing actions of these muscles above and below, if the cord be injured in the neck below the origin of the phrenic nerve, the inferior muscles becoming paralysed, the antagonism of muscular forces is thereby interrupted, and the pelvic organs are, under such circumstances, equally withdrawn from the sphere of volition. The antagonism of the abdominal muscles to the diaphragm being necessary, in order that the pelvic viscera may be acted upon, if the cord be injured in the lower dorsal region, so as to paralyse the abdominal walls and the perinaeal muscles, the downward pressure of the diaphragm alone could not evacuate the pelvic organs voluntarily, for the abdominal muscles are now incapable of deflecting the line of force backwards and downwards through the pelvic axis; and the perinaeal muscles being also unable to act in agreement, the contents of the viscera pass involuntarily. Again, as the muscular apparatus which occupies the pelvic outlet acts antagonistic to the abdomen and thorax, when by an injury to the cord in the sacral spine the perinaeal apparatus alone becomes paralysed, its relaxation allows the thoracic and abdominal force to evacuate the pelvic organs involuntarily. It would appear, therefore, that the term "paralysis" of the bladder or rectum, when following spinal injuries, &c. &c. means, or should mean, only a paralytic state of the abdomino-pelvic muscular apparatus, entirely or in part. For, in fact, neither the bladder nor rectum ever acts voluntarily *per se* any more than the stomach does, and therefore the name "detrusor" urinae, as applied to the muscular coat investing the bladder, is as much a misnomer (if it be meant that the act of voiding the organ at will be dependent upon it) as would be the name "detrusor" applied to the muscular coat of the stomach, under the meaning that this were the agent in the spasmodic effort of vomiting.

The urinary bladder, G, Plate 49, (in the adult body,) occupies the true pelvic region when the organ is collapsed, or only partly distended. It is situated behind the pubic symphysis and in front of the rectum, C,--the latter lies between it and the sacrum, A. In early infancy, when the pelvis is comparatively small, the bladder is situated in the hypogastric region, with its summit pointing towards the umbilicus; as the bladder varies in shape, according to whether it be empty or full, its relations to neighbouring parts, especially to those in connexion with its summit, vary also considerably. When empty, the back and upper surface of the bladder collapse against its forepart, and in this state the organ lies flattened against the pubic symphysis. Whether the bladder be distended or not, the small intestines lie in contact with its upper surface, and compress it in the manner of a soft elastic cushion. When distended largely, its summit is raised above the pubic symphysis, the small intestines having yielded place to it, and in this state it can be felt by the hand laid upon the hypogastrium.

The shape of the bladder varies in different individuals. In some it is rounded, in others pyriform, in others peaked towards its summit. Its capacity varies also considerably at different ages and in different sexes. When distended, its long axis will be found to coincide with a line passing from a point midway between the navel and pubes to the point of the coccyx, the obliquity of this direction being greatest when the body is in the erect posture, for the intestines now gravitate upon it. When the body is recumbent, the bladder recedes somewhat from the pubes, and as the intestines do not now press upon it from above, it allows of being distended to a much greater degree without causing uneasiness, and a desire to void its contents.

The manner in which the bladder is connected to neighbouring parts is such as to admit of its full distension. Its summit, back, and upper sides are free and covered by the elastic peritonaeum, whilst its front, lower sides, and base are adherent to adjacent parts, and divested of the serous membrane. On tracing the peritonaeum from the front wall of the abdomen to its point of reflexion over the summit of the bladder, we find the membrane to be in this part so loosely adherent, that the bladder when much distended, raises the peritonaeum above the level of the upper margin of the pubic symphysis. In this state the organ may be punctured immediately above the pubic symphysis without endangering the serous sac. When the bladder is collapsed, the peritonaeum follows its summit below the level of the pubes, and in this position of the organ such an operation would be inadmissible, if indeed the necessity for it can now be conceived.

By removing the os innominatum, A D, Plate 48, together with the internal obturator, and levator ani muscles, which arise from its inner side, we obtain a lateral view, Plate 49, of the pelvic viscera, and of the vessels &c. connected with them. Those parts of the bladder, G, and the rectum, C, which are invested by the peritonaeum, are also now fully displayed. On tracing this membrane from before backwards, over the summit of the bladder, G, we find it descending deeply upon the posterior surface of the organ, before it becomes reflected so as to ascend over the forepart of the rectum. This duplicature of the serous membrane, H H, is named the recto-vesical pouch, and it is required to ascertain with all the exactness possible the level to which it descends, so as to avoid it in the operation of puncturing the bladder through the rectum. The serous pouch descends lower in some bodies than in others; but in all there exists a space, of greater or less dimensions, between it and the prostate, V, whereat the base of the bladder is in direct apposition with the rectum, W, the serous membrane not intervening.

When the peritonaeum is traced from one iliac fossa to the other, we find it sinking deeply into the hollow of the pelvis behind the bladder, so as to form the sides of the recto-vesical pouch; but when traced over the summit of the bladder, this organ is seen to have the membrane reflected upon it, almost immediately below the pelvic brim. At the situations where the peritonaeum becomes reflected in front, laterally, and behind, upon the sides of the bladder, the membrane is thrown into folds, which are named "false ligaments." The pelvic fascia, in being reflected to the bladder from the front and sides of the pelvis, at a lower level than that of the peritonaeum, forms the "true ligaments." In addition to these ligaments, which serve to keep the base and front of the bladder fixed in the pelvis, other structures, such as the ureters, K, the vasa deferentia, I, the hypogastric cords, the urachus, and the bloodvessels, embrace the organ in various directions, and act as bridles, to limit its expansion more or less in all directions, but least so towards its summit, which is always comparatively free.

The neck and outlet of the bladder, V, are situated at the anterior part of its base, and point towards the subpubic space. The prostate gland, V, surrounds its neck, and occupies a position behind and below the pubic arch, D, and in front of the rectum, W. The gland, V, being of a rounded form and dense structure, can be felt in this situation by the finger, passed upwards through the bowel. The prostate is suspended from the back of the pubic arch by the anterior true ligament of the bladder, and at its forepart, where the membranous portion of the urethra commences, this passes through the deep perinaeal fascia, X. The anterior fibres of the levator ani muscle embrace the prostate on both its sides. Behind the base of the prostate, the ureter, K, is seen to enter the coats of the bladder obliquely, whilst the vas deferens, I, joined by the vesicula seminalis, L, penetrates the substance of the prostate, V, at its lower and back part, which lies in apposition with the rectum.

The rectum, W C, at its middle and upper parts, occupies the hollow of the sacrum, A Q, and is behind the bladder. The lower third of the rectum, W, not being covered by the peritonaeum, is that part on which the various surgical operations are performed. At its upper three-fifths, the rectum describes a curve corresponding to that of the sacrum; and if the bladder be full, its convex back part presses the bowel against the bone, causing its curve to be greater than if the bladder were empty and collapsed. This fact requires to be borne in mind, for, in order to introduce a bougie, or to allow a large injection to pass with freedom into the bowel, the bladder should be first evacuated. The coccygeal bones, Q, continuing in the curve of the sacrum, bear the rectum, W, forwards against the base of the bladder, and give to this part a degree of obliquity upwards and backwards, in respect to the perinaeum and anus. From the point where the prostate, V, lies in contact with the rectum, W, this latter curves downwards, and slightly backwards, to the anus, P. The prostate is situated at a distance of about an inch and a half or two inches from the anus--the distance varying according to whether the bladder and bowel be distended or not. [Footnote]

[Footnote: The distance between any two given parts is found to vary in different cases. "In subjects of an advanced age," Mr. Stanley remarks, "a deep perinaeum, as it is termed, is frequently met with. This may be occasioned either by an unusual quantity of fat in the perinaeum, or by an enlarged prostate, or by the dilatation of that part of the rectum which is contiguous to the prostate and bladder. Under either of these circumstances, the prostate and bladder become situated higher in the pelvis than naturally, and consequently at a greater distance from the perinaeum."--*On the Lateral Operation of Lithotomy*.]

The arteries of the bladder are derived from the branches of the internal iliac, S. The rectum receives its arteries from the inferior mesenteric and pudic. The veins which course upwards from the rectum are large and numerous, and devoid of valves. When these veins become varicose, owing to a stagnation of their circulation, produced from whatever cause, the bowel is liable to be affected with haemorrhoids or to assume a haemorrhagic tendency.

The pudic artery, S *s,* is a branch of the internal iliac. It passes from the pelvis by the great sciatic foramen, below the pyriformis muscle, and in company with the sciatic artery. The pudic artery and vein wind around the spine, E, of the ischium, where they are joined by the pudic nerve, derived from, T, the sacral plexus. The artery, in company with the nerve and vein, re-enters the pelvis by the small sciatic foramen, and gets under cover of a dense fibrous membrane (obturator fascia), between which and the obturator muscle, it courses obliquely downwards and forwards to the forepart of the perinaeum. At the place where the vessel re-enters the pelvis, it lies removed at an interval of an inch and a half from the perinaeum, but becomes more superficial as it approaches the subpubic space, N. The levator ani muscle separates the pudic vessels and nerves from the sides of the rectum and bladder. The principal branches given off from the pudic artery of either side, are (1st), the inferior hemorrhoidal, to supply the lower end of the rectum; (2nd), the transverse and superficial perinaeal; (3rd), the artery of the bulb; (4th), that which enters the corpus cavernosum of the penis, N; and (5th), the dorsal artery of the penis. [Footnote] The branches given off from the pudic nerve correspond in number and place to those of the artery. Having now considered the relations of the pelvic organs in a lateral view, we are better prepared to understand these relations when seen at their perinaeal aspect.

[Footnote: The pudic artery, or some one of its branches, occasionally undergoes marked deviations from the ordinary course. In Mr. Quain's work, ("Anatomy of the Arteries,") a case is represented in which the artery of the bulb arose from the pudic as far back as the tuber ischii, and crossed the line of incision made in the lateral operation of lithotomy. In another figure is seen a vessel ("accessory pudic"), which, passing between the base of the bladder and the levator ani muscle, crosses in contact with the left lobe of the prostate.]

DESCRIPTION OF PLATES 48 & 49.

PLATE 48.

A. The anterior superior iliac spine.

B. The anterior inferior iliac spine.

C. The acetabulum; *c,* the ligamentum teres.

D. The tuber ischii.

E. The spine of the ischium.

F. The pubic horizontal ramus.

G. The summit of the bladder covered by the peritonaeum.

H. The femoral artery.

I. The femoral vein.

K. The anterior crural nerve.

L. The thyroid ligament.

M. The spermatic cord.

N. The corpus cavernosum penis; *n*, its artery.

O. The urethra; *o*, the bulbus urethrae.

P. The sphincter ani muscle.

Q. The coccyx.

R. The sacro-sciatic ligament.

S. The pudic artery and nerve.

T. The sacral nerves.

U. The pyriformis muscle, cut.

V. The gluteal artery.

W. The small gluteus muscle.

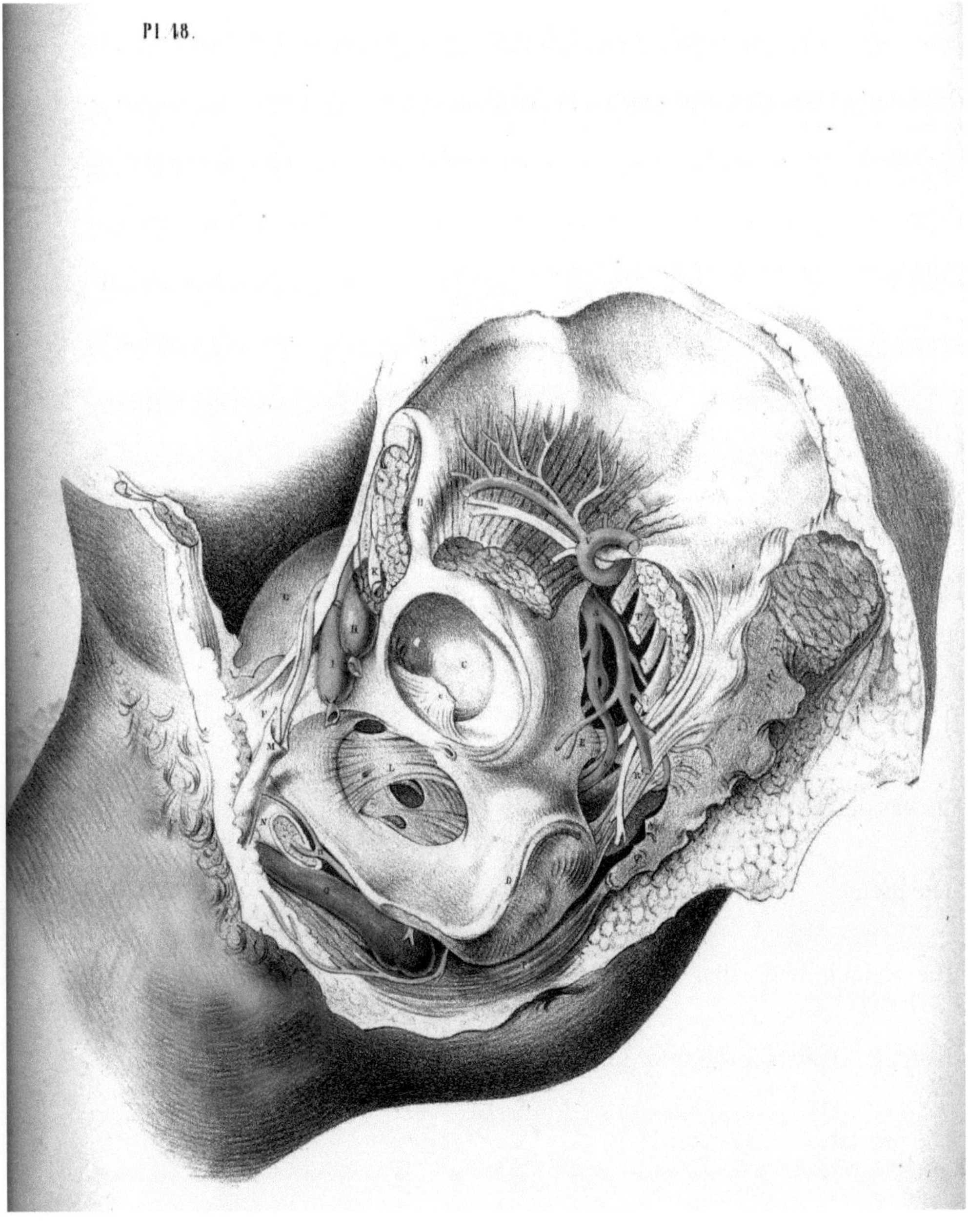

Plate 48

PLATE 49.

A. The part of the sacrum which joins the ilium.

B. The external iliac artery, cut across.

C. The upper part of the rectum.

D. The ascending pubic ramus.

E. The spine of the ischium, cut.

F. The horizontal pubic ramus, cut.

G. The summit of the bladder covered by the peritonaeum; G *, its side, not covered by the membrane.

H H. The recto-vesical peritonaeal pouch,

I. The vas deferens.

K. The ureter.

L. The vesicula seminalis.

M, N, O, P, Q, R, S, T, U, refer to the same parts as in Plate 48.

V. The prostate.

W. The lower part of the rectum.

X. The deep perinaeal fascia.

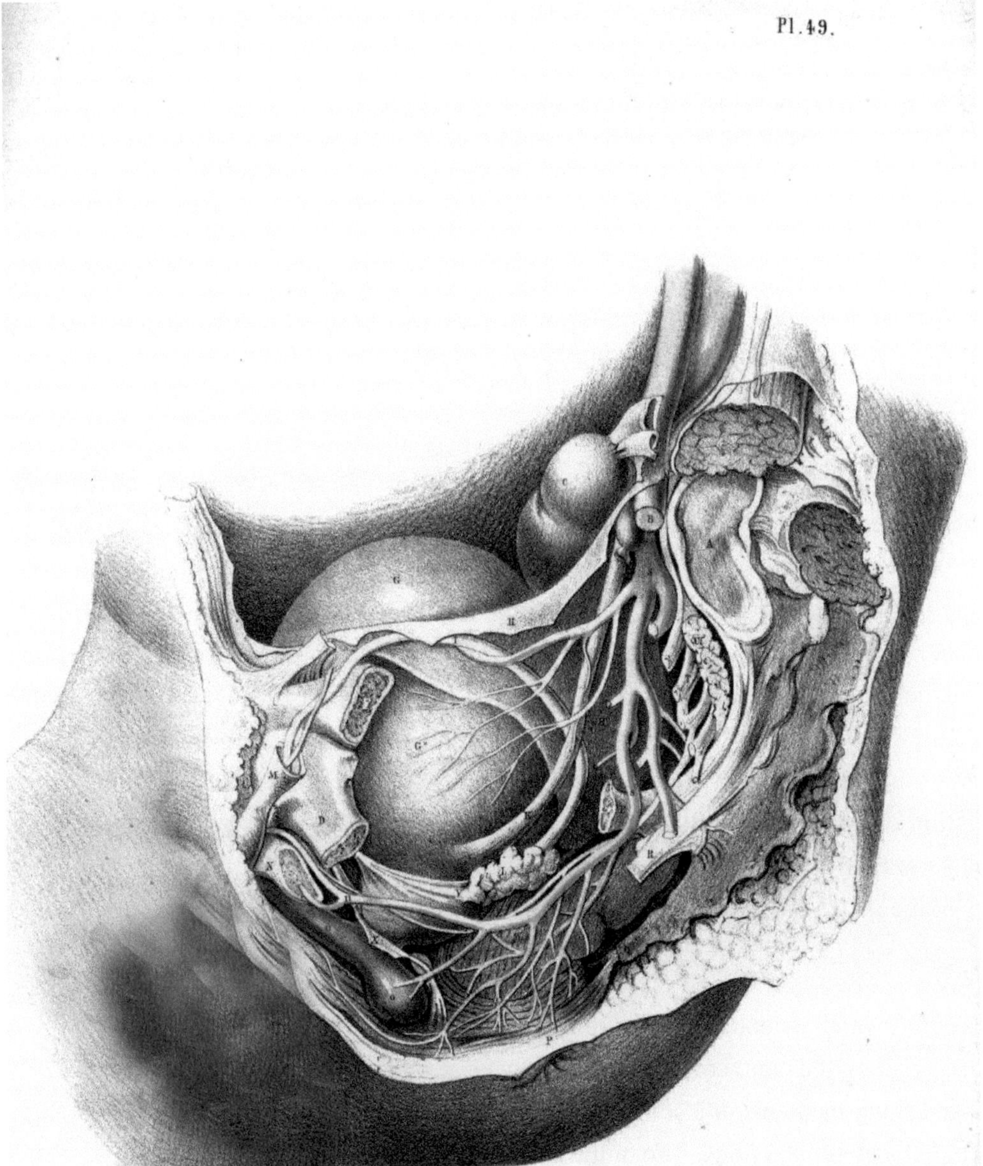

Plate 49

THE SURGICAL DISSECTION OF THE SUPERFICIAL STRUCTURES OF THE MALE PERINAEUM.

Remarks on the median line. Congenital malformations. Extravasation of urine into the sac of the superficial fascia. Symmetry of the parts. Surgical boundaries of the perinaeum. Median and lateral important parts to be avoided in lithotomy, and the operation for fistula in ano.

The median line of the body is marked as the situation where the opposite halves unite and constitute a perfect symmetrical figure. Every structure--superficial as well as deep--which occupies the median line is either single, by the union of halves, or dual, by the cleavage and partition of halves. The two sides of the body being absolutely similar, the median line at which they unite is therefore common to both. Union along the median line is an occlusion taking place by the junction of sides; and every hiatus or opening, whether normal or abnormal, which happens at this line, signifies an omission in the process of central union. The sexual peculiarities are the results of the operation of this law, and all forms which are anomalous to either sex, may be interpreted as gradations in the same process of development; a few of these latter occasionally come under the notice of the surgeon.

The region which extends from the umbilicus to the point of the coccyx is marked upon the cutaneous surface by a central raphe dividing the hypogastrium, the penis, the scrotum, and the perinaeum respectively into equal and similar sides. The umbilicus is a cicatrix formed after the metamorphosis of a median foetal structure--the placental cord, &c. In the normal form, the meatus urinarius and the anus coincide with the line of the median raphe, and signify omissions at stated intervals along the line of central union. When between these intervals the process of union happens likewise to be arrested, malformations are the result; and of these the following are examples:--Extrusion of the bladder at the hypogastrium is caused by a congenital hiatus at the lower part of the linea alba, which is in the median line; Epispadias, which is an urethral opening on the dorsum of the penis; and

Hypospadias, which is a similar opening on its under surface, are of the same nature--namely, omissions in median union. Hermaphrodism may be interpreted simply as a structural defect, compared to the normal form of the male, and as a structural excess compared to that of the female. Spina bifida is a congenital malformation or hiatus in union along the median line of the sacrum or loins. As the process of union along the median line may err by a defect or omission, so may it, on the other hand, err by an excess of fulfilment, as, for example, when the urethra, the vagina, or the anus are found to be imperforate. As the median line of union thus seems to influence the form of the hypogastrium, the genitals, and the perinaeum, the dissection of these parts has been conducted accordingly.

By removing the skin and subjacent adipose membrane from the hypogastrium, we expose the superficial fascia. This membrane, E E E*, Fig. 1, Plate 50, is, in the middle line, adherent to B, the linea alba, and thereby contributes to form the central depression which extends from the navel to the pubes. The adipose tissue, which in some subjects accumulates on either side of the linea alba, renders this depression more marked in them. At the folds of the groin the fascia is found adherent to Poupart's ligament, and this also accounts for the depressions in both these localities. From the central linea alba to which the fascia adheres, outwards on either side to the folds of both groins, the membrane forms two distinct sacs, which droop down in front, so as to invest the testicles, E**, and penis in a manner similar to that of the skin covering these parts. As the two sacs of the superficial fascia join each other at the line B, coinciding with the linea alba, they form by that union the suspensory ligament of the penis, which is a structure precisely median.

The superficial fascia having invested the testicles each in a distinct sac, the adjacent sides of both these sacs, by joining together, form the median septum scroti, E, Fig. 2, Plate 50. In the perinaeum, Fig. 1, Plate 51, the fascia, A, may be traced from the back of the scrotum to the anus. In this region the membrane is found to adhere laterally to the rami of the ischium and pubes; whilst along the median perinaeal line the two sacs of which the membrane is composed unite, as in the scrotum, and form an imperfect septum. In front of the anus, beneath the sphincter ani, the fascia degenerates into cellular membrane, one layer of which is spread over the adipose tissue in the ischio-rectal space, whilst its deeper and stronger layer unites with the deep perinaeal fascia, and by this connexion separates the urethral from the anal spaces. The superficial fascia of the hypogastrium, the scrotum, and the perinaeum forming a continuous membrane, and being adherent to the several parts above noticed, may be regarded as a general double sac, which isolates the inguino-perinaeal region from the femoral and anal regions, and hence it happens that when the urethra becomes ruptured, the urine which is extravasated in the perinaeum, is allowed to pass over the scrotum and the abdomen, involving these parts in consequent inflammation, whilst the thighs and anal space are exempt. The tunicae vaginales, which form the immediate coverings of the testicles, cannot be entered by the urine, as they are distinct sacs originally protruded from the abdomen. It is in consequence of the imperfect state of the inguino-perinaeal septum of the fascia, that urine effused into one of the sacs is allowed to enter the other.

Like all the other structures which join on either side of the median line, the penis appears as a symmetrical organ, D D, Fig. 2, Plate 50. While viewed in section, its two corpora cavernosa are seen to unite anteriorly, and by this union to form a septum "pectiniforme;" posteriorly they remain distinct and lateral, F F, Fig. 2, Plate 51, being attached to the ischio-pubic rami as the crura penis. The urethra, B, Fig. 2, Plate 50, is also composed of two sides, united along the median line, but forming between them a canal by the cleavage and partition of the urethral septum. All the other structures of the perinaeum will be seen to be either double and lateral, or single and median, according as they stand apart from, or approach, or occupy the central line.

The perinaeum, Figs. 1, 2, Plate 51, is that space which is bounded above by the arch of the pubes, behind by C, the os coccygis, and the lower borders of, I I, the glutaei muscles and sacro-sciatic ligaments, and laterally by D D, the ischiatic tuberosities. The osseous boundaries can be felt through the integuments. Between the back of the scrotum and the anus the perinaeum swells on both sides of the raphe, A B, Fig. 3, Plate 50, and assumes a form corresponding with the bag of the superficial fascia which encloses the structures connected with the urethra. The anus is centrally situated in the depression formed between D D, the ischiatic tuberosities, and the double folds of the nates.

The perinaeum, Fig. 3, Plate 50, is, for surgical purposes, described as divisible into two spaces (anterior and posterior) by a transverse line drawn from one tuber ischii, D, to the other, D, and crossing in front of the anus. The anterior space, A D D, contains the urethra; the posterior space, D D C, contains the rectum. The central raphe, A B C, traverses both these spaces. The anterior or urethral space is (while viewed in reference to its osseous boundaries) triangular in shape, the apex being formed by the pubic symphysis beneath A, whilst two lines drawn from A to D D, would coincide with the ischio-pubic rami which form its sides. The raphe in the anterior space indicates the central position of the urethra, as may be ascertained by passing a sound into the bladder, when the shaft of the instrument will be felt prominently between the points A B. Behind the point B, the sound or staff sinks deeper in the perinaeum as it follows the curve of the urethra towards the bladder, and becomes overlaid by the bulb, &c.

The ischiatic tuberosities, D D, Fig. 3, Plate 50, are, in all subjects, sufficiently prominent to be felt through the integuments, &c.; and the line which, when drawn from one to the other, serves to divide the two perinaeal spaces, forms the base of the anterior one. In well-formed subjects, the anterior space is equiangular, the base being equal to each side; but according as the tuberosities approach the median line, the base becomes narrowed, and the triangle is thereby rendered acute. These circumstances influence the direction in which the first incision in the lateral operation of lithotomy should be made. When the tuberosity of the left ischium stands well apart from the perinaeal centre, the line of incision, B E, Fig. 3, Plate 50, is carried obliquely from above downwards and outwards; but in cases where the tuberosity approaches the centre, the incision must necessarily be made more vertical. The posterior perinaeal space may be described on the surface by two lines drawn from D D, the ischiatic tuberosities, to C, the point of the coccyx, whilst the transverse line between D and D bounds it above.

By removing the integument and superficial fascia, we expose the superficial vessels and nerves, together with the muscles in the neighbourhood of the urethra and the anus. The accelerator urinae, E, Fig. 2, Plate 51, which embraces the urethra, and the sphincter ani, B C, which surrounds the anus, H, occupy the median line, and are divided each into halves by a central tendon, E B C, which traverses the perinaeum from before backwards, to the point of the coccyx. On either side of the anus, in the ischio-rectal space, D D, Fig. 1, Plate 51, is found a considerable quantity of granular adipose tissue, traversed by the inferior haemorrhoidal arteries and nerves-branches of the pudic artery and nerve.

In front of the anus are seen two small muscles (transversae perinaei), G G, Fig. 2, Plate 51, each arising from the tuber ischii of its own side, and the two becoming inserted into, B, the central tendon. These transverse muscles serve to mark the boundary between the anterior and posterior perinaeal spaces. Behind each muscle is found a small artery, crossing to the median line. The left transverse muscle and artery are always divided in the lateral operation of lithotomy. On the outer sides of the anterior perinaeal space are seen the erectores penis muscles, F F, overlaying the crura penis. Between each muscle and the accelerator urinae, the superficialis perinaei artery and nerve course forwards to the scrotum, &c.

The perinaeal muscles having been brought fully into view, Plate 52, Fig. 1, their symmetrical arrangement on both sides of the median line at once strikes the attention. On either side of the anterior space appears a small angular interval, L, formed between B, the accelerator urinae, D, the erector penis, and E, the transverse muscle. Along the surface of this interval, the superficial perinaeal artery and nerve are seen to pass forwards; and deep in it, beneath these, may also be observed, L, the artery of the bulb, arising from the pudic, and crossing inwards, under cover of the anterior layer of the membrane named the deep perinaeal fascia. The first incision in the lateral operation of lithotomy is commenced over the inferior inner angle of this interval.

The muscles occupying the anterior perinaeal space require to be removed, Fig. 1, Plate 53, in order to expose the urethra, B M, the crus penis, D, and the deep perinaeal fascia. The fascia will be now seen stretched across the subpubic triangular space, reaching from one ischio-pubic ramus to the other, whilst by its lower border, corresponding with the line of the transversae perinaei muscles, it becomes continuous with the superficial fascia, in the manner before described. The deep perinaeal fascia (triangular ligament) encloses between its two layers, C E, on either side of the urethra, the pudic artery, the artery of the bulb, Cowper's glands, and some muscular fibres occasionally to be met with, to which the name "Compressor urethrae" has been assigned. At this stage of the dissection, as the principal vessels and parts composed of erectile tissue are now in view, their relative situations should be well noticed, so as to avoid wounding them in the several cutting operations required to be performed in their vicinity.

Along the median line (marked by the raphe) from the scrotum to the coccyx, and close to this line on either side, the vessels are unimportant as to size. The urethra lies along the middle line in the anterior perinaeal space; the rectum occupies the middle in the posterior space. When either of these parts specially requires to be incised--the urethra for impassable stricture, &c., and the lower part of the rectum for fistula in ano--the operation may be performed without fear of inducing dangerous arterial haemorrhage. With the object of preserving from injury these important parts, deep incisions at, or approaching to, the middle line must be avoided. The outer (ischio-pubic) boundary of the perinaeum is the line along which the pudic artery passes. The anterior half of this boundary supports also the crus penis; hence, therefore, in order to avoid these, all deep incisions should be made parallel to, but removed to a proper distance from this situation. The structures placed at the middle line, B M F, Fig. 2, Plate 52, and those in connexion with the left perinaeal boundary, D G L, require (in order to insure the safety of these parts) that the line of incision necessary to gain access to the neck of the bladder in lithotomy should be made through the left side of the perinaeum from a point midway between M, the bulb, and D, crus penis above, to a point, K, midway between the anus, F, and tuber ischii, G, below. As the upper end of this incision is commenced over the situation of the superficial perinaeal artery and the artery of the bulb, the knife at this place should only divide the skin and superficial fascia. The lower end, K, just clears the outer side of the dilated lower part of the rectum. The middle of the incision is over the left lobe of the prostate gland and neck of the bladder, which parts, together with the membranous portion of the urethra, are still concealed by the deep perinaeal fascia, the structures between its layers, and the anterior fibres of K, the levator ani muscle. The incision, if made in due reference to the relative situation of the parts above noticed, will leave them untouched; but when the pudic artery, or some one of its branches, deviates from its ordinary course and crosses the line of incision, a serious haemorrhage will ensue, despite the anatomical knowledge of the most experienced operator. When it is requisite to divide the superficial and deep sphincter ani as in the operation for complete fistula in ano, if the incision be made transversely in the ischio-rectal fossa, the haemorrhoidal arteries and nerves converging towards the anus will be the more likely to escape being wounded.

DESCRIPTION OF THE FIGURES OF PLATES 50 & 51.

PLATE 50.

FIGURE 1.

A. The umbilicus.

B. The linea alba.

C. The suspensory ligament of the penis.

D D. The two corpora cavernosa penis.

E E**. The hypogastric and scrotal superficial fascia.

F F. The spermatic cords.

FIGURE 2.

A. The umbilicus.

B. The urethra.

C*. The tunica vaginalis; *c,* the testicle invested by the tunic.

D D. The corpora cavernosa seen in section.

E. The scrotal raphe and septum scroti.

FIGURE 3.

A B. The perinaeal raphe.

C. The place of the coccyx.

D D. The projections of the ischiatic tuberosities.

BE. The line of section in lithotomy.

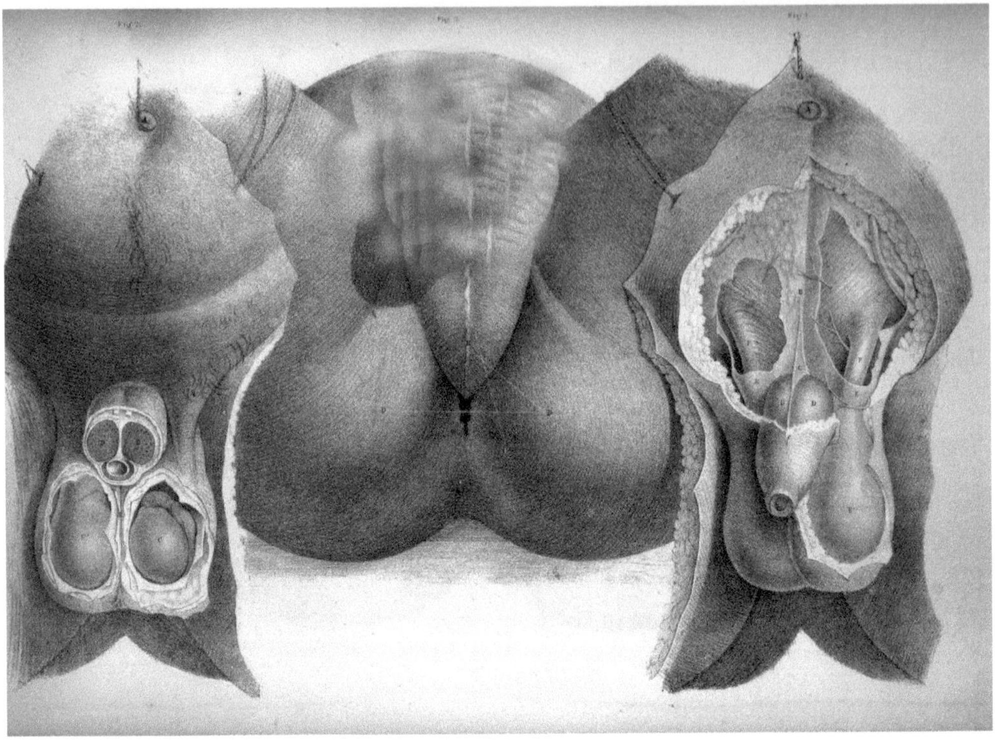

Figure 2-3-1

Plate 50

PLATE 51.

FIGURE 1.

A. The superficial fascia covering the urethral space.

B. The sphincter ani.

C. The coccyx.

D D. The right and left ischiatic tuberosities.

H. The anus.

I I. The glutei muscles.

FIGURE 2.

A, B, C, D, H, I. The same parts as in Fig. 1.

E. The accelerator urinae muscle.

F F. Right and left erector penis muscle.

G G. Right and left transverse muscle.

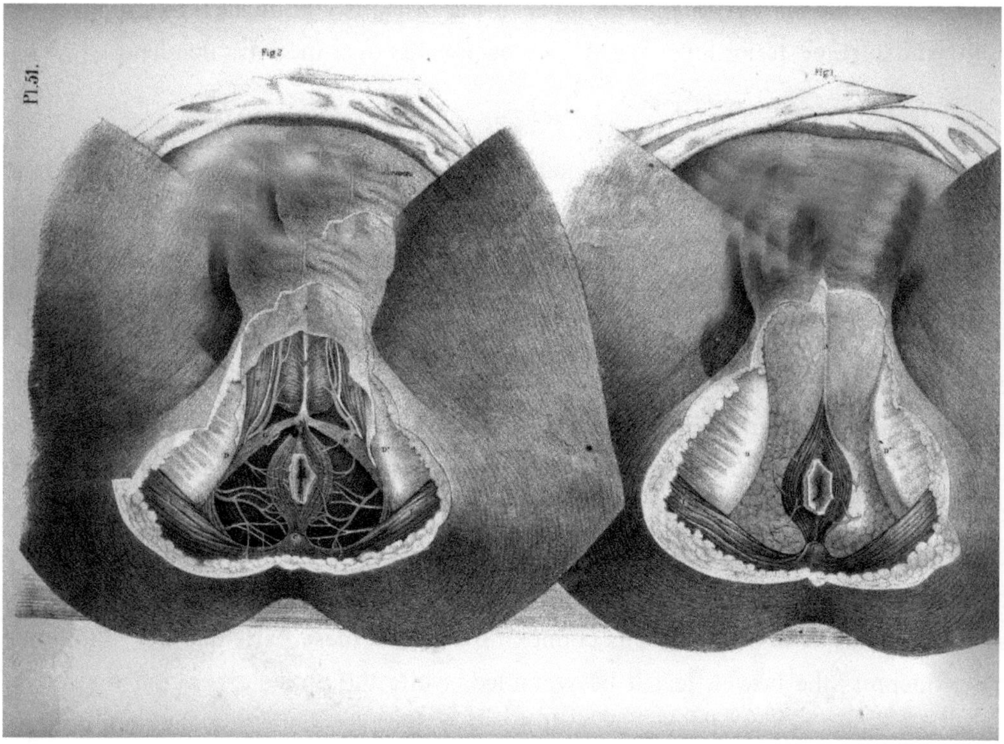

Figure 2-1

Plate 51

THE SURGICAL DISSECTION OF THE DEEP STRUCTURES OF THE MALE PERINAEUM; THE LATERAL OPERATION OF LITHOTOMY.

Relative position of the parts at the base of the bladder. Puncture of the bladder through the rectum and of the urethra in the perinaeum. General rules for lithotomy.

The urethra, at its membranous part, M, Fig. 1, Plate 53, which commences behind the bulb, perforates the centre of the deep perinaeal fascia, E E, at about an inch and a half in front of F, the anus. The anterior layer of the fascia is continued forwards over the bulb, whilst the posterior layer is reflected backwards over the prostate gland.

Behind the deep perinaeal fascia, the anterior fibres of K, the levator ani muscle, arise from either side of the pubic symphysis posteriorly, and descend obliquely down wards and forwards, to be inserted into the sides of N N, the rectum above the anus. These fibres of the muscle, and the lower border of the fascia which covers them, lie immediately in front of the prostate, C C, Fig. 2, Plate 53, and must necessarily be divided in the operation of lithotomy. Previously to disturbing the lower end of the rectum from its natural position in the perinaeum, its close relation to the prostate and base of the bladder should be noticed. While the anus remains connected with the deep perinaeal fascia in front, the fibres of the levator ani muscle of the left side may be divided; and by now inserting the finger between them and the rectum, the left lobe of the prostate can be felt in apposition with the forepart of the bowel, an inch or two above the anus. It is owing to this connexion between these parts that the lithotomist has to depress the bowel, lest it be wounded, while the prostate is being incised. If either the bowel or the bladder, or both together, be over-distended, they are brought into closer apposition, and the rectum is consequently more exposed to danger during the latter stages of the operation. The prostate being in contact with the rectum, the surgeon is enabled to examine by the touch, *per anum*, the state of the gland. If the prostate be diseased and irregularly enlarged, the urethra, which passes through it, becomes, in general, so distorted, that the surgeon, after passing the catheter along the urethra as far as the prostate, will find it necessary to guide the point of the instrument into the bladder, by the finger introduced into the bowel. The middle or third lobe of the prostate being enlarged, bends the prostatic part of the urethra upwards. But when either of the lateral lobes is enlarged, the urethra becomes bent towards the opposite side.

By dividing the levator ani muscle on both sides of the rectum, F, Fig. 2, Plate 53, and detaching and depressing this from the perinaeal centre, the prostate, C C, and base of the bladder, P, are brought into view. The pelvic fascia may be now felt reflected from the inner surface of the levator ani muscle to the bladder at a level corresponding with the base of the prostate, and the neck of the bladder in front, and the vesiculae seminales, N N, laterally. In this manner the pelvic fascia serves to insulate the perinaeal space from the pelvic cavity. The prostate occupies the centre of the perinaeum. If the perinaeum were to be penetrated at a point midway between the bulb of the urethra and the anus, and to the depth of two inches straight backwards, the instrument would transfix the apex of the gland. Its left lobe lies directly under the middle of the line of incision which the lithotomist makes through the surface; a fibrous membrane forms a capsule for the gland, and renders its surface tough and unyielding, but its proper substance is friable, and may be lacerated or dilated with ease, after having partly incised its fibrous envelope. The membranous part of the urethra, M, Fig. 2, Plate 53, enters the apex of the prostate, and traverses this part in a line, nearer to the upper than to the under surface; and that portion of the canal which the gland surrounds, is named prostatic. The prostate is separated from the pudic artery by the levator ani muscle, and from the artery of the bulb, by the deep perinaeal fascia and the muscular fibres enclosed between its two layers.

The prostate being a median structure, is formed of two lobes, united at the median line. The bulbus urethrae being also a median structure, is occasionally found notched in the centre, and presenting a bifid appearance. On the base of the bladder, P, Fig. 2, Plate 53, the two vasa deferentia, Q Q, are seen to converge from behind forwards, and enter the base of the gland; a triangular interval is thus formed between the vasa, narrower before than behind, and at the middle of this place the point of the trocar is to be passed (through the rectum,) for the purpose of evacuating the contents of the bladder, when other measures fail. When this operation is required to be performed, the situation of the prostate is first to be ascertained through the bowel; and at a distance of an inch behind the posterior border of the gland, precisely in the median line, the distended base of the bladder may be safely punctured. If the trocar pierce the bladder at this point, the seminal vessels converging to the prostate from either side, and the recto-vesical serous pouch behind, will escape being wounded. If the prostate happen to be much enlarged, the relative position of the neighbouring parts will be found disturbed, and in such case the bladder can be punctured above the pubes with greater ease and safety. In cases of *impassable* stricture, when extravasation of urine is threatened, or has already occurred, the urethra should be opened in the perinaeum behind the place where the stricture is situated, and this (in the present instance) certainly seems to be the more effectual measure, for at the same time that the stricture is divided, the contents of the bladder may be evacuated through the perinaeum. If the membranous part of the urethra be that where the stricture exists, a staff with a central groove is to be passed as far as the strictured part, and having ascertained the position of the instrument by the finger in the bowel, the perinaeum should be incised, at the middle line, between the bulb of the urethra and the anus. The urethra in this situation will be found to curve backwards at the depth of an inch or more from the surface. The point of the staff is now to be felt for, and the urethra is to be incised upon it. The bistoury is next to be carried backwards through the stricture till it enters that part of the urethra (usually dilated in such cases) which intervenes between the seat of obstruction and the neck of the bladder.

The lateral operation of lithotomy is to be performed according to the above described anatomical relations of the parts concerned. The bowel being empty and the bladder moderately full, a staff with a groove in its left side is to be passed by the urethra into the bladder. The position and size of the prostate is next to be ascertained by the left fore-finger in the rectum. Having now explored the surface of the perinaeum in order to determine the situation of the left tuberosity and ischio-pubic ramus, in relation to the perinaeal middle line, the staff being held steadily against the symphysis pubis, the operator proceeds to divide the skin and superficial fascia on the left side of the perinaeum, commencing the incision on the left of the raphe about an inch in front of the anus, and carrying it downwards and outwards midway between the anus and ischiatic tuberosity, to a point below these parts. The left fore-finger is then to be passed along the incision for the purpose of parting the loose cellular tissue; and any of the more resisting structures, such as the transverse and levator ani muscles, are to be divided by the knife. Deep in the forepart of the wound, the position of the staff is now to be felt for, and the structures which cover the membranous portion of the urethra are to be cautiously divided. Recollecting now that the artery of the bulb passes anterior to the staff in the urethra on a level with the bulb, the vessel is to be avoided by inserting the point of the knife in the groove of the staff as far backwards--that is, as near the apex of the prostate--as possible. The point of the knife having been inserted in the groove of the staff, the bowel is then to be depressed by the left fore-finger; and now the knife, with its back to the staff, and its edge lateralized (towards the lower part of the left tuber ischii), is to be pushed steadily along the groove in the direction of the staff, and made to divide the membranous part of the urethra and the anterior two-thirds of the left lobe of the prostate. The gland must necessarily be divided to this extent if the part of the urethra which it surrounds be traversed by the knife. The extent to which the prostate is divided depends upon the degree of the angle which the knife, passing along the urethra, makes with the staff. The greater this angle is, the greater the extent to which the gland will be incised. The knife being next withdrawn, the left fore-finger is to be passed through the opening into the bladder, and the parts are to be dilated by the finger as it proceeds, guided by the staff. The staff is now to be removed while the point of the finger is in the neck of the bladder, and the forceps is to be passed into the bladder along the finger as a guide. The calculus, now in the grip of the forceps, is to be extracted by a slow undulating motion.

The general rules to be remembered and adopted in performing the operation of lithotomy are as follow:--1st, The incision through the skin and sub-cutaneous cellular membrane should be freely made, in order that the stone may be easily extracted and the urine have ready egress. The incision which (judging from the anatomical relations of the parts) appears to be best calculated to effect these objects, is one which would extend from a point an inch above the anus to a point in the posterior perinaeal space an inch or more below the anus. The wound thus made would *depend in relation to the neck of the bladder;* the important parts, vessels, &c., in the anterior perinaeal space would be avoided where the incision, if extended upwards, would have no effect whatever in facilitating the extraction of the stone or the egress of the urine; and what is also of prime importance, the external opening would directly correspond with the incision through the prostate and neck of the bladder. 2nd, After the incision through the skin and superficial fascia is made, the operator should separate as many of the deeper structures as will admit of it, by the finger rather than by the knife; and especially use the knife cautiously towards the extremities of the wound, so as to avoid the artery of the bulb, and the bulb itself in the upper part, and the rectum below. The pudic artery will not be endangered if the deeper parts be divided by the knife, with its edge directed downwards and outwards, while its point slides securely along the staff in the prostate. 3rd, The prostate should be incised sparingly, for, in addition to the known fact that the gland when only partly cut admits of dilatation to a degree sufficient to admit the passage of even a stone of large size, it is also stated upon high authority that by incising the prostate and neck of the bladder to a length equal to the diameter of the stone, such a proceeding is more frequently followed with disastrous results, owing to the circumstance that the pelvic fascia being divided at the place where it is reflected upon the base of the gland and the side and neck of the bladder, allows the urine to infiltrate the cellular tissue of the pelvis. [Footnote]

[Footnote: "The object in following this method," Mr. Liston observes, "is to avoid all interference with the reflexion of the ilio-vesical fascia from the sides of the pelvic cavity over the base of the gland and side of the bladder. If this natural boundary betwixt the external and internal cellular tissue is broken up, there is scarcely a possibility of preventing infiltration of the urine, which must almost certainly prove fatal. The prostate and other parts around the neck of the bladder are very elastic and yielding, so that without much solution of their continuity, and without the least laceration, the opening can be so dilated as to admit the fore-finger readily through the same wound; the forceps can be introduced upon this as a guide, and they can also be removed along with a stone of considerable dimensions, say from three to nearly five inches in circumference, in one direction, and from four to six in the largest."--*Practical Surgery*, page 510. This doctrine (founded, no doubt, on Mr. Liston's own great experience) coincides with that first expressed by Scarpa, Le Cat, and others. Sir Benjamin Brodie, Mr. Stanley, and Mr. Syme are also advocates for limited incisions, extending no farther than a partial division of the prostate, the rest being effected by dilatation. The experience, however, of Cheselden, Martineau, and Mr. S. Cooper, inclined them in favour of a rather free incision of the prostate and neck of the bladder proportioned to the size of the calculus, so that this may be extracted freely, without lacerating or contusing the parts, "and," says the distinguished lithotomist Klein, "upon this basis rests the success of my operations; and hence I invariably make it a rule to let the incision be rather too large than too small, and never to dilate it with any blunt instrument when it happens to be too diminutive, but to enlarge it with a knife, introduced, if necessary, several times."--*Practische Ansichten der Bedeutendsten Chirurgische Operationen*. Opinions of the highest authority being thus opposed, in reference to the question whether free or limited incisions in the neck of the bladder are followed respectively by the greater number of fatal or favourable results, and these being thought mainly to depend upon whether the pelvic fascia be opened or not, one need not hesitate to conclude, that since facts seem to be noticed in support of both modes of practice equally, the issue of the cases themselves must really be dependent upon other circumstances, such as the state of the constitution, the state of the bladder, and the relative position of the internal and external incisions. "Some individuals (observes Sir B. Brodie) are good subjects for the operation, and recover perhaps without a bad symptom, although the operation may have been very indifferently performed. Others may be truly said to be bad subjects, and die, even though the operation be performed in the most perfect manner. What is it that constitutes the essential difference between these two classes of cases? It is, according to my experience, the presence or absence of organic disease."--*Diseases of the Urinary Organs*.]

The position in which the staff is held while the membranous urethra and prostate are being divided should be regulated by the operator himself. If he requires the perinaeum to be protruded and the urethra directed towards the place of the incision, he can effect this by depressing the handle of the instrument a little towards the right groin, taking care at the same time that the point is kept beyond the prostate in the interior of the bladder.

DESCRIPTION OF THE FIGURES OF PLATES 52 & 53.

PLATE 52.

FIGURE 1.

A. The urethra.

B. Accelerator urinae muscle.

C. Central perinaeal tendon.

D D. Right and left erector penis muscle.

E E. The transverse muscles.

F. The anus.

G G. The ischiatic tuberosities.
H. The coccyx.
I I. The glutei muscles.
K K. The levator ani muscle.
L. The left artery of the bulb.

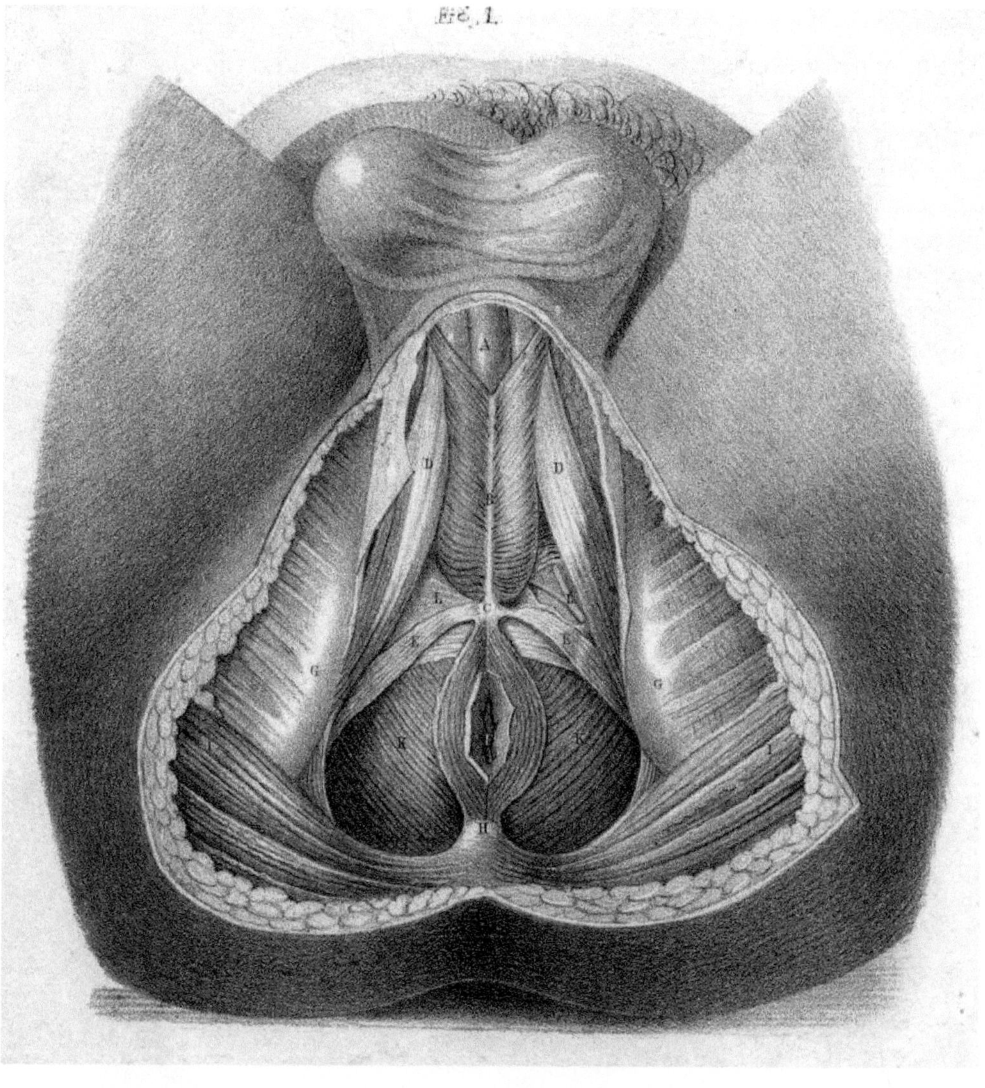

Plate 52.--Figure 1

FIGURE 2.

A, D, F, G, H, I, K, L refer to the same parts as in Fig. 1, Plate 52.

B. The urethra.

C. Cowper's glands between the two layers of--

E. The deep perinaeal fascia.

M. The bulb of the urethra.

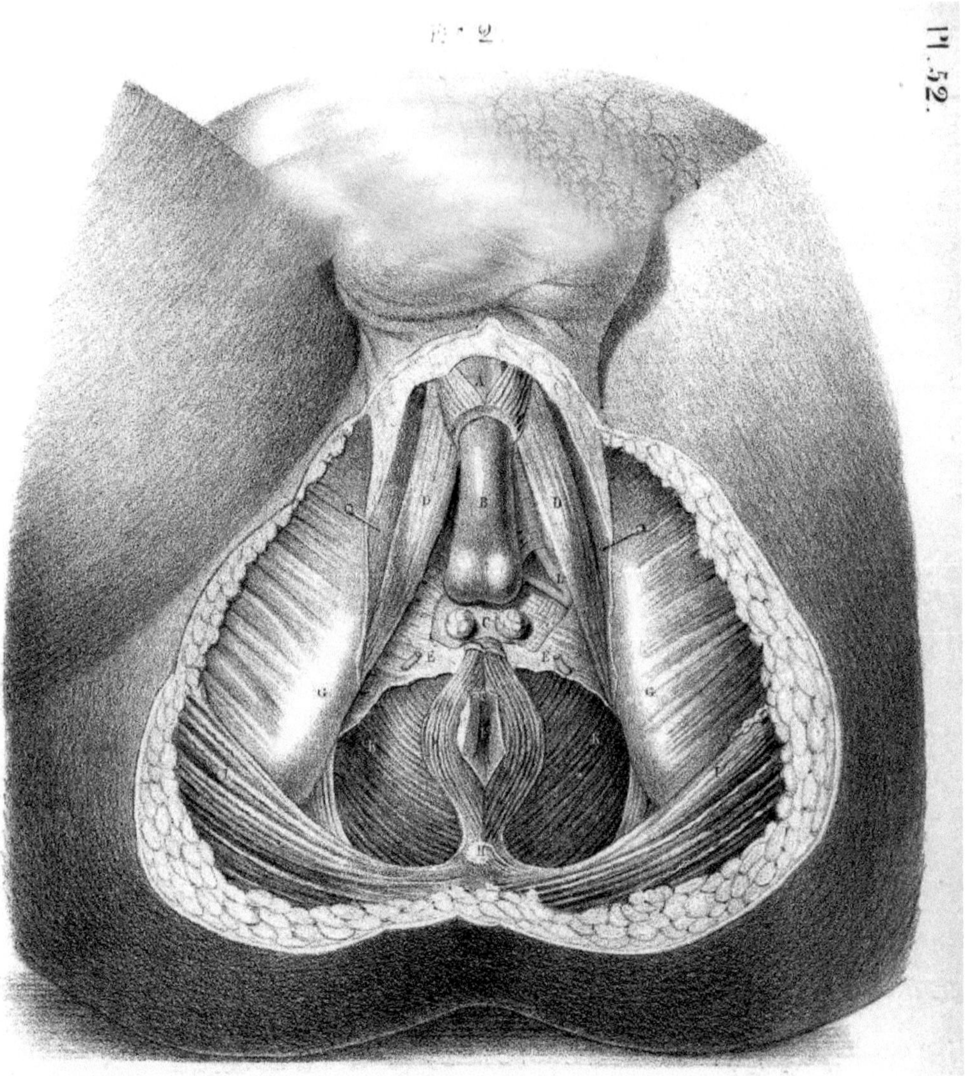

Plate 52.--Figure 2

PLATE 53.

FIGURE 1.

A, B, C, E, F, G, H, I, K, L refer to the same parts as in Fig. 2, Plate 52.

D D. The two crura penis.

M. The urethra in section

N N. The rectum.

O. The sacro-sciatic ligament.

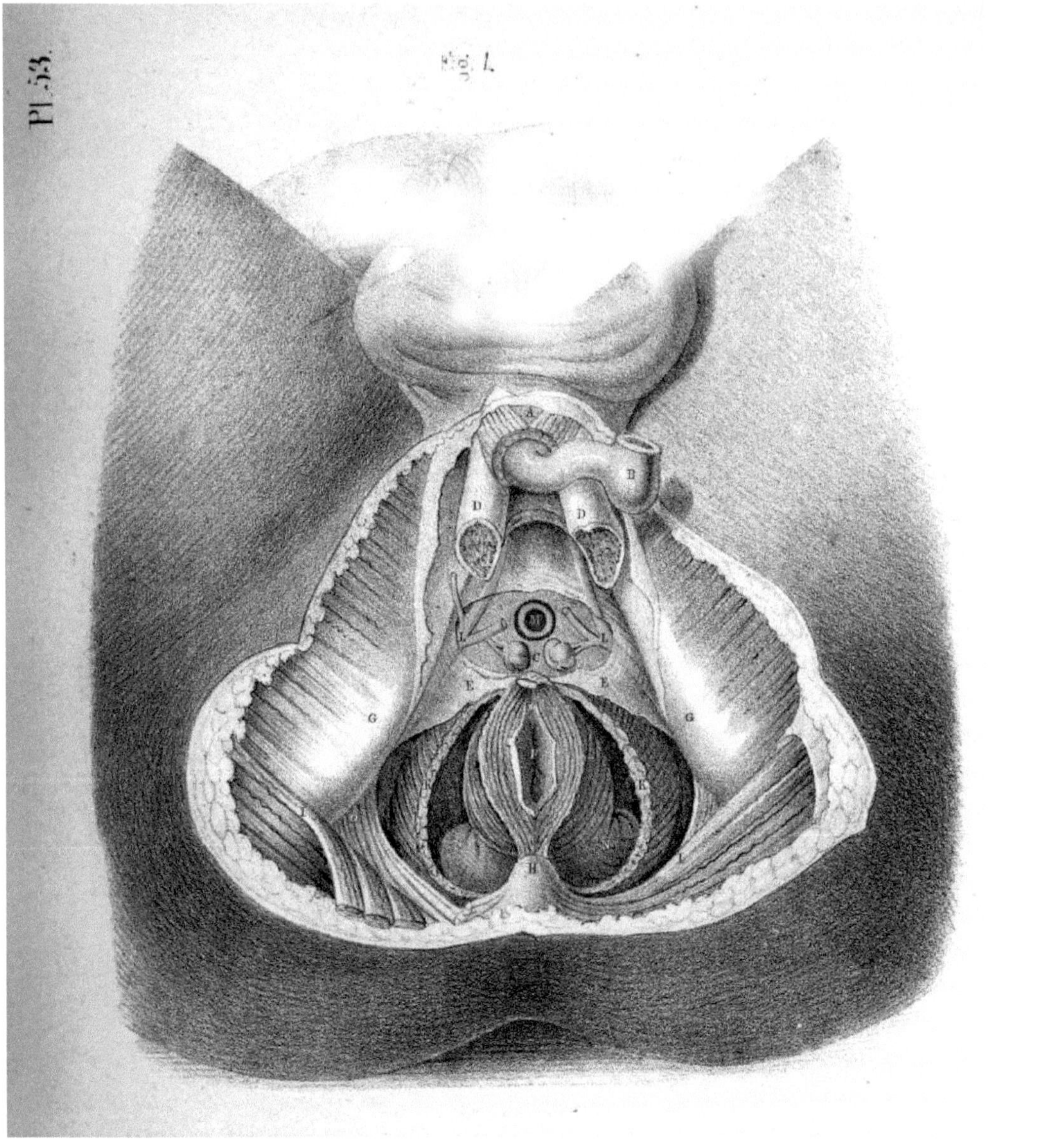

Plate 53.--Figure 1

FIGURE 2.

A, B, D, G, H, I, K, L, O refer to the same parts as in Fig. 1, Plate 53.

C C. The two lobes of the prostate.

F. The rectum turned down.

M. The membranous part of the urethra.

N N. The vesiculae seminales.

P. The base of the bladder.

Q Q. The two vasa deferentia.

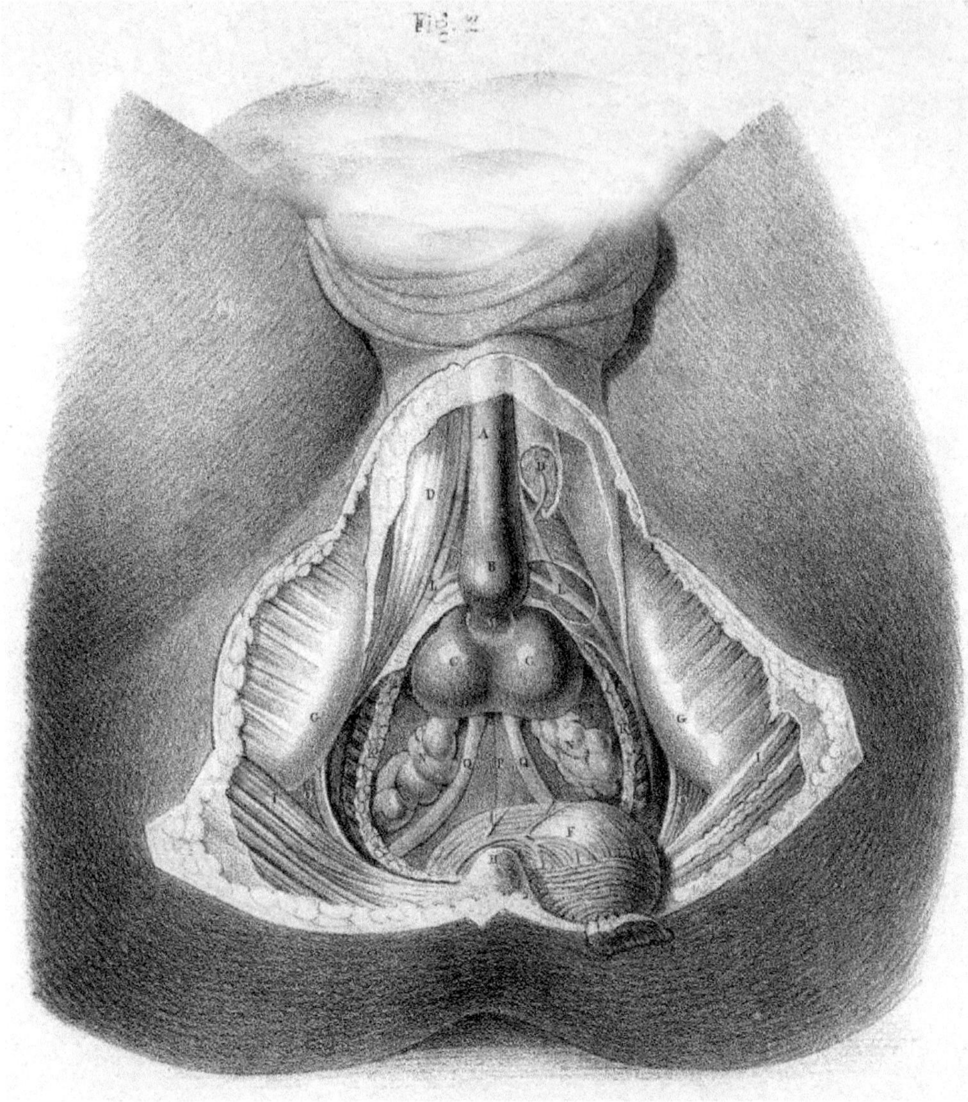

Plate 53.--Figure 2.

THE SURGICAL DISSECTION OF THE MALE BLADDER AND URETHRA; LATERAL AND BILATERAL LITHOTOMY COMPARED.

Lines of incision in both operations. Urethral muscles--their analogies and significations. Direction, form, length, structure, &c., of the urethra at different ages. Third lobe of the prostate. Physiological remarks. Trigone vesical. Bas fond of the bladder. Natural form of the prostate at different ages.

Having examined the surgical relations of the bladder and adjacent structures, in reference to the lateral operation of lithotomy, it remains to reconsider these same parts as they are concerned in the bilateral operation and in catheterism.

Fig. 1, Plate 54, represents the normal relations of the more important parts concerned in lithotomy as performed at the perinaeal region. The median line, AA, drawn from the symphysis pubis above, to the point of the coccyx below, is seen to traverse vertically the centres of the urethra, the prostate, the base of the bladder, the anus, and the rectum. These several parts are situated at different depths from the perinaeal surface. The bulb of the urethra and the lower end of the bowel are on the same plane comparatively superficial. The prostate lies between these two parts, and on a plane deeper than they. The base of the bladder is still more deeply situated than the prostate; and hence it is that the end of the bowel is allowed to advance so near the pendent bulb, that those parts are in a great measure concealed by these. As the apex of the prostate lies an inch (more or less) deeper than the bulb, so the direction of the membranous urethra, which intervenes between the two, is according to the axis of the pelvic outlet; the prostatic end of the membranous urethra being deeper than the part near the bulb. The scalpel of the lithotomist, guided by the staff in this part of the urethra, is made to enter the neck of the bladder *deeply* in the same direction. On comparing the course of the pudic arteries with the median line, A A, we find that they are removed from it at a wider interval below than above; and also that where the vessels first enter the perinaeal space, winding around the spines of the ischia, they are much deeper in this situation (on a level with the base of the bladder) than they are when arrived opposite the bulb of the urethra. The transverse line B B, drawn in front of the anus from one tuber ischii to the other, is seen to divide the perinaeum into the anterior and posterior spaces, and to intersect at right angles the median line A A. In the same way the line B B divides transversely both pudic arteries, the front of the bowel, the base of the prostate, and the sides of the neck of the bladder. Lateral lithotomy is performed in reference to the line A A; the bilateral operation in regard to the line B B. In order to avoid the bulb and rectum at the median line, and the pudic artery at the outer side of the perinaeum, the lateral incisions are made obliquely in the direction of the lines CD. In the bilateral operation the incision necessary to avoid the bulb of the urethra in front, the rectum behind, and the pudic arteries laterally, is required to be made of a semicircular form, corresponding with the forepart of the bowel; the cornua of the incision being directed behind. In the lateral operation, the incision C through the integument, crosses at an acute angle the deeper incision D, which divides the neck of the bladder, the prostate, &c. The left lobe of the prostate is divided obliquely in the lateral operation; both lobes transversely in the bilateral.

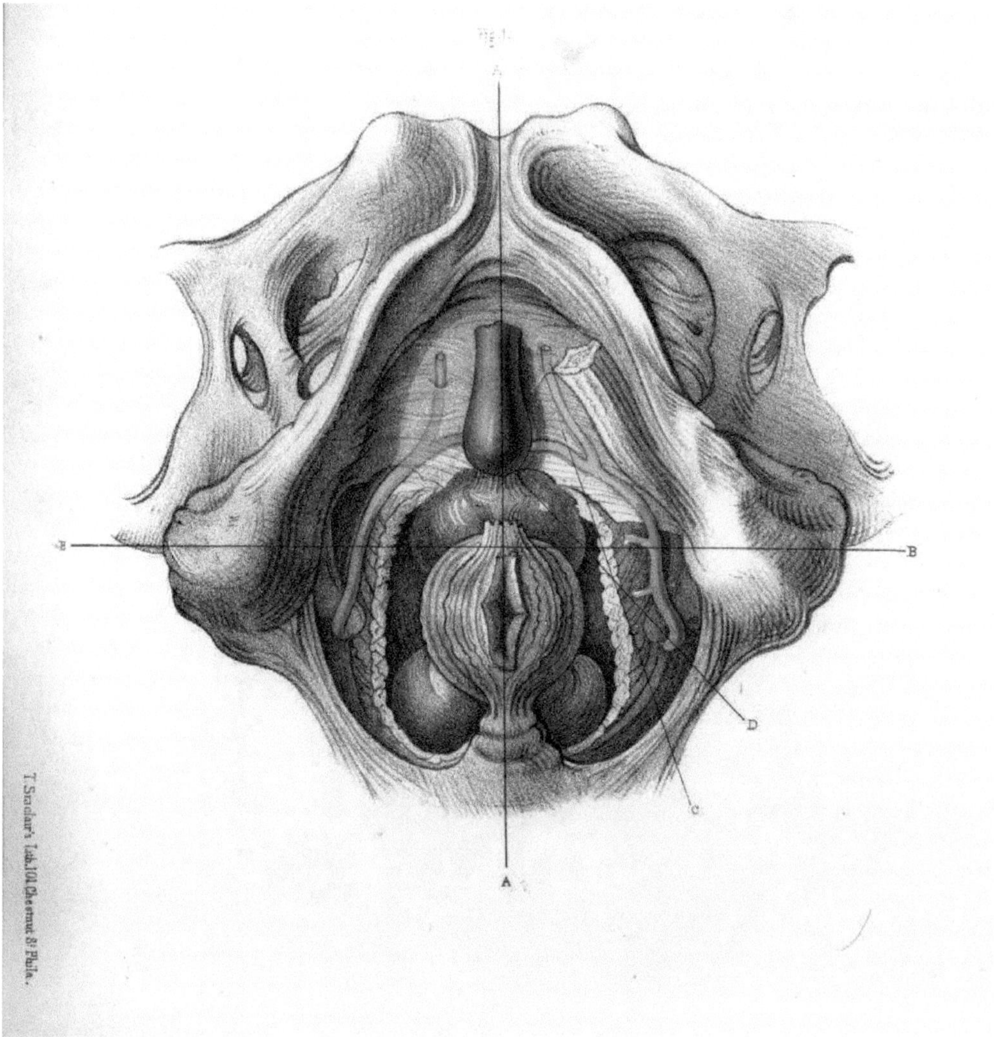

Plate 54, Figure 1.

Fig. 2, Plate 54.--If the artery of the bulb happen to arise from the pudic opposite the tuber ischii, or if the inferior hemorrhoidal arteries be larger than usual, these vessels crossing the lines of incision in both operations will be divided. If the superficial lateral incision C, Fig. 1, be made too deeply at its forepart, the artery of the bulb, even when in its usual place, will be wounded; and if the deep lateral incision D be carried too far outwards, the trunk of the pudic artery will be severed. These accidents are incidental in the bilateral operation also, in performing which it should be remembered that the bulb is in some instances so large and pendulous, as to lie in contact with the front of the rectum.

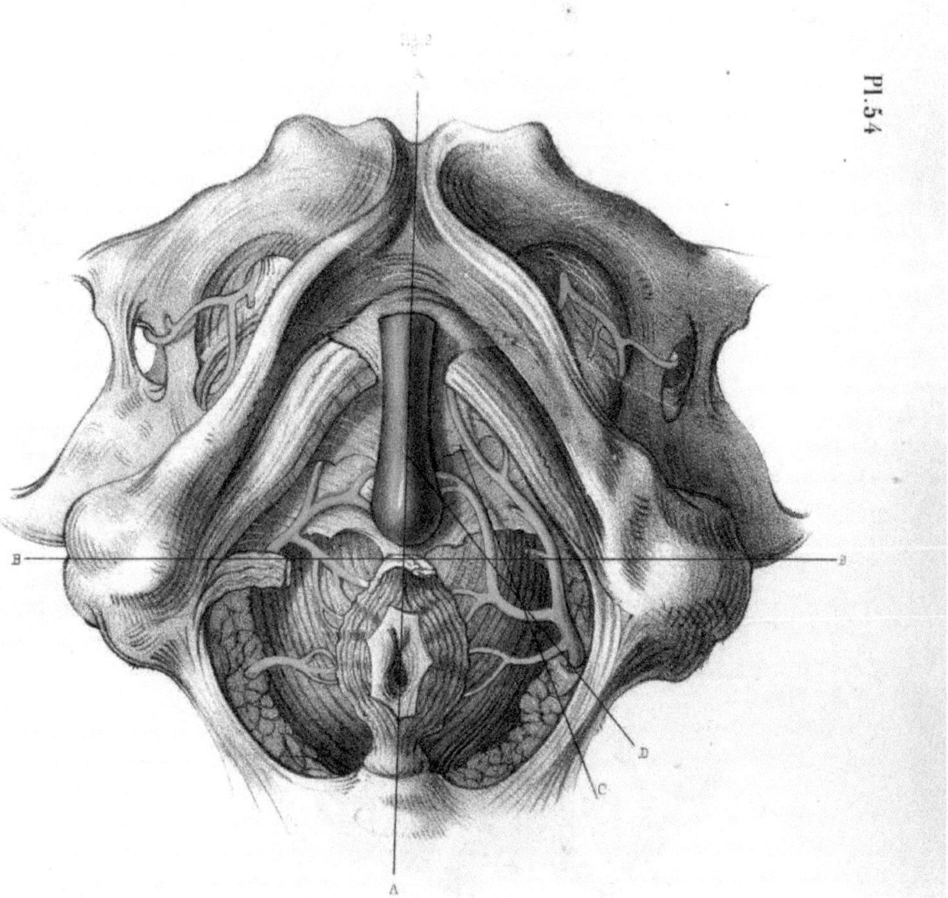

Plate 54, Figure 2.

Fig. 1, Plate 55.--When the pudic artery crosses in contact with the prostate, F, it must inevitably be divided in either mode of operation. Judging from the shape of the prostate, I am of opinion that this part, whether incised transversely in the line B B, or laterally in the line D, will exhibit a wound in the neck of the bladder of equal dimensions. When the calculus is large, it is recommended to divide the neck of the bladder by an incision, combined of the transverse and the lateral. The advantages gained by such a combination are, that while the surface of the section made in the line D is increased by "notching" the right lobe of the prostate in the direction of the line B, the sides of both sections are thereby rendered more readily separable, so as to suit with the rounded form of the calculus to be extracted. These remarks are equally applicable as to the mode in which the superficial perinaeal incision should be made under the like necessity. If the prostate be *wholly* divided in either line of section, the pelvic fascia adhering to the base of this body will be equally subject to danger. By incising the prostate transversely, B B, the seminal ducts, G H, which enter the base of this body, are likewise divided; but by the simple lateral incision D being made through the forepart of the left lobe, F, these ducts will escape injury. [Footnote] On the whole, therefore, the lateral operation appears preferable to the bilateral one.

[Footnote: As to the mode in which the superficial and deep incisions in lateral lithotomy should be made, a very eminent operating surgeon remarks--"a free incision of the skin I consider a most important feature in the operation; but beyond this the application of the knife should, in my opinion, be extremely limited. In so far as I can perceive, there should be no hesitation in cutting any part of the gland which seems to offer resistance, with the exception, perhaps, of its under surface, where the position of the seminal ducts, and other circumstances, should deter the surgeon from using a cutting instrument."--Wm. Fergusson, *Practical Surgery*, 3d Am. Ed., p. 610.]

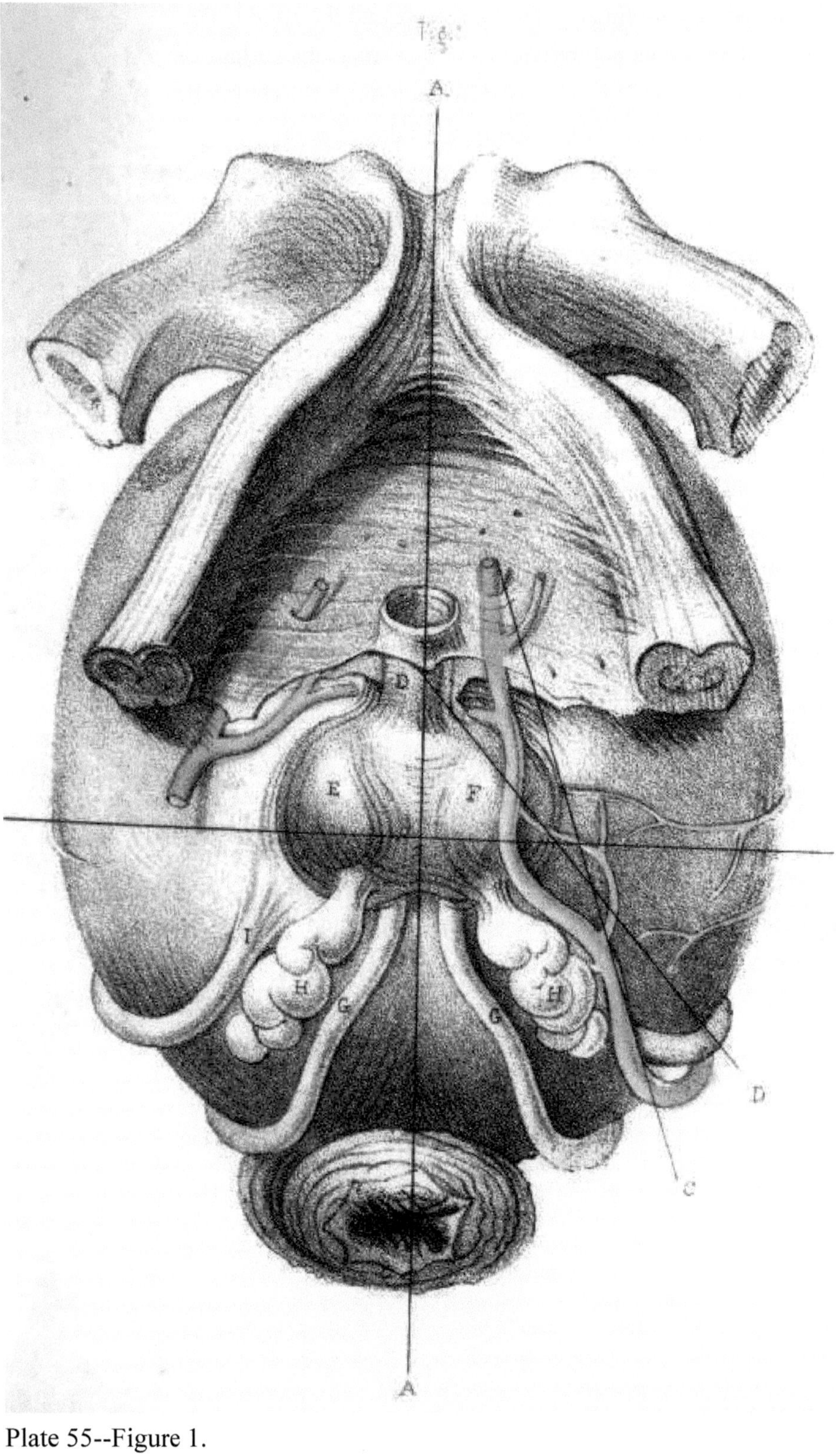

Plate 55--Figure 1.

Fig. 2, Plate 55.--The muscular structures surrounding the membranous urethra and the neck of the bladder, and which are divided in lithotomy, have been examined from time to time by anatomists with more than ordinary painstaking, owing to the circumstance that they are found occasionally to offer, by spasmodic contraction, an obstacle to the passage of the catheter along the urethral canal. These muscles do not appear to exist in all subjects alike. In some, they are altogether wanting; in others, a few of them only appear; in others, they seem to be not naturally separable from the larger muscles which are always present. Hence it is that the opinions of anatomists respecting their form, character, and even their actual existence, are so conflicting, not only against each other, but against nature. In Fig. 2, Plate 55, I have summed together all the facts recorded concerning them, [Footnote] and on comparing these facts with what I have myself observed, the muscles seem to me to assume originally the form and relative position of the parts B C D E F viewed in their totality. Each of these parts of muscular structure arises from the ischio-pubic ramus, and is inserted at the median line A A. They appear to me, therefore, to be muscles of the same category, which, if all were present, would assume the serial order of B C D E F. When one or more of them are omitted from the series, there occurs anatomical variety, which of course occasions variety in opinion, fruitless though never ending. By that interpretation of the parts which I here venture to offer, and to which I am guided by considerations of a higher law of formation, I encompass and bind together, as with a belt, all the dismembered parts of variety, and of these I construct a uniform whole. Forms become, when not viewed under comparison, as meaningless hieroglyphics, as the algebraic symbols $a + c - d = 11$ are when the mind is devoid of the power of calculation.

[Footnote: The part C is that alone described by Santorini, who named it "elevator urethrae," as passing beneath the urethra. The part B is that first observed and described by Mr. Guthrie as passing above the urethra. The part F represents the well-known "transversalis perinaei," between which and the part C there occasionally appears the part E, supposed to be the "transversalis alter" of Albinus, and also the part D, which is the "ischio bulbosus" of Cruveilhier. It is possible that I may not have given one or other of these parts its proper name, but this will not affect their anatomy.]

Fig. 2.

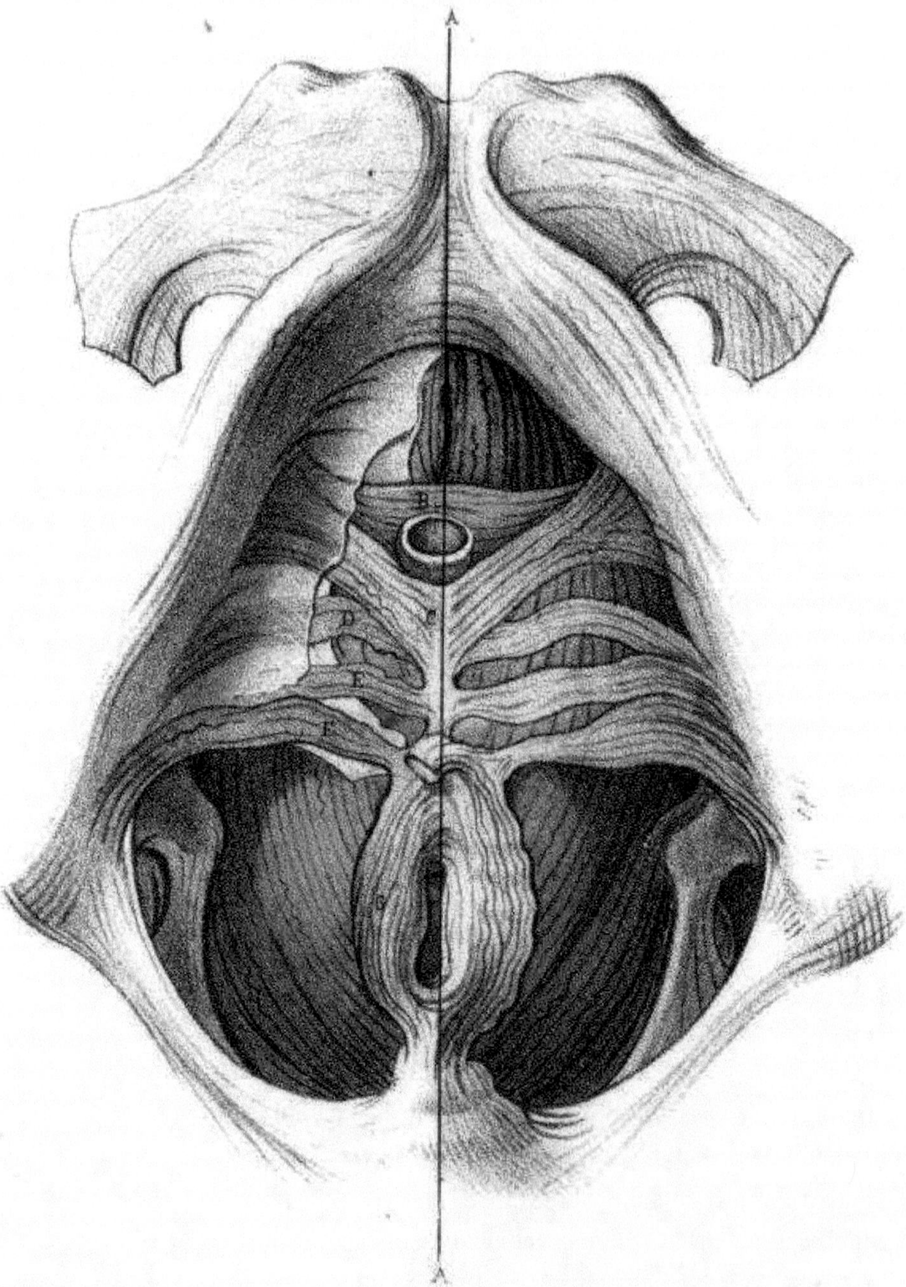

Plate 55--Figure 2

Fig. 3, Plate 55.--The membranous urethra A is also in some instances embraced by two symmetrical fasciculi of muscular fibres B B, which arising from the posterior and lower part of the symphysis pubis, descend on either side of the canal and join beneath it. The muscles B C, Fig. 2, Plate 55, are between the two layers of the deep perinaeal fascia, while the muscle B B, Fig. 3, Plate 55, lies like the forepart of the levator ani, C C, behind this structure and between it and the anterior ligaments of the bladder. [Footnote] As to the interpretation of the muscle, I, myself, am inclined to believe that it is simply a part of the levator ani, and for these reasons--1st, it arises from the pubic symphysis, and is inserted into the perinaeal median line with the levator ani; 2nd, the fibres of both muscles overlie the forepart of the prostate, and present the same arrangement in parallel order; 3rd, the one is not naturally separable from the other.

[Footnote: This is the muscle, B B, which is described by Santorini as the "levator prostatae;" by Winslow as "le prostatique superieur;" by Wilson as the "pubo-urethrales;" by Muller as not existing; by Mr. Guthrie as forming (when existing), with the parts B C, Fig. 2, Plate 55, his "compressor isthmi urethrae;" and by M. Cruveilhier as being *part of the levator ani muscle*. "As in one case," (observes Mr. Quain,) "I myself saw a few vertical muscular fibres connected with the transverse compressor, it has been thought best to retain the muscle in the text."--Dr. Quain's *Anat.*, Am. Ed. vol. ii. p. 539.]

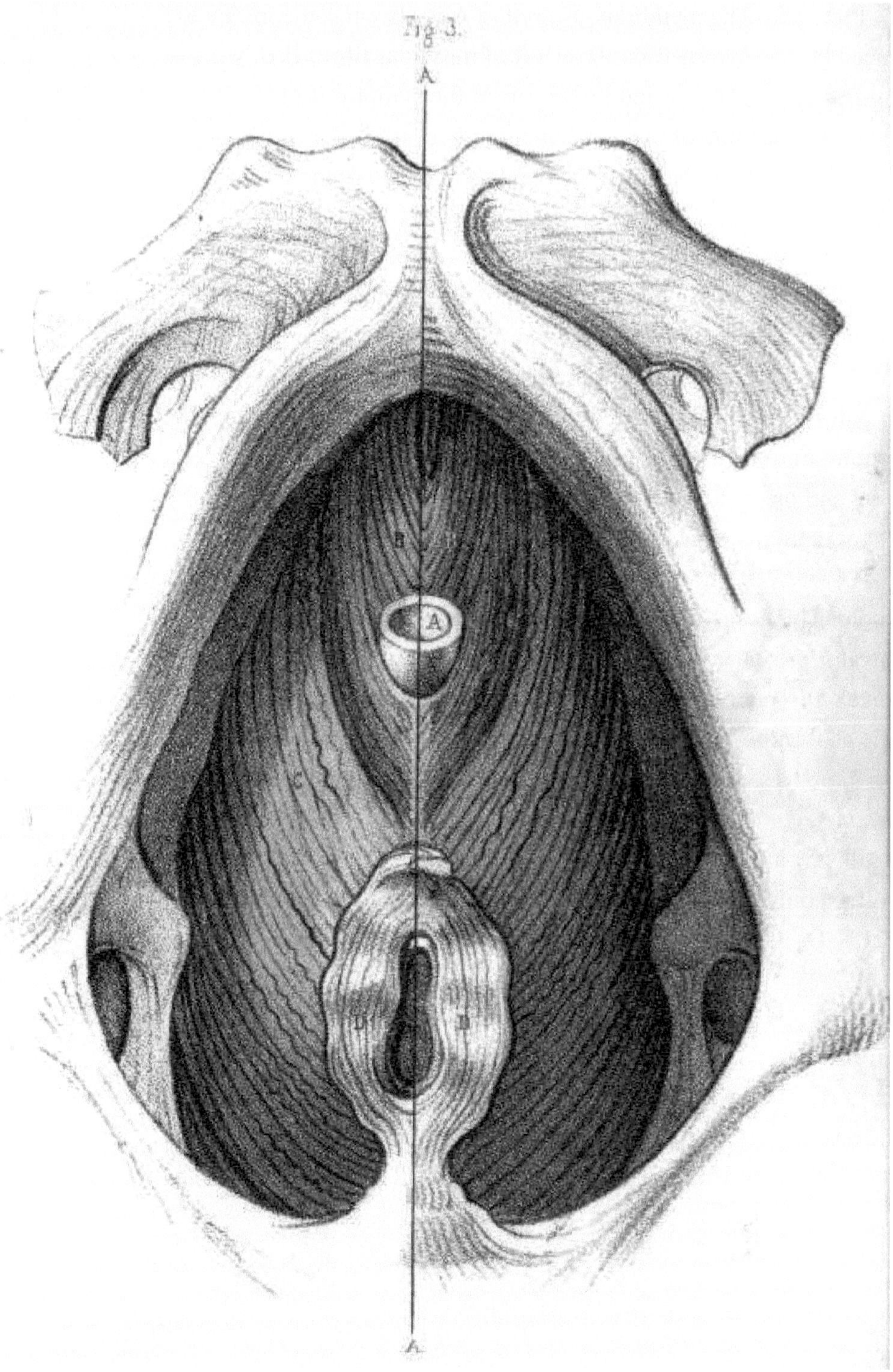

Plate 55--Figure 3

Fig. 1, Plate 56, represents by section the natural forms of the urethra and bladder. The general direction of the urethra measured during its relaxed state from the vesical orifice to the glans is usually described as having the form of the letter S laid procumbent to the right side [capital S rotated 90 degrees right] or to the left [capital S rotated 90 degrees left]. But as the anterior half of the canal is moveable, and liable thereby to obliterate the general form, while the posterior half is fixed, I shall direct attention to the latter half chiefly, since upon its peculiar form and relative position depends most of the difficulty in the performance of catheterism. The portion of the urethra which intervenes between the neck of the bladder, K, and the point E, where the penis is suspended from the front of the symphysis pubis by the suspensory ligament, assumes very nearly the form of a semicircle, whose anterior half looks towards the forepart, and whose posterior half is turned to the back of the pubis. The pubic arch, A, spans crossways, the middle of this part of the urethra, G, opposite the bulb H. The two extremes, F K, of this curve, and the lower part of the symphysis pubis, occupy in the adult the same antero-posterior level; and it follows, therefore, that the distance to which the urethra near its bulb, H, is removed from the pubic symphysis above must equal the depth of its own curve, which measures about an inch perpendicularly. The urethral aperture of the triangular ligament appears removed at this distance below the pubic symphysis, and that portion of the canal which lies behind the ligament, and ascends obliquely backwards and upwards to the vesical orifice on a level with the symphysis pubis in the adult should be remembered, as varying both in direction and length in individuals of the extremes of age. In the young, this variation is owing to the usual high position of the bladder in the pelvis, whilst in the old it may be caused by an enlarged state of the prostate. The curve of the urethra now described is permanent in all positions of the body, while that portion of the canal anterior to the point F, which is free, relaxed, and moveable, can by traction towards the umbilicus be made to continue in the direction of the fixed curve F K, and this is the general form which the urethra assumes when a bent catheter of ordinary shape is passed along the canal into the bladder. The length of the urethra varies at different ages and in different individuals, and its structure in the relaxed state is so very dilatable that it is not possible to estimate the width of its canal with fixed accuracy. As a general rule, the urethra is much more dilatable, and capable consequently of receiving an instrument of much larger bore in the aged than in the adult.

The three portions into which the urethra is described as being divisible, are the spongy, the membranous, and the prostatic. These names indicate the difference in the structure of each part. The spongy portion is the longest of the three, and extending from the glans to the bulb may be said on a rough, but for practical purposes, a sufficiently accurate estimate to comprise seven parts of the whole urethra, which measures nine. The membranous and prostatic portions measure respectively one part of the whole. These relative proportions of the three parts are maintained in different individuals of the same age, and in the same individual at different ages. The spongy part occupies the inferior groove formed between the two united corpora cavernosa of the penis, and is subcutaneous as far back as the scrotum under the pubes, between which point and the bulb it becomes embraced by the accelerator urinae muscle. The bulb and glans are expansions or enlargements of the spongy texture, and do not affect the calibre of the canal. When the spongy texture becomes injected with blood, the canal is rendered much narrower than otherwise. The canal of the urethra is uniform-cylindrical. The meatus is the narrowest part of it, and the prostatic part is the widest. At the point of junction between the membranous and spongy portions behind the bulb, the canal is described as being naturally constricted. Behind the meatus exists a dilatation (fossa navicularis), and opposite the bulb another (sinus of the bulb). Muscular fibres are said to enter into the structure of the urethra, but whether such be the case or not, it is at least very certain that they never prove an obstacle to the passage of instruments, or form the variety of stricture known as spasmodic. The urethra is lined by a delicate mucous membrane presenting longitudinal folds, which become obliterated by distention; and its entire surface is numerously studded with the orifices of mucous cells (lacunae), one of which, larger than the rest, appears on the upper side of the canal near the meatus. Some of these lacunae are nearly an inch long, and all of them open in an oblique direction forwards. Instruments having very narrow apices are liable to enter these ducts and to make false passages. The ducts of Cowper's glands open by very minute orifices on the sides of the spongy urethra anterior to and near the bulb. On the floor of the prostatic urethra appears the crest of the veru montanum, upon which the two seminal ducts open by orifices directed forwards. On either side of the veru montanum the floor of the prostate may be seen perforated by the "excretory ducts" of this so-called *gland.* The part K, which is here represented as projecting from the floor of the bladder, near its neck, is named the "uvula vesicae," (Lieutaud.) It is the same as that which is named the "third lobe of the prostate," (Home.) The part does not appear as proper to the bladder in the healthy condition, Fig. 2, Plate 56. On either side of the point K may be seen the orifices, M M, of the ureters, opening upon two ridges of fibrous substance directed towards the uvula. These are the fibres which have been named by Sir Charles Bell as "the muscles of the ureters;" but as they do not appear in the bladder when in a state of health, I do not believe that nature ever intended them to perform the function assigned to them by this anatomist. And the same may be said of the fibres, which surrounding the vesical orifice, are supposed to act as the "sphincter vesicae." The form of that portion of the base of the bladder which is named "trigone vesical" constitutes an equilateral triangle, and may be described by two lines drawn from the vesical orifice to both openings of the ureters, and another line reaching transversely between the latter. Behind the trigone a depression called "bas fond" is formed in the base of the bladder. Fig. 2, Plate 56, represents the prostate of a boy nine years of age. Fig. 3, Plate 56, represents that of a man aged forty years. A difference as to form and size, &c., is observable between both.

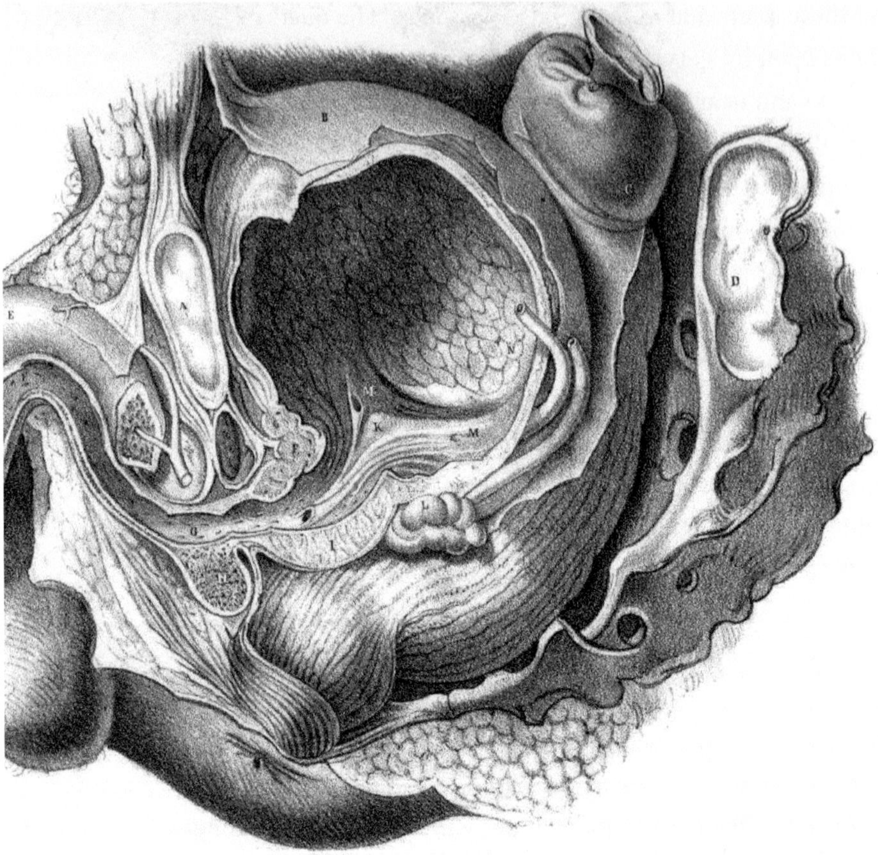

Fig. 2.

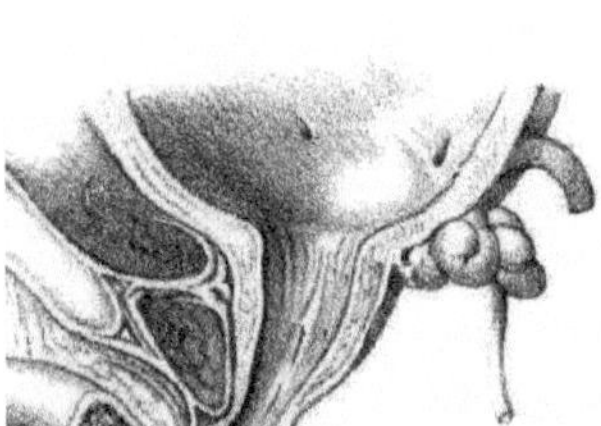

Fig. 3.

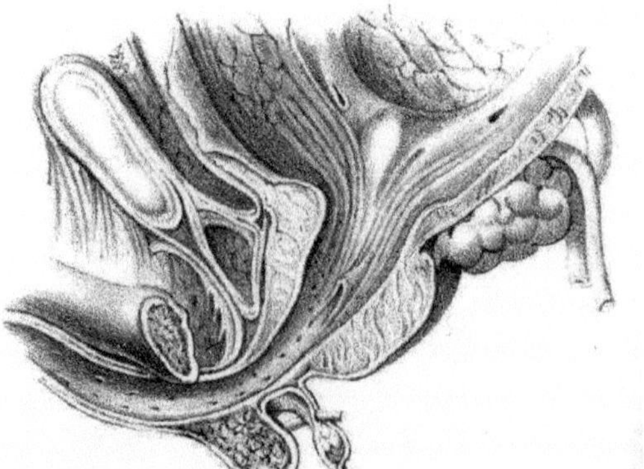

Plate 56--Figure 1, 2, 3

CONGENITAL AND PATHOLOGICAL DEFORMITIES OF THE PREPUCE AND URETHRA; STRICTURES AND MECHANICAL OBSTRUCTIONS OF THE URETHRA.

General remarks. Congenital phymosis. Gonorrhoeal paraphymosis and phymosis. Effect of circumcision. Protrusion of the glans through an ulcerated opening in the prepuce. Congenital hypospadias. Ulcerated perforations of the urethra. Congenital epispadias. Urethral fistula, stricture, and catheterism. Sacculated urethra. Stricture opposite the bulb and the membranous portion of the urethra. Observations respecting the frequency of stricture in these parts. Calculus at the bulb. Polypus of the urethra. Calculus in its membranous portion. Stricture midway between the meatus and bulb. Old callous stricture, its form, &c. Spasmodic stricture of the urethra by the urethral muscles. Organic stricture. Surgical observations.

When any of the central organs of the body presents in a form differing from that which we term natural, or structurally perfect and efficient, if the deformity be one which results as a malformation, ascribable to an error in the law of development, it is always characterized as an excess or defect of the substance of the organ at, and in reference to, the median line. And when any of the canals which naturally open upon the external surface at the median line happens to deviate from its proper position, such deviation, if it be the result of an error in the law of development, always occurs, by an actual necessity, at the median line. On the contrary, though deformities which are the results of diseased action in a central organ may and do, in some instances, simulate those which occur by an error in the process of development, the former cannot bear a like interpretation with the latter, for those are the effects of ever-varying circumstances, whereas these are the effects of certain deviations in a natural process--a law, whose course is serial, gradational, and in the sequent order of a continuous chain of cause and effect.

Fig. 1, represents the prepuce in a state of congenital phymosis. The part hypertrophied and pendent projects nearly an inch in front of the meatus, and forms a canal, continued forwards from this orifice. As the prepuce in such a state becomes devoid of its proper function, and hence must be regarded, not only as a mere superfluity, but as a cause of impediment to the generative function of the whole organ, it should be removed by an operation.

Fig. 2, represents the prepuce in the condition of paraphymosis following gonorrhoeal inflammation. The part appears constricting the penis and urethra behind the corona glandis. This state of the organ is produced in the following-mentioned way:--the prepuce, naturally very extensible, becomes, while covering the glans, inflamed, thickened, and its orifice contracted. It is during this state withdrawn forcibly backwards over the glans, and in this situation, while being itself the first cause of constriction, it induces another--namely, an arrest to the venous circulation, which is followed by a turgescence of the glans. In the treatment of such a case, the indication is, first, to reduce by gradual pressure the size of the glans, so that the prepuce may be replaced over it; secondly, to lessen the inflammation by the ordinary means.

Fig. 3 exhibits the form of a gonorrhoeal phymosis. The orifice of the prepuce is contracted, and the tissue of it infiltrated. If in this state of the part, consequent upon diseased action, or in that of Fig. 1, which is congenital, the foreskin be retracted over the glans, a paraphymosis, like Fig. 2, will be produced.

Fig. 4, shows a form of phymosis in which the prepuce during inflammation has become adherent to the whole surface of the glans. The orifice of the prepuce being directly opposite the meatus, and the parts offering no obstruction to the flow of urine, an operation for separating the prepuce from the glans would not be required.

Fig. 1, Plate 58.--In this figure the urethra appears communicating with a sac like a scrotum. A bougie is represented entering by the meatus, traversing the upper part of the sac, and passing into the membranous part of the urethra beyond. This case which was owing to a congenital malformation of the urethra, exhibits a dilatation of the canal such as might be produced behind a stricture wherever situated. The urine impelled forcibly by the whole action of the abdominal muscles against the obstructing part dilates the urethra behind the stricture, and by a repetition of such force the part gradually yields more and more, till it attains a very large size, and protrudes at the perinaeum as a distinct fluctuating tumour, every time that an effort is made to void the bladder. If the stricture in such a case happen to cause a complete retention of urine, and that a catheter cannot be passed into the bladder, the tumour should be punctured prior to taking measures for the removal of the stricture. (Sir B. Brodie.)

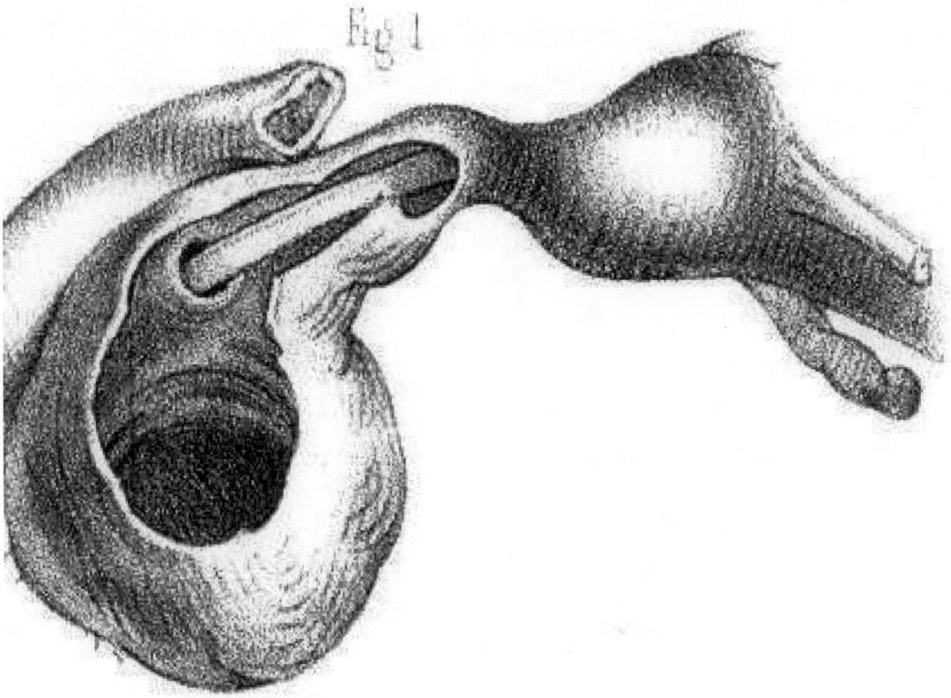

Plate 58.--Figure 1.

Fig. 2, Plate 58, represents two close strictures of the urethra, one of which is situated at the bulb, and the other at the adjoining membranous part. These are the two situations in which strictures of the organic kind are said most frequently to occur, (Hunter, Home, Cooper, Brodie, Phillips, Velpeau.) False passages likewise are mentioned as more liable to be made in these places than elsewhere in the urethral canal. These occurrences--the disease and the accident--would seem to follow each other closely, like cause and consequence. The frequency with which false passages occur in this situation appears to me to be chiefly owing to the anatomical fact, that the urethra at and close to the bulb is the most dependent part of the curve, F K, Fig. 1, Plate 56; and hence, that instruments descending to this part from before push forcibly against the urethra, and are more apt to protrude through it than to have their points turned so as to ascend the curve towards the neck of the bladder. If it be also true that strictures happen here more frequently than elsewhere, this circumstance will of course favour the accident. An additional cause why the catheter happens to be frequently arrested at this situation and to perforate the canal, is owing to the fact, that the triangular ligament is liable to oppose it, the urethral opening in this structure not happening to coincide with the direction of the point of the instrument. In the figure, part of a bougie traverses the urethra through both strictures and lodges upon the enlarged prostate. Another instrument, after entering the first stricture, occupies a false passage which was made in the canal between the two constricted parts.

Fig. 2

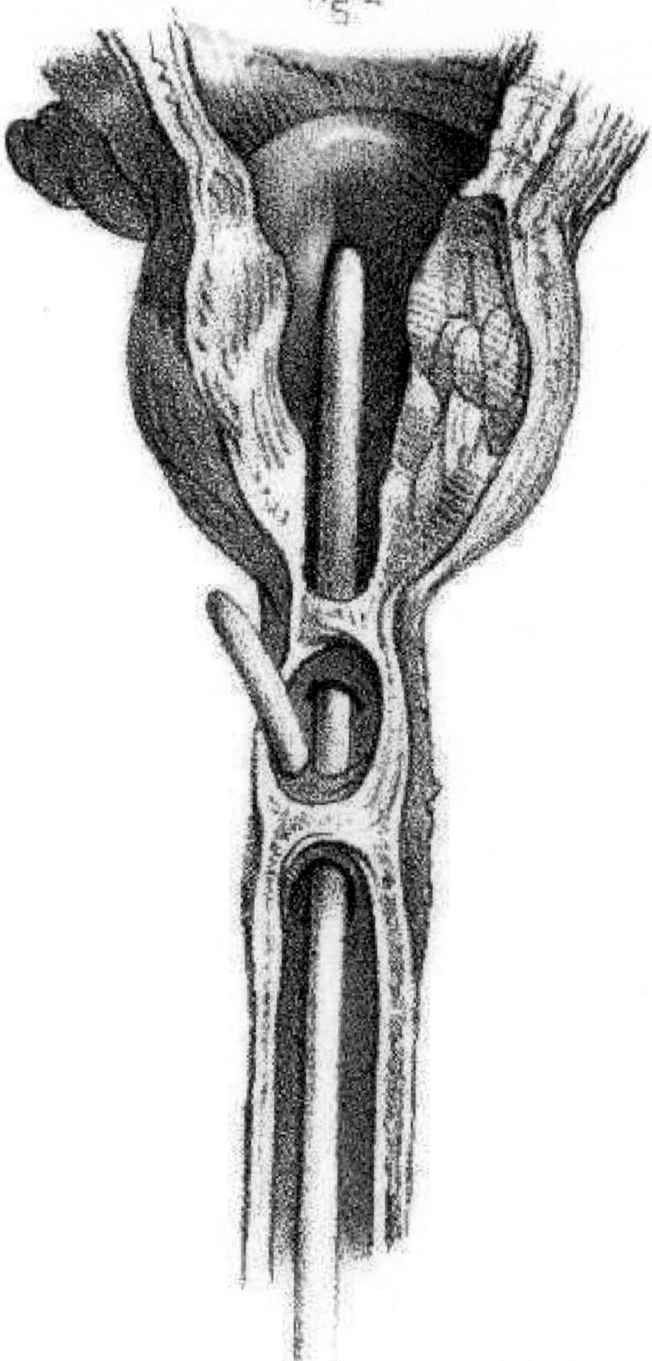

Plate 58.--Figure 2.

Fig. 3, Plate 58.--A calculus is here represented lodging in the urethra at the bulb. The walls of the urethra around the calculus appear thickened. Behind the obstructing body the canal has become dilated, and, in front of it, contracted. In some instances the calculus presents a perforation through its centre, by which the urine escapes. In others, the urine makes its exit between the calculus and the side of the urethra, which it dilates. In this latter way the foreign body becomes loosened in the canal and gradually pushed forwards as far as the meatus, within which, owing to the narrowness of this aperture, it lodges permanently. If the calculus forms a complete obstruction to the passage of the urine, and its removal cannot be effected by other means, an incision should be made to effect this object.

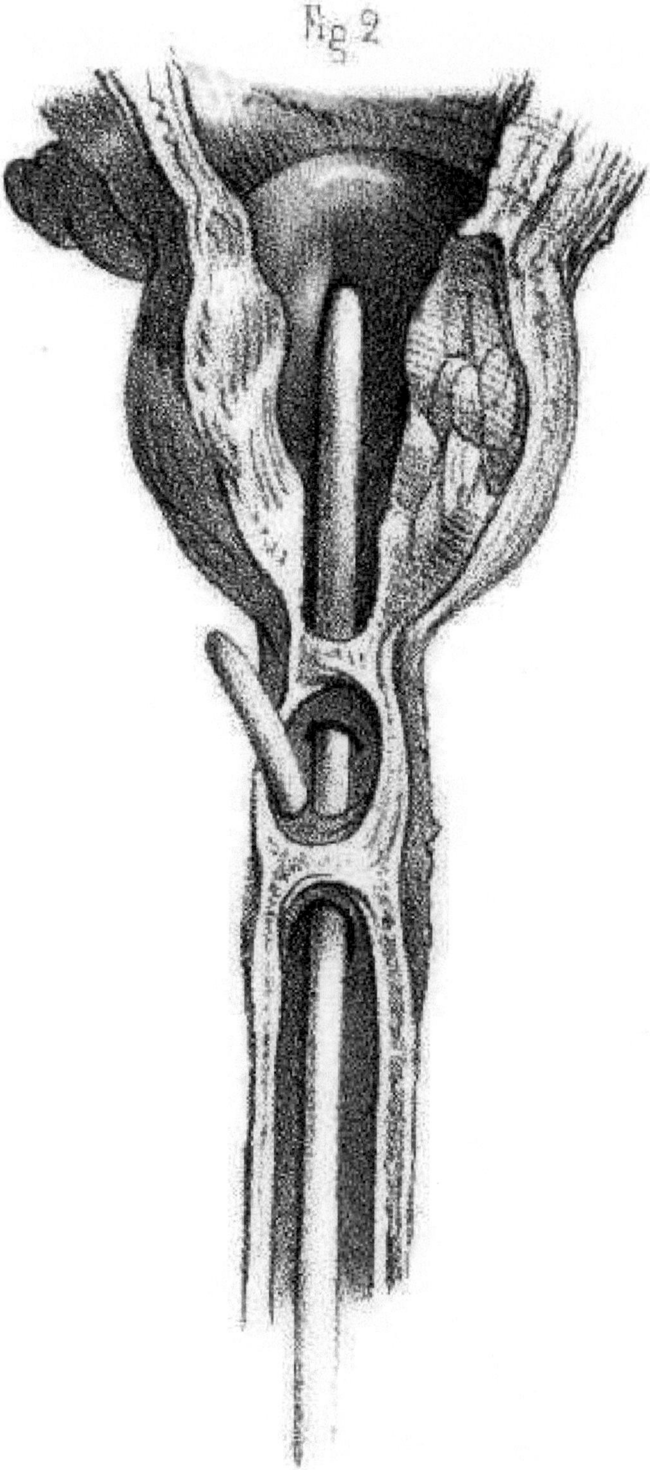

Plate 58.--Figure 3.

Fig. 4, Plate 58, represents the neck of the bladder and neighbouring part of the urethra of an ox, in which a polypous growth is seen attached by a long pedicle to the veru montanum and blocking up the neck of the bladder. Small irregular tubercles of organized lymph, and tumours formed by the lacunae distended by their own secretion, their orifices being closed by inflammation, are also found to obstruct the urethral canal.

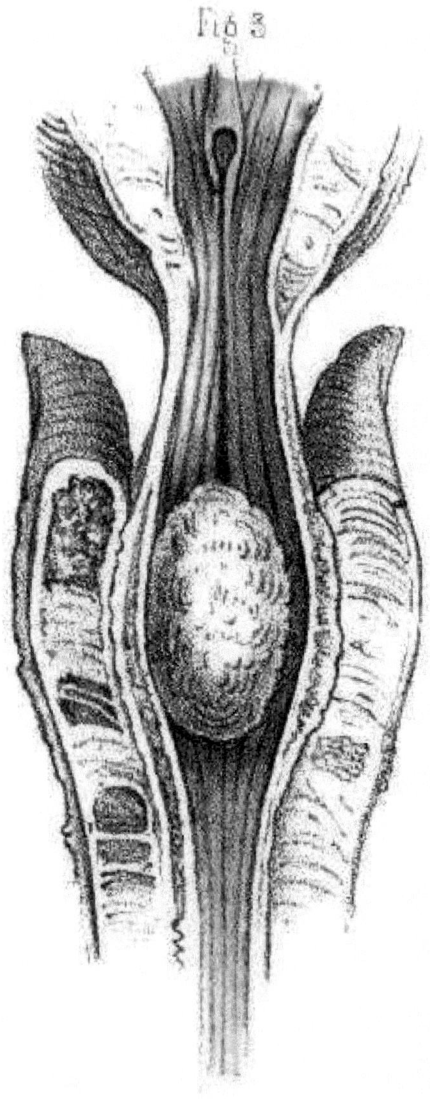

Plate 58.--Figure 4.

Fig. 5, Plate 58.--In this figure is represented a small calculus impacted in and dilating the membranous part of the urethra.

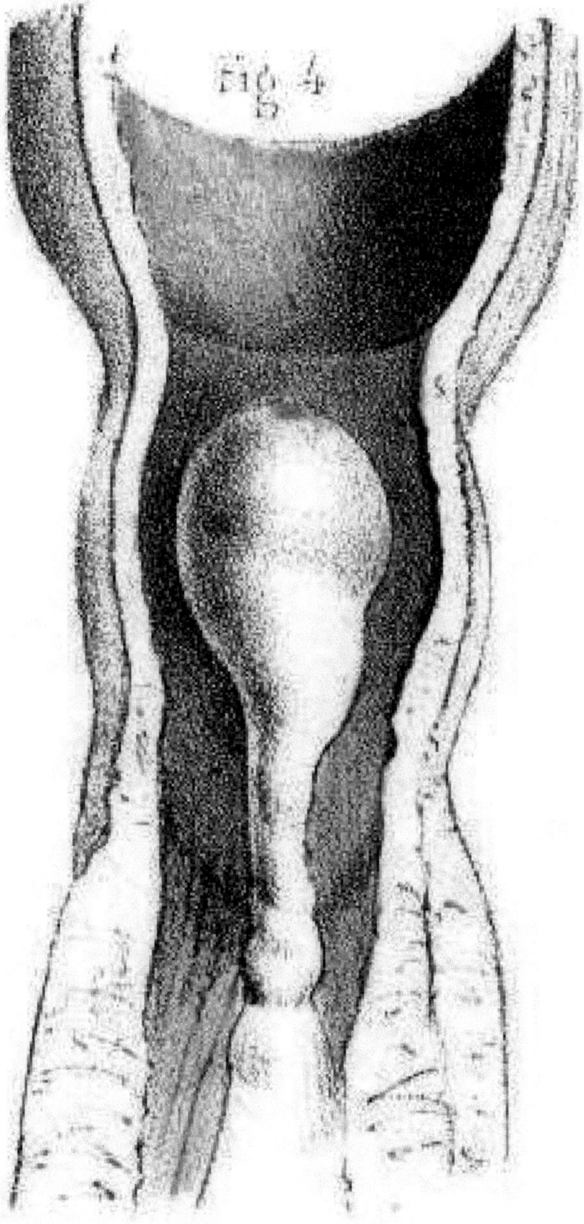

Plate 58.--Figure 5.

Fig. 6, Plate 58.--Two strictures are here shown to exist in the urethra, one of which is situated immediately in front of the bulb, and the other at a point midway between the bulb and the meatus.

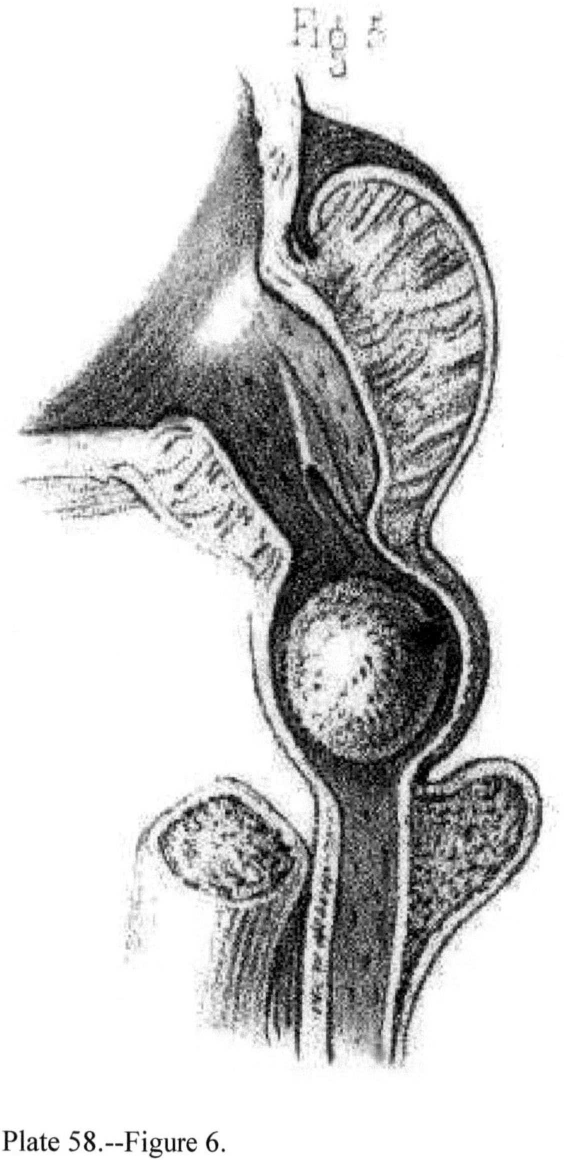

Plate 58.--Figure 6.

Fig. 7, Plate 58.--A stricture is here shown situated at the bulb.

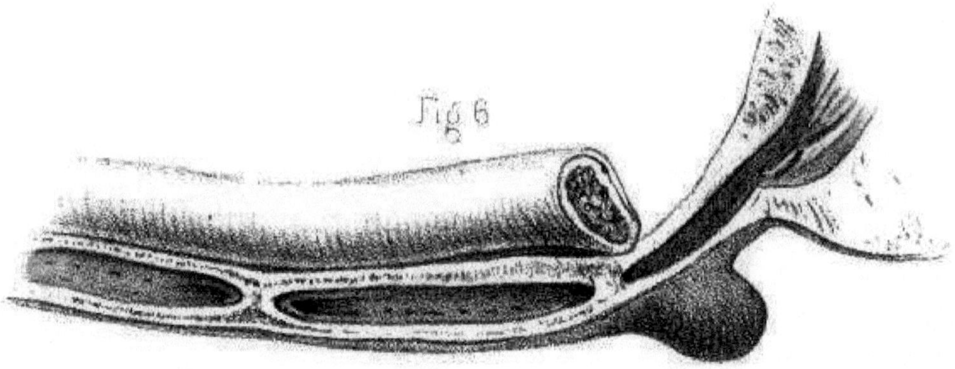

Plate 58.--Figure 7.

Fig. 8, Plate 58, represents a stricture of the canal in front of the bulb.

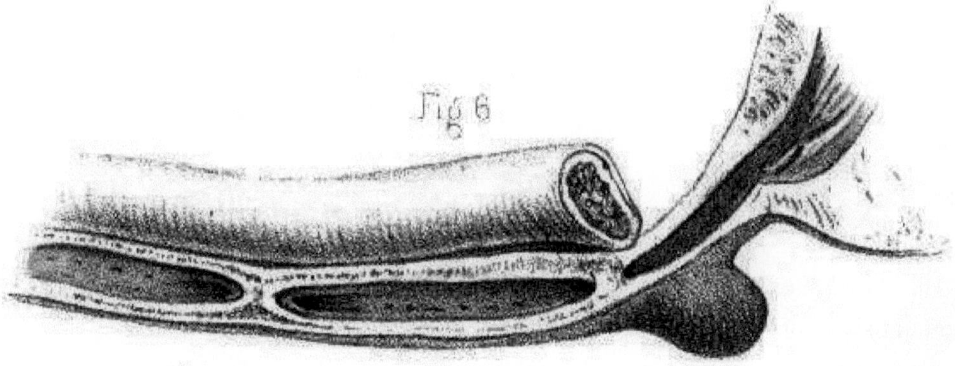

Plate 58.--Figure 8.

Fig. 9, Plate 58, represents the form of an old callous stricture half an inch long, situated midway between the bulb and the meatus. This is perhaps the most common site in which a stricture of this kind is found to exist. In some instances of old neglected cases the corpus spongiosum appears converted into a thick gristly cartilaginous mass, *several inches in extent*, the passage here being very much contracted, and chiefly so at the middle of the stricture. When it becomes impossible to dilate or pass the canal of such a stricture by the ordinary means, it is recommended to divide the part by the lancetted stilette. (Stafford.) Division of the stricture, by any means, is no doubt the readiest and most effectual measure that can be adopted, provided we know clearly that the cutting instrument engages fairly the part to be divided. But this is a knowledge less likely to be attained if the stricture be situated behind than in front of the triangular ligament.

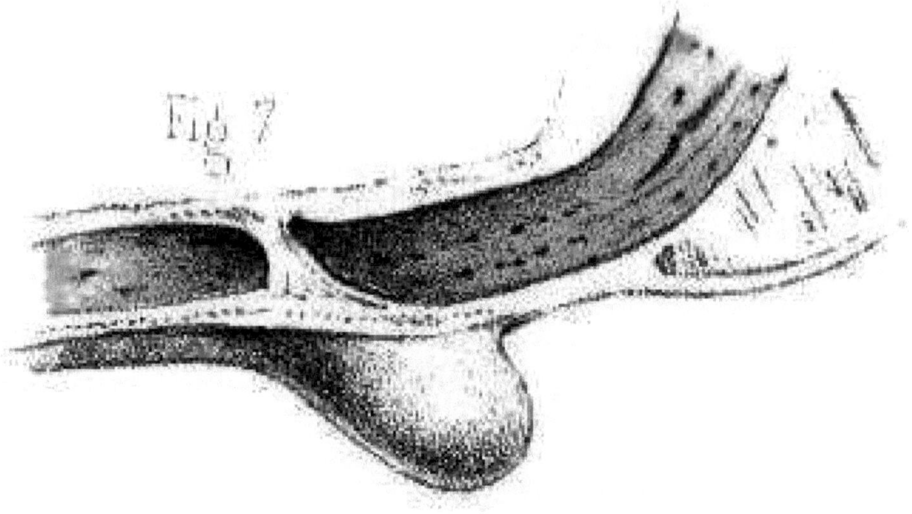

Plate 58.--Figure 9.

Fig. 10, Plate 58, exhibits a lateral view of the muscular parts which surround the membranous portion of the urethra and the prostate; *a*, the membranous urethra embraced by the compressor urethrae muscle; *b*, the levator prostatae muscle; *c*, the prostate; *d*, the anterior ligament of the bladder.

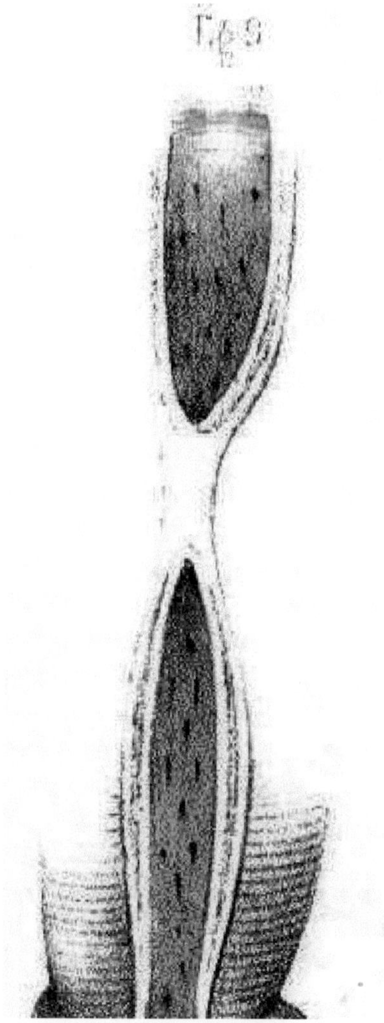

Plate 58.--Figure 10.

Fig. 11, Plate 58.--A posterior view of the parts seen in Fig. 10; *a,* the urethra divided in front of the prostate; *b b*, the levator prostatae muscle; *c c,* the compressor urethrae; *d d*, parts of the obturator muscles; *e e*, the anterior fibres of the levator ani muscle; *f g*, the triangular ligament enclosing between its layers the artery of the bulb, Cowper's glands, the membranous urethra, and the muscular parts surrounding this portion of the canal. The fact that the flow of urine through the urethra happens occasionally to be *suddenly* arrested, and this circumstance contrasted with the opposite fact that the organic stricture is of *slow formation,* originated the idea that the former occurrence arose from a spasmodic muscular contraction. By many this spasm was *supposed* to be due to the urethra being itself muscular. By others, it was *demonstrated* as being dependent upon the muscles which surround the membranous part of the urethra, and which act upon this part and constrict it. From my own observations I have formed the settled opinion that the urethra itself is not muscular. And though, on the one hand, I believe that this canal, *per se,* never causes by active contraction the spasmodic form of stricture, I am far from supposing, on the other, that *all* sudden arrests to the passage of urine through the urethra are solely attributable to spasm of the muscles which embrace this canal.

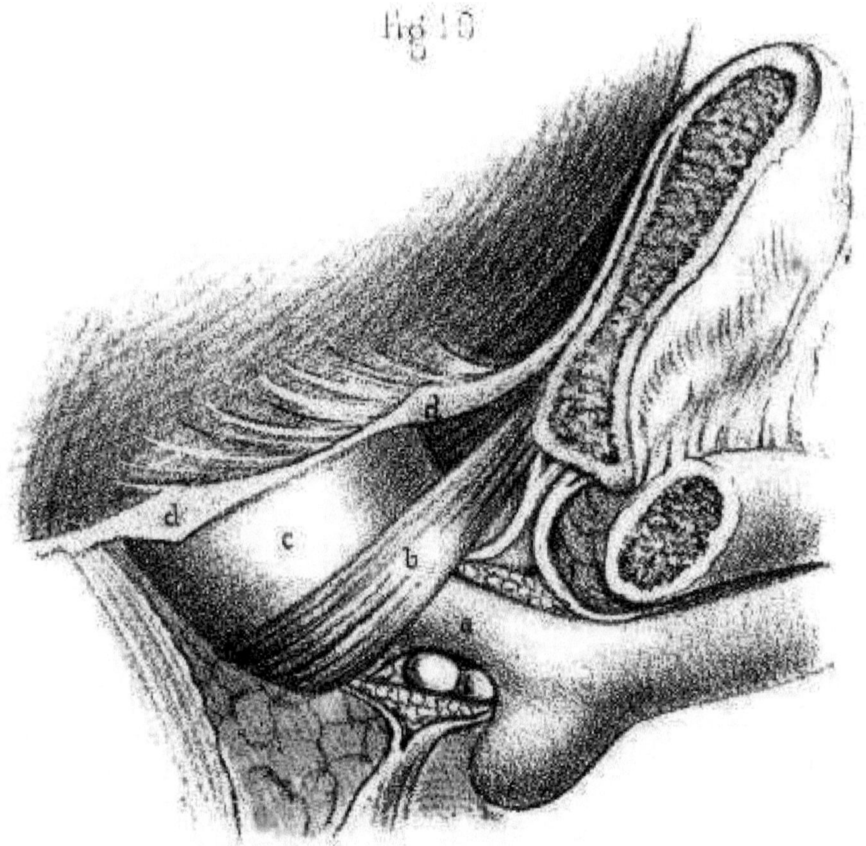

Plate 58.--Figure 11.

THE VARIOUS FORMS AND POSITIONS OF STRICTURES AND OTHER OBSTRUCTIONS OF THE URETHRA; FALSE PASSAGES; ENLARGEMENTS AND DEFORMITIES OF THE PROSTATE.

General remarks. Different forms of the organic stricture. Coexistence of several. Prostatic abscess distorting and constricting the urethra. Perforation of the prostate by catheters. Series of gradual enlargements of the third lobe of the prostate. Distortion of the canal by the enlarged third lobe--by the irregular enlargement of the three lobes—by a nipple-shaped excrescence at the vesical orifice.

Impediments to the passage of the urine through the urethra may arise from different causes, such as the impaction of a small calculus in the canal, or any morbid growth (a polypus, &c.) being situated therein, or from an abscess which, though forming externally to the urethra, may press upon this tube so as either to obstruct it partially, by bending one of its sides towards the other, or completely, by surrounding the anal on all sides. These causes of obstruction may happen in any part of the urethra, but there are two others (the prostatic and the spasmodic) which are, owing to anatomical circumstances, necessarily confined to the posterior two-thirds of the urethra. The portion of the urethra surrounded by the prostate can alone be obstructed by this body when it has become irregularly enlarged, while the spasmodic stricture can only happen to the membranous portion of the urethra, and to an inch or two of the canal anterior to the bulb, these being the parts which are embraced by muscular structures. The urethra itself not being muscular, cannot give rise to the spasmodic form of stricture. But that kind of obstruction which is common to all parts of the urethra, and which is dependent, as well upon the structures of which the canal is uniformly composed, as upon the circumstance that inflammation may attack these in any situation and produce the same effect, is the permanent or organic stricture. Of this disease the forms are as various as the situations are, for as certainly as it may reasonably be supposed that the plastic lymph, effused in an inflamed state of the urethra from any cause, does not give rise to stricture of any special or particular form, exclusive of all others; so as certainly may it be inferred that, in a structurally uniform canal, inflammation points to no one particular place of it, whereat by preference to establish the organic stricture.

The membranous part of the canal is, however, mentioned as being the situation most prone to the disease; but I have little doubt, nevertheless, that owing to general rules of this kind being taken for granted, upon imposing authority, many more serious evils (false passages, &c.) have been effected by catheterism than existed previous to the performance of this operation.[Footnote]

[Footnote: Home describes "a natural constriction of the urethra, directly behind the bulb, which is probably formed with a power of contraction to prevent," &c. This is the part which he says is "most liable to the disease of stricture." (*Strictures of the Urethra.*) Now, if anyone, even among the acute observing microscopists, can discern the structure to which Home alludes, he will certainly prove this anatomist to be a marked exception amongst those who, for the enforcement of any doctrine, can see any thing or phenomenon they wish to see. And, if Hunter were as the mirror from which Home's mind was reflected, then the observation must be imputed to the Great Original. Upon the question, however, as to which is the most frequent seat of stricture, I find that both these anatomists do not agree, Hunter stating that its usual seat is just in front of the bulb, while Home regrets, as it were, to be obliged to differ from "his immortal friend," and avers its seat to be an infinitesimal degree behind the bulb. Sir A. Cooper again, though arguing that the most usual situation of stricture is that mentioned by Hunter, names, as next in order of frequency, strictures of the membranous and prostatic parts of the urethra. Does it not appear strange now, how questions of this import should have occupied so much of the serious attention of our great predecessors, and of those, too, who at the present time form the vanguard of the ranks of science? Upon what circumstance, either anatomical or pathological, can one part of the urethra be more liable to the organic stricture than another?]

Figs. 1 and 2, Plate 59.--In these figures are presented seven forms of organic stricture occurring, in different parts of the urethra. In *a*, Fig. 1, the mucous membrane is thrown into a sharp circular fold, in the centre of which the canal, appears much contracted; a section of this stricture appears in *b,* Fig. 2. In *b*, Fig. 1, the canal is contracted laterally by a prominent fold of the mucous membrane at the opposite side. In c, Fig. 1, an organized band of lymph is stretched across the canal; this stricture is seen in section in c, Fig. 2. In *e*, Fig. 1, a stellate band of organized lymph, attached by pedicles to three sides of the urethra, divides the canal into three passages. In *d*, Fig. 1, the canal is seen to be much contracted towards the left side by a crescentic fold of the lining membrane projecting from the right. In *f*, the canal appears contracted by a circular membrane, perforated in the centre; a section of which is seen at *a*, Fig. 2. The form of the organic stricture varies therefore according to the three following circumstances:--1st. When lymph becomes effused within the canal upon the surface of the lining mucous membrane, and contracts adhesions across the canal. 2ndly. When lymph is effused external to the lining membrane, and projects this inwards, thereby narrowing the diameter of the canal. 3rdly. When the outer and inner walls of a part of the urethra are involved in the effused organizable matter, and on contracting towards each other, encroach at the same time upon the area of the canal. This latter state presents the form, which is known as the old callous tough stricture, extending in many instances for an inch or more along the canal. In cases where the urethra becomes obstructed by tough bands of substance, *c e,* which cross the canal directly, the points of flexible catheters, especially if these be of slender shape, are apt to be bent upon the resisting part, and on pressure being continued, the operator may be led to suppose that the instrument traverses the stricture, while it is most probably perforating the wall of the urethra. But in those cases where the diameter of the canal is circularly contracted, the stricture generally presents a conical depression in front, which, receiving the point of the instrument, allows this to enter the central passage unerringly. A stricture formed by a crescentic septum, such as is seen in *b d*, Fig. 1, offers a more effectual obstacle to the passage of a catheter than the circular septum like *a f*.

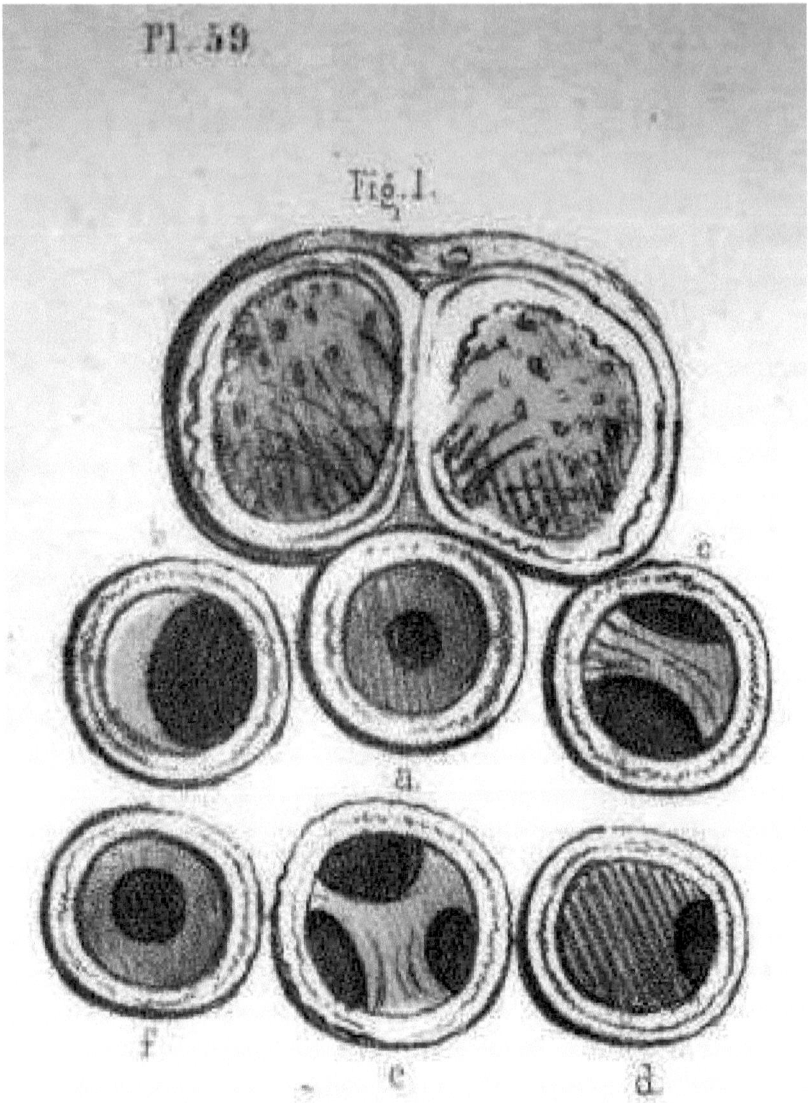
Pl. 59
Fig. I.
b
c
a.
f
e
d

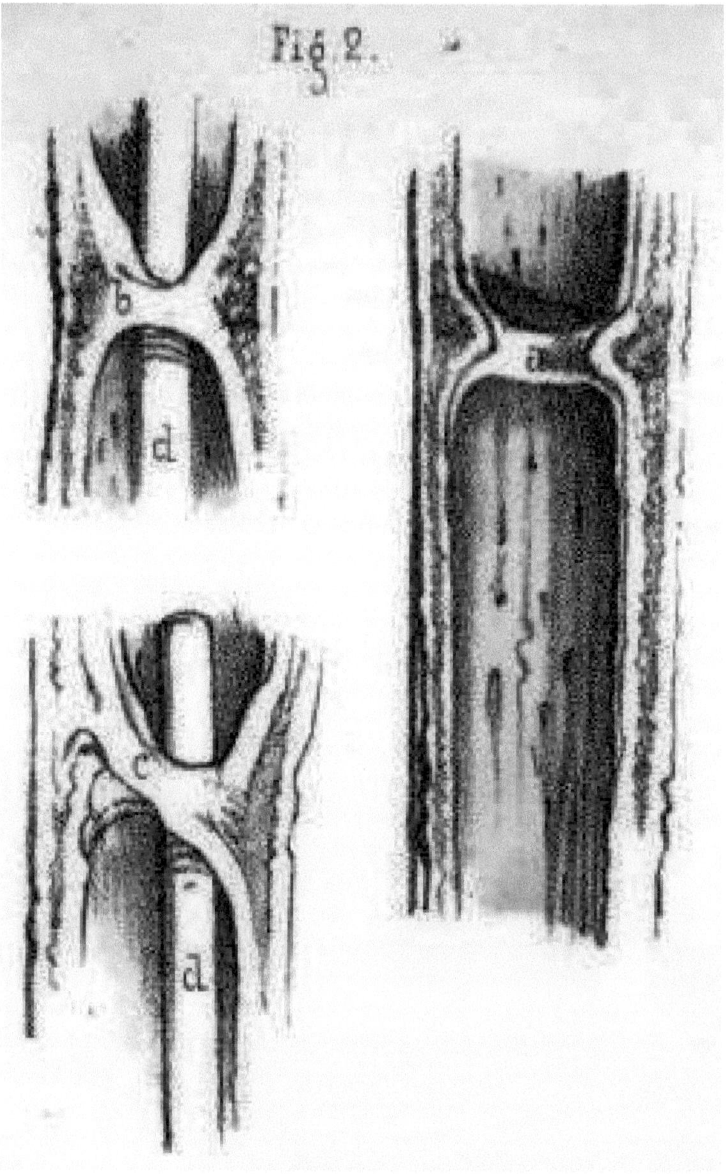

Fig. 3, Plate 59.--In this there are seen three separate strictures, *a, b, c*, situated in the urethra, anterior to the bulb. In some cases there are many more strictures (even to the number of six or seven) situated in various parts of the urethra; and it is observed that when one stricture exists, other slight tightnesses in different parts of the canal frequently attend it. (Hunter.) When several strictures occur in various parts of the urethra, they may occasion as much difficulty in passing an instrument as if the whole canal between the extreme constrictions were uniformly narrowed.

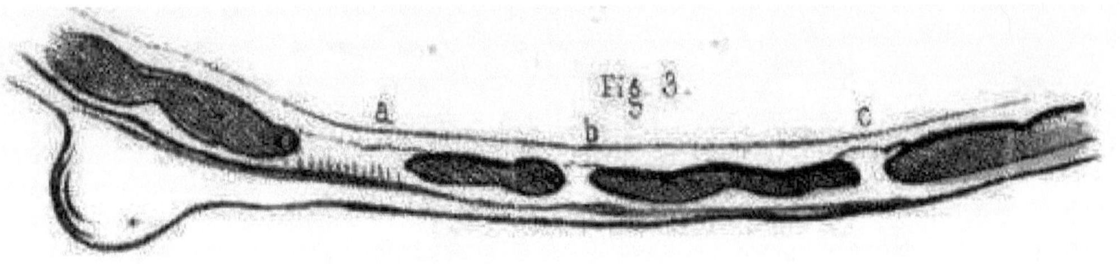

Plate 59.--Figure 3.

Fig. 4, Plate 59.--In this the canal is constricted at the point *a,* midway between the bulb and glans. A false passage has been made under the urethra by an instrument which passed out of the canal at the point *f,* anterior to the stricture *a*, and re-entered the canal at the point *c,* anterior to the bulb. When a false passage of this kind happens to be made, it will become a permanent outlet for the urine, so long as the stricture remains. For it can be of no avail that we avoid re-opening the anterior perforation by the catheter, so long as the urine prevented from flowing by the natural canal enters the posterior perforation. Measures should be at once taken to remove the stricture.

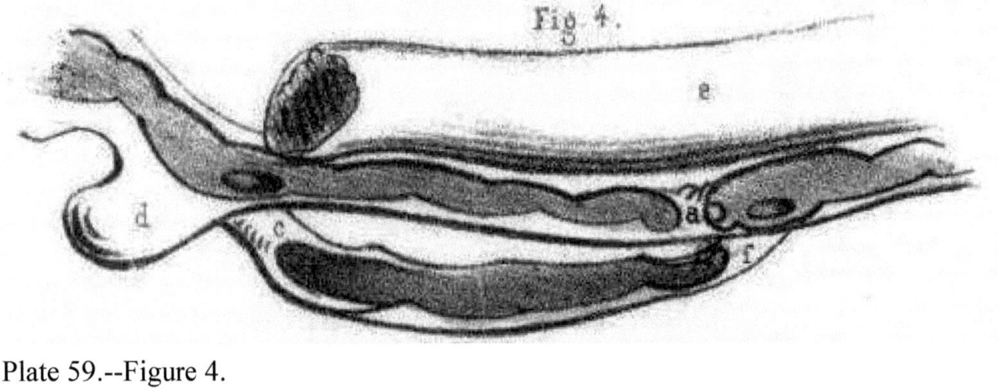

Plate 59.--Figure 4.

Fig. 5, Plate 59.--The stricture *a* appears midway between the bulb and glans, the area of the passage through the stricture being sufficient only to admit a bristle to pass. It would seem almost impossible to pass a catheter through a stricture so close as this, unless by a laceration of the part, combined with dilatation.

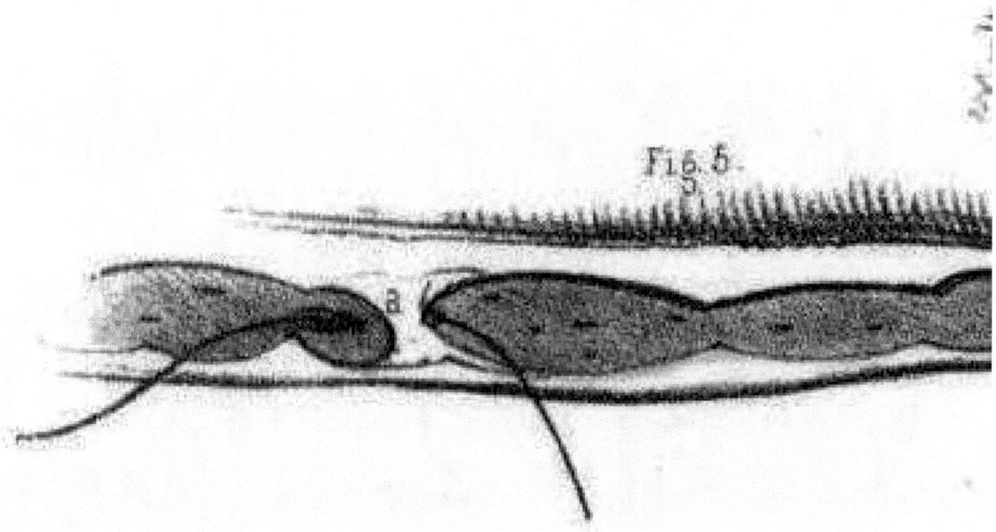

Plate 59.--Figure 5.

Fig. 6, Plate 59.--Two instruments, *a, b,* have made false passages beneath the mucous membrane, in a case where no stricture at all existed. The resistance which the instruments encountered in passing out of the canal having been mistaken, no doubt, for that of passing through a close stricture.

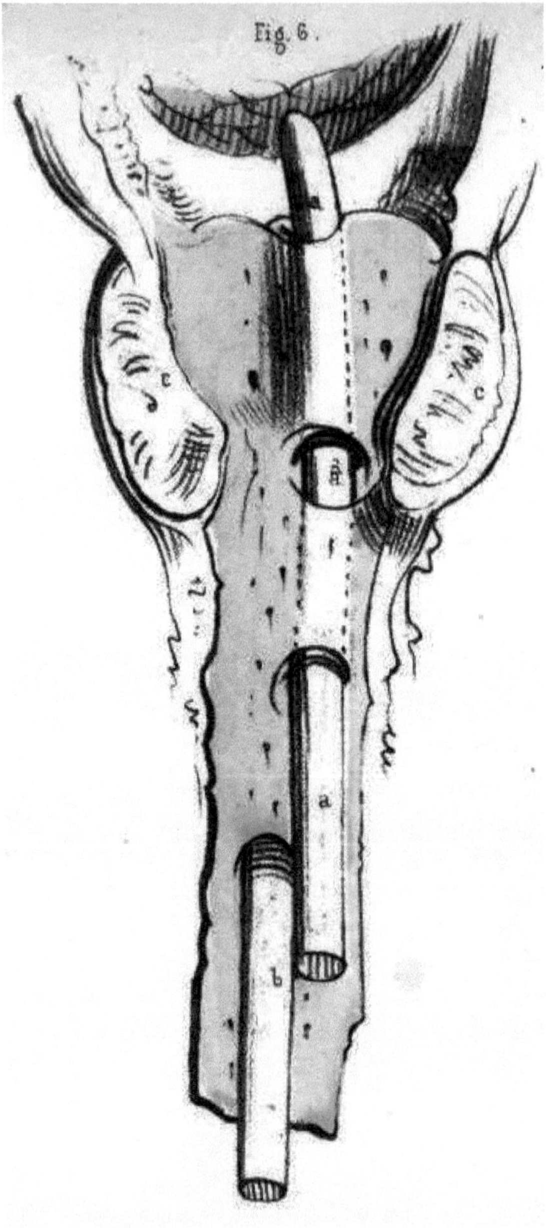

Plate 59.--Figure 6.

Fig. 7, Plate 59.--A bougie, *b b,* is seen to perforate the urethra anterior to the stricture *c,* situated an inch behind the glans, and after traversing the substance of the right corpus cavernosum *d*, for its whole length, re-enters the neck of the bladder through the body of the prostate.

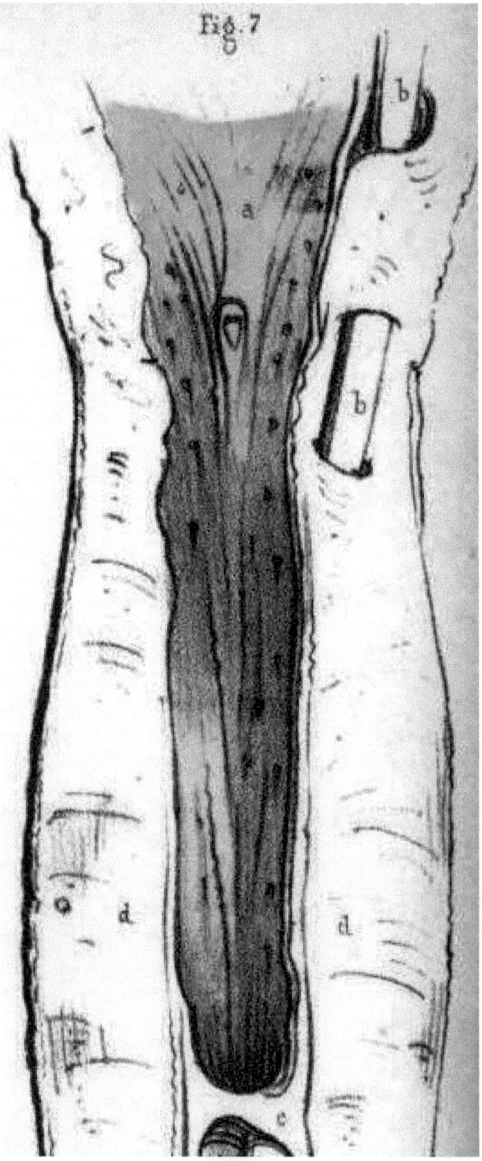

Plate 59.--Figure 7.

Fig. 8, Plate 59.--A bougie, *c c,* appears tearing and passing beneath the lining membrane, *d d*, of the prostatic urethra. It is remarked that the origin of a false passage is in general anterior to the stricture. It may, however, occur at any part of the canal in which no stricture exists, if the hand that impels the instrument be not guided by a true knowledge of the form of the urethra; and perhaps the accident happening from this cause is the more general rule of the two.

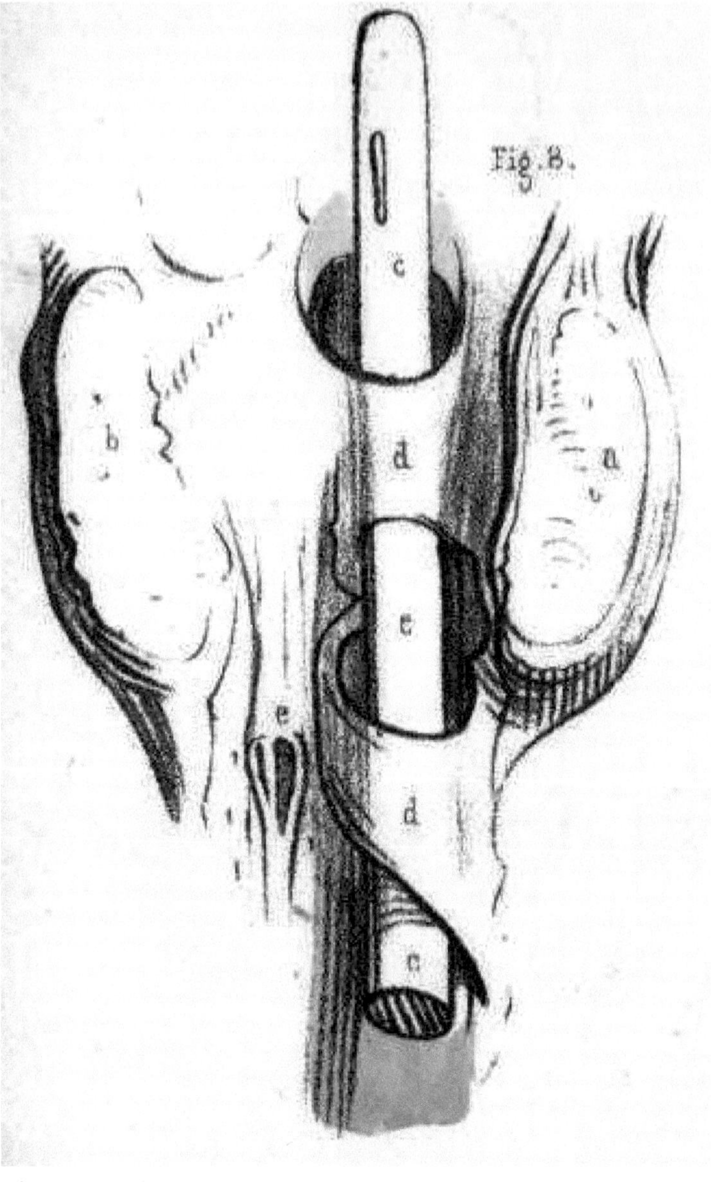

Plate 59.--Figure 8.

Fig. 9, Plate 59.--Two strictures are represented here, the one, *e*, close to the bulb *d,* the other, *f*, an inch anterior to this part. In the prostate, *a b*, are seen irregularly shaped abscess pits, communicating with each other, and projecting upwards the floor of this body to such a degree, that the prostatic canal appears nearly obliterated.

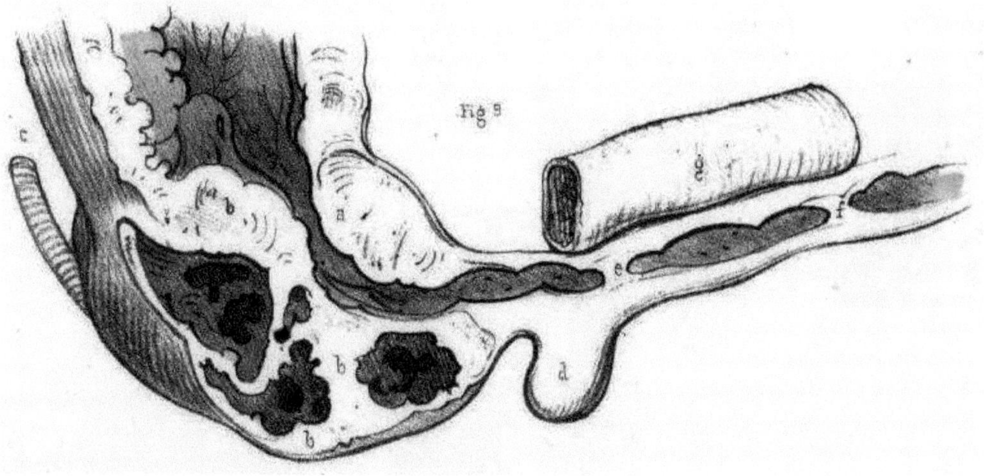

Plate 59.--Figure 9.

Fig. 10, Plate 59.--Two bougies, *d e,* are seen to enter the upper wall of the urethra, *c*, anterior to the prostate, *a b.* This accident happens when the handle of a rigid instrument is depressed too soon, with the object of raising its point over the enlarged third lobe of the prostate.

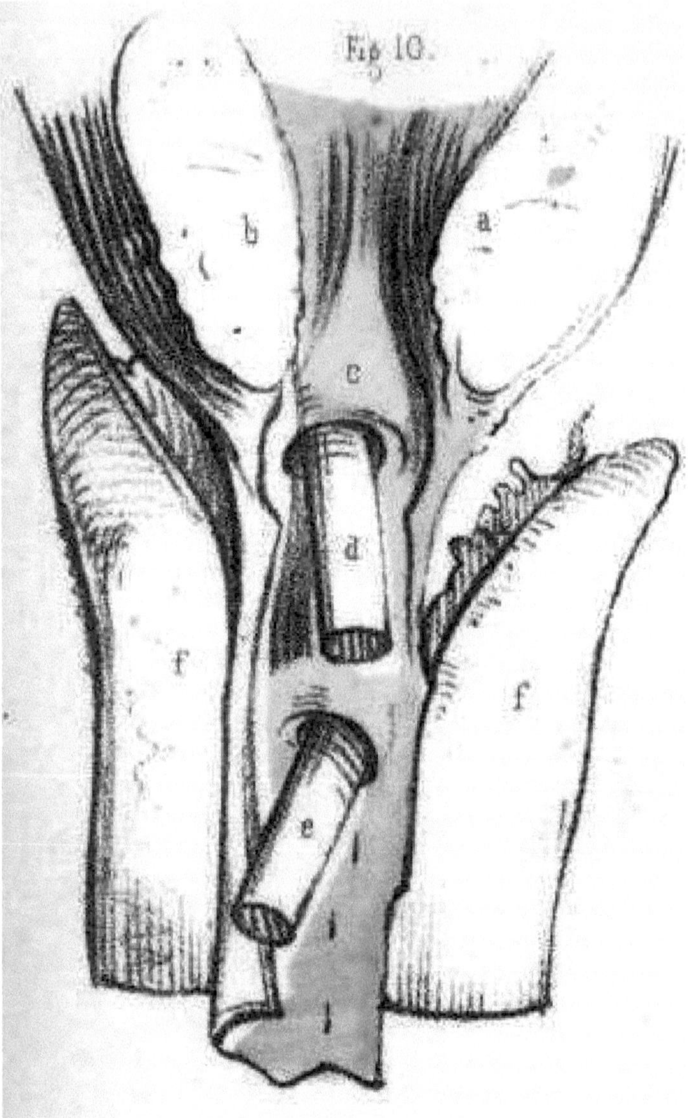

Plate 59.--Figure 10.

Fig. 11, Plate 59.--Two instruments appear transfixing the prostate, of which body the three lobes, *a, b, c*, are much enlarged. The instrument *d* perforates the third lobe, *a*; while the instrument *e* penetrates the right lobe, *c*, and the third lobe, *a.* This accident occurs when instruments not possessing the proper prostatic bend are forcibly pushed forwards against the resistance at the neck of the bladder.

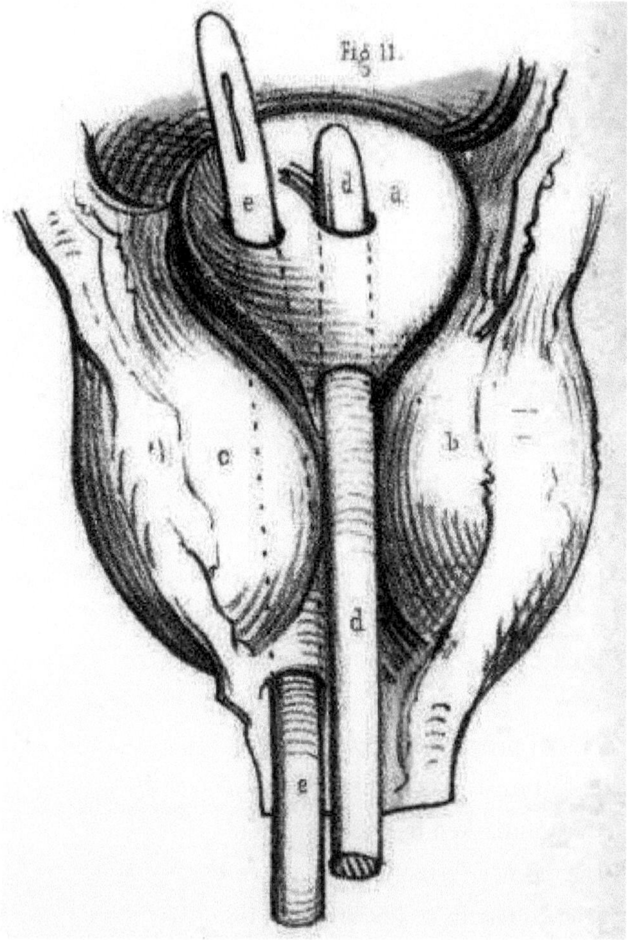

Plate 59.--Figure 11.

Fig. 12, Plate 59.--In this case an instrument, *d d,* after passing beneath part of the lining membrane, *e e,* anterior to the bulb, penetrates *b,* the right lobe of the prostate. A second instrument, *c c,* penetrates the left lobe. A third smaller instrument, *f f,* is seen to pass out of the urethra anterior to the prostate, and after transfixing the right vesicula seminalis external to the neck of the bladder, enters this viscus at a point behind the prostate. The resistance which the two larger instruments met with in penetrating the prostate, made it seem, perhaps, that a tight stricture existed in this situation, to match which the smaller instrument, *f f,* was afterwards passed in the course marked out.

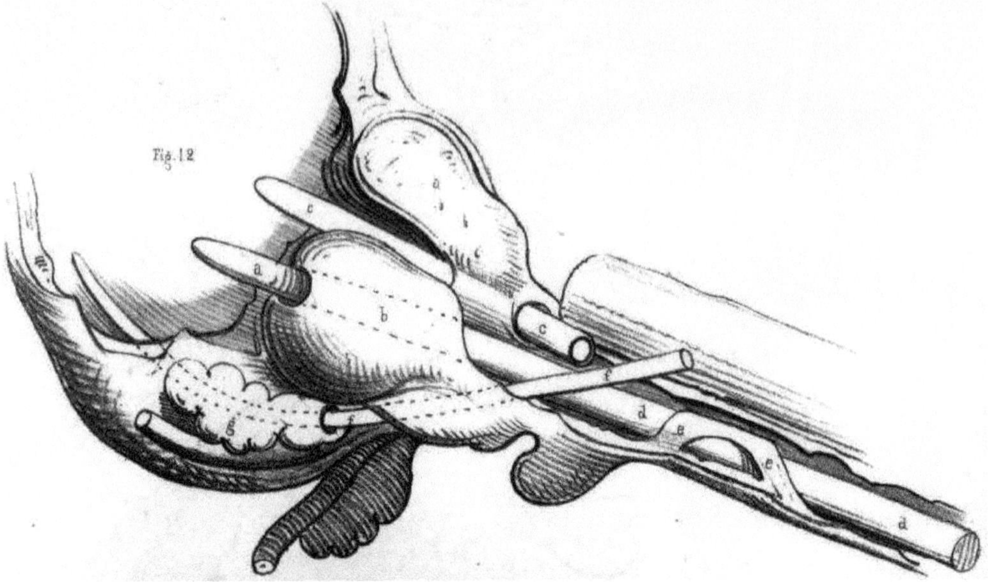

Plate 59.--Figure 12.

Figs. 1 to 5, Plate 60, represent a series of prostates, in which the third lobe gradually increases in size. In Fig. 1, which shows the healthy state of the neck of the bladder, unmarked by the prominent lines which are said to bound the space named "trigone vesical," or by those which indicate the position of the "muscles of the ureters," the third lobe does not exist. In Fig. 2 it appears as the uvula vesicae, *a*. In Fig. 3 the part *a* is increased, and under the name now of third lobe is seen to contract and bend upwards the prostatic canal. In Fig. 4 the effect which the growth of the lobe, *a*, produces upon the form of the neck of the bladder becomes more marked, and the part presenting perforations, *e e*, produced by instruments, indicates that by its shape it became an obstacle to the egress of the urine as well as to the entrance of instruments. A calculus of irregular form is seen to lodge behind the third lobe, and to be out of the reach of the point of a sound, supposing this to enter the bladder over the apex of the lobe. In Fig. 5 the three lobes are enlarged, but the third is most so, and while standing on a narrow pedicle attached to the floor of the prostate, completely blocks up the neck of the bladder. [Footnote]

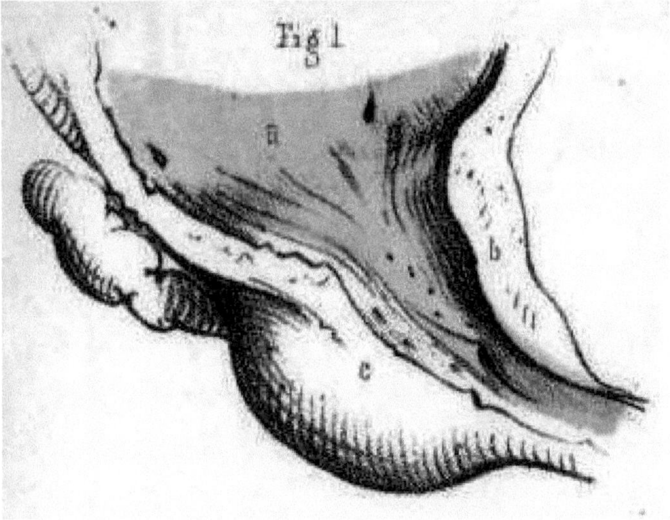

Plate 60.--Figure 1

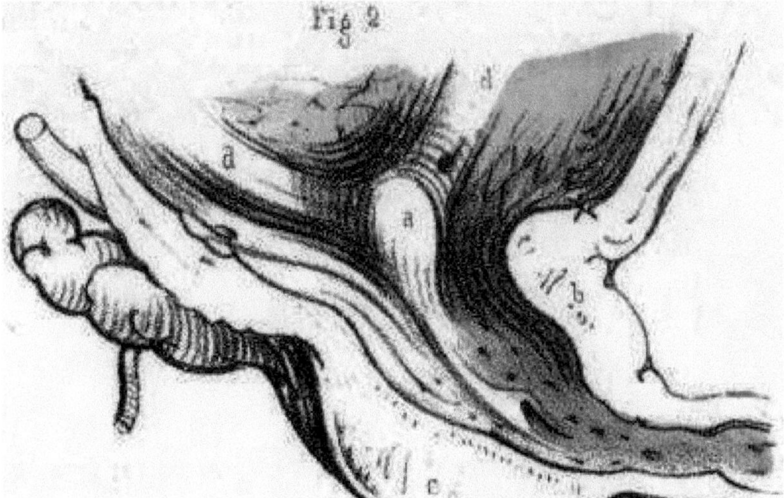

Plate 60.--Figure 2

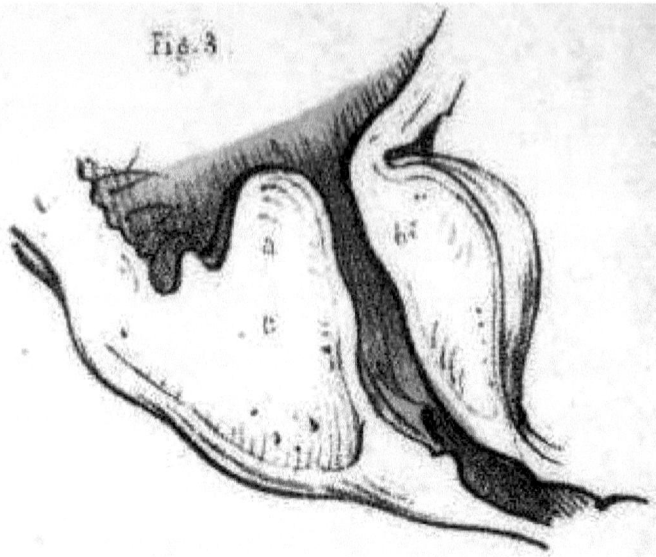

Plate 60.--Figure 3

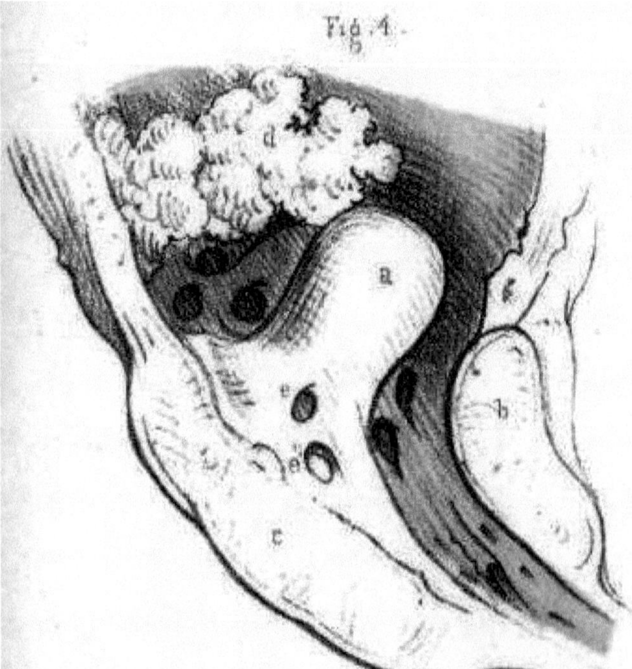

Plate 60.--Figure 4

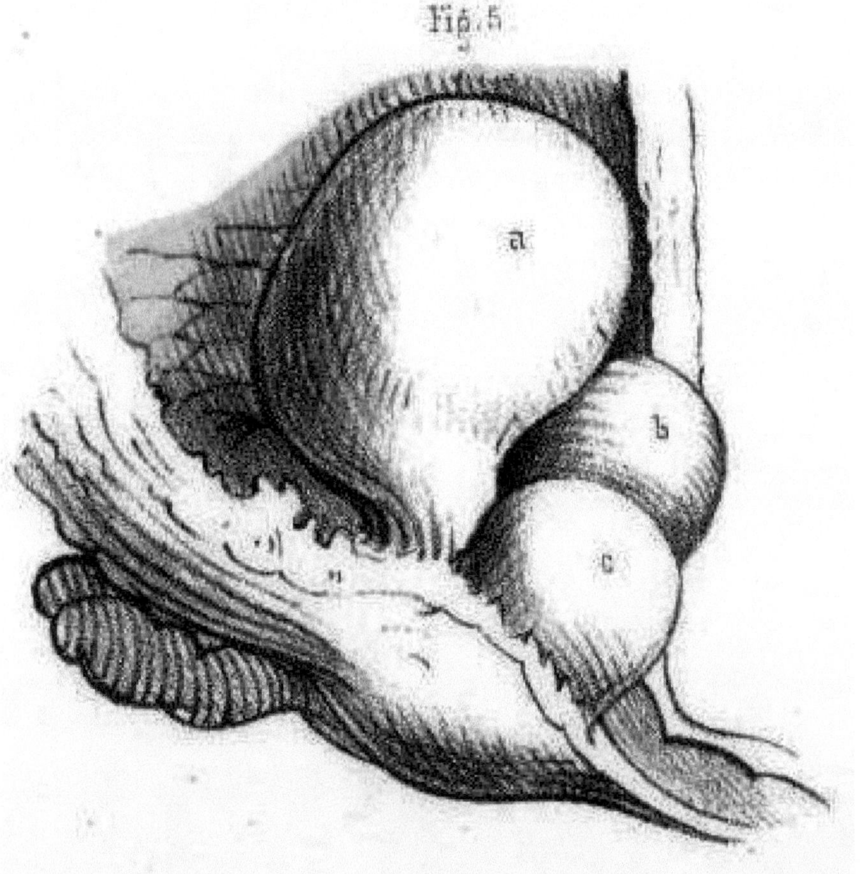

Plate 60.--Figure 5

[Footnote: On comparing this series of figures, it must appear that the third lobe of the prostate is the product of diseased action, in so far at least as an unnatural hypertrophy of a part may be so designated. It is not proper to the bladder in the healthy state of this organ, and where it does manifest itself by increase it performs no healthy function in the economy. When Home, therefore, described this part as a new fact in anatomy, he had in reality as little reason for so doing as he would have had in naming any other tumour, a thing unknown to normal anatomy. Langenbeck (Neue Bibl. b. i. p. 360) denies its existence in the healthy state. Cruveilhier (Anat. Pathog. liv. xxvii.) deems it incorrect to reckon a third lobe as proper to the healthy bladder.]

Fig. 6, Plate 60.--The prostatic canal is bent upwards by the enlarged third lobe to such a degree as to form a right angle with the membranous part of the canal. A bougie is seen to perforate the third lobe, and this is the most frequent mode in which, under such circumstances, and with instruments of the usual imperfect form, access may be gained to the bladder for the relief of retention of urine. "The new passage may in every respect be as efficient as one formed by puncture or incision in any other way." (Fergusson.)

Fig. 6.

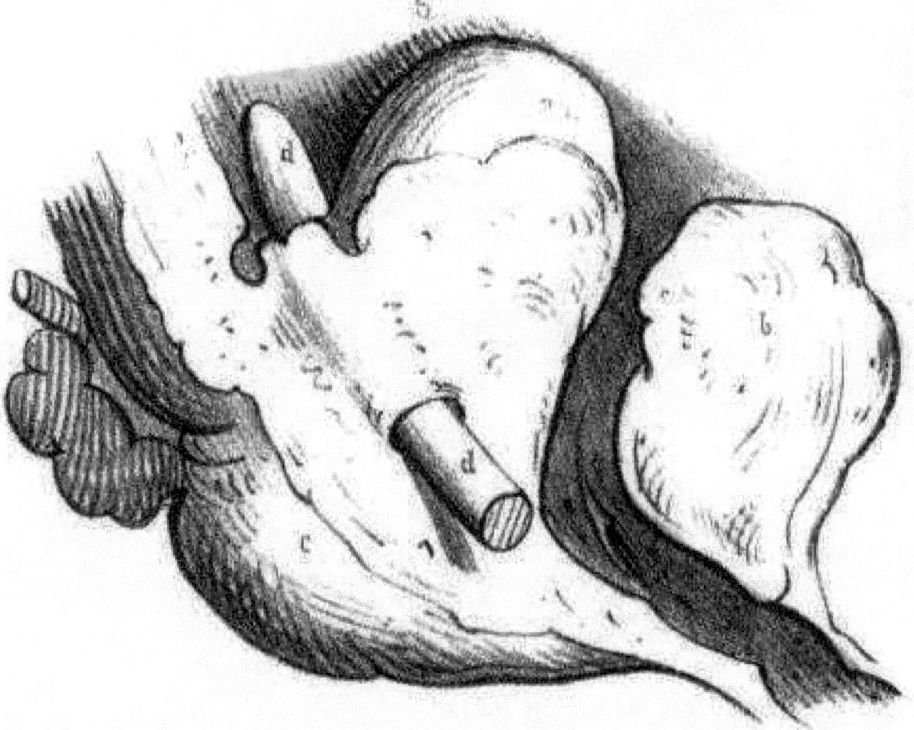

Plate 60.--Figure 6

Fig. 7, Plate 60.--The three lobes of the prostate, *a, b, c,* are equally enlarged. The prostatic canal is consequently much contracted and distorted, so that an instrument on being passed into the bladder has made a false passage through the third lobe. When a catheter is suspected to have entered the bladder by perforating the prostate, the instrument should be retained in the newly made passage till such time as this has assumed the cylindrical form of the instrument. If this be done, the new passage will be the more likely to become permanent. It is ascertained that all false passages and fistulae by which the urine escapes, become after a time lined with a membrane similar to that of the urethra. (Stafford.)

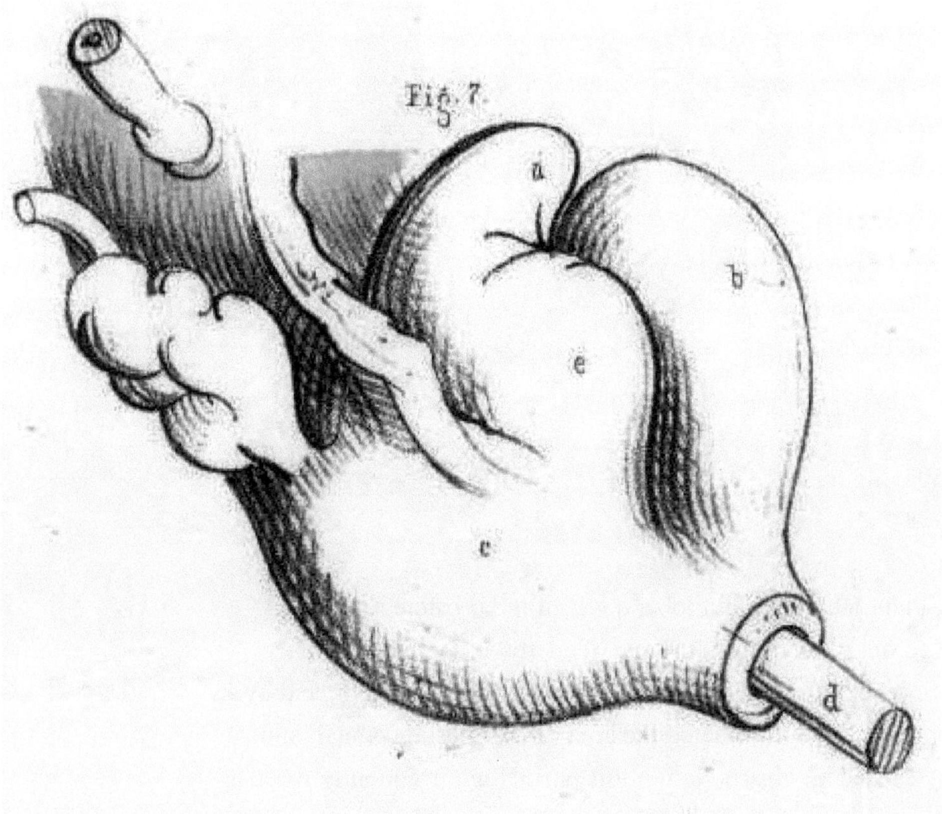

Plate 60.--Figure 7

Fig. 8, Plate 60.--The three lobes, *a, b, c,* of the prostate are irregularly enlarged. The third lobe, *a a*, projecting from below, distorts the prostatic canal upwards and to the right side.

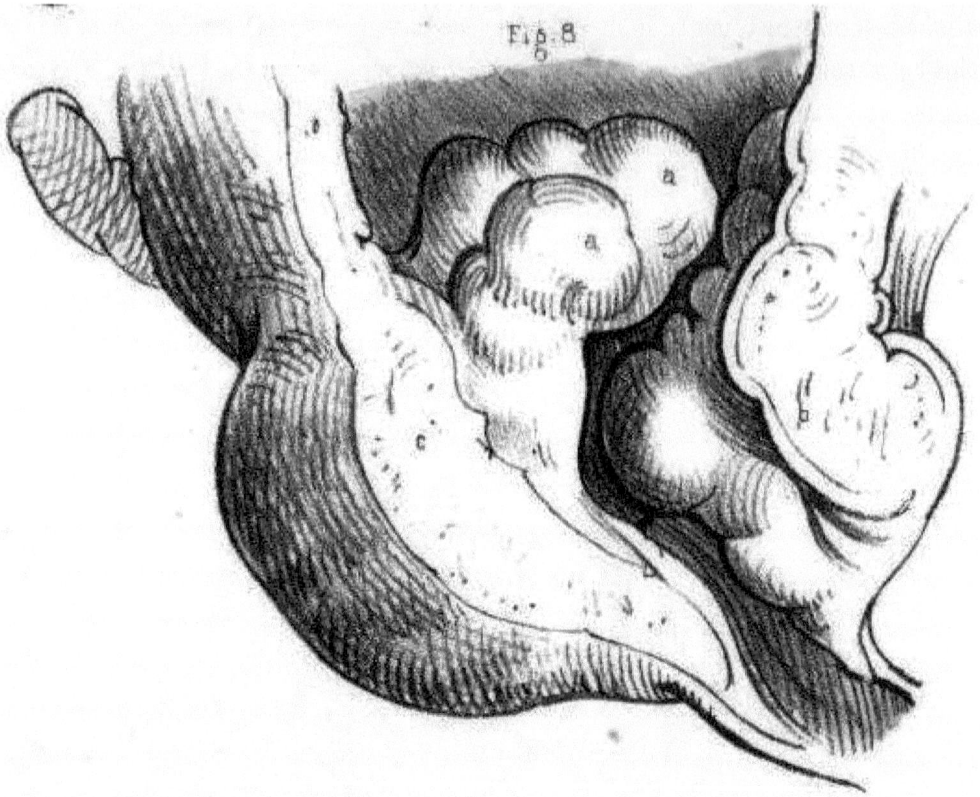

Plate 60.--Figure 8.

Fig. 9, Plate 60.--The right lobe, *a c c,* of the prostate appears hollowed out so as to form the sac of an abscess which, by its projection behind, pressed upon the forepart of the rectum, and by its projection in front, contracted the area of the prostatic canal, and thereby caused an obstruction in this part. Not unfrequently when a catheter is passed along the urethra, for the relief of a retention of urine caused by the swell of an abscess in this situation, the sac becomes penetrated by the instrument, and, instead of urine, pus flows. The sac of a prostatic abscess frequently opens of its own accord into the neighbouring part of the urethra, and when this occurs it becomes necessary to retain a catheter in the neck of the bladder, so as to prevent the urine entering the sac.

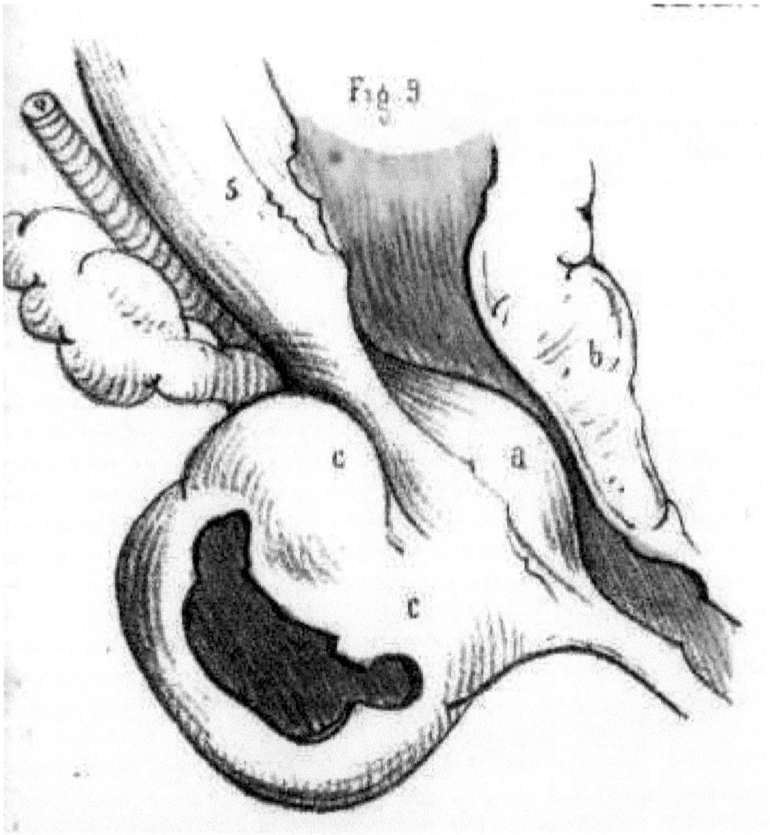

Plate 60.--Figure 9.

Fig. 10, Plate 60.--The prostate presents four lobes of equal size, and all projecting largely around the neck of the bladder. The prostatic canal is almost completely obstructed, and an instrument has made a false passage through the lobe *a*.

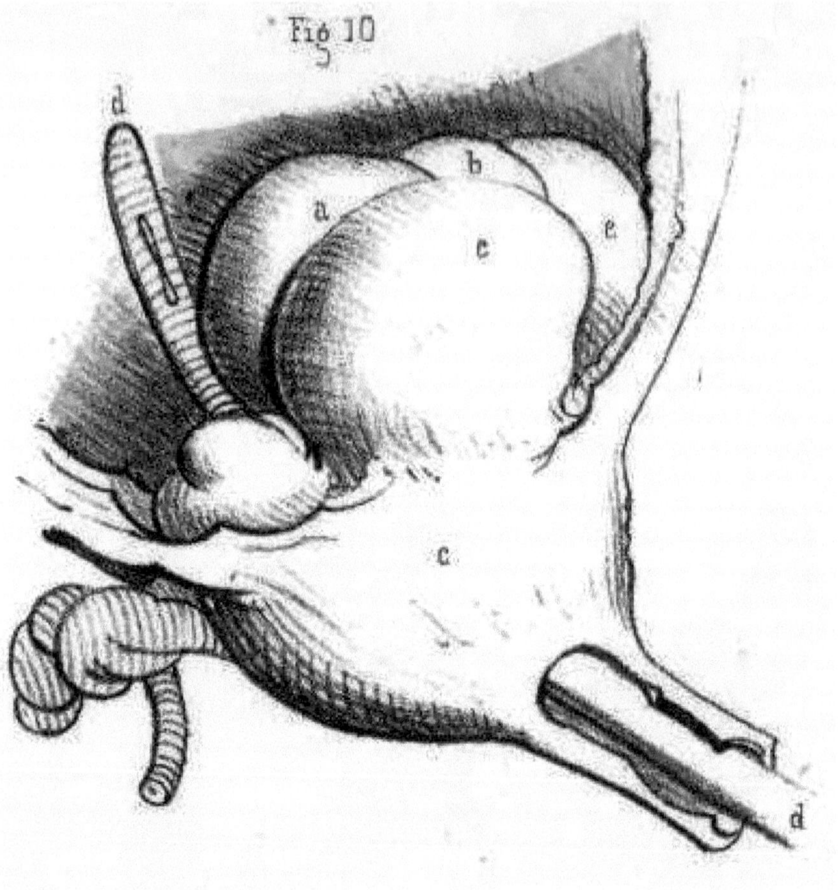

Plate 60.--Figure 10.

Fig. 11, Plate 60.--The third lobe of the prostate is viewed in section, and shows the track of the false passage made by the catheter, *d*, through it, from its apex to its base. The proper canal is bent upwards from its usual position, which is that at present marked by the instrument in the false passage.

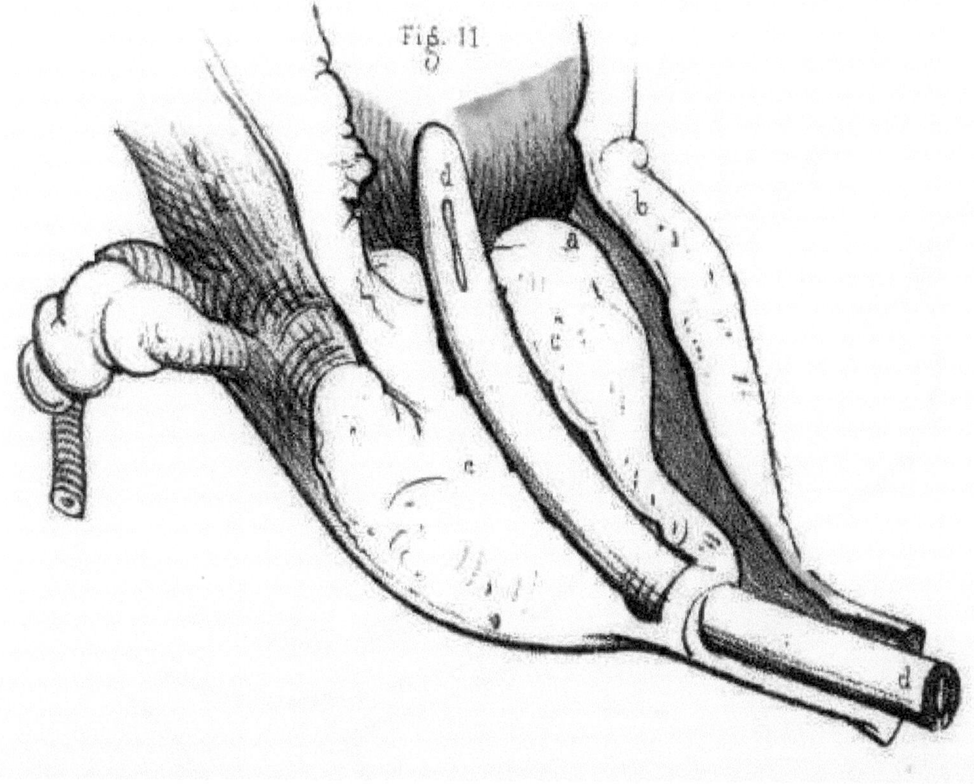

Plate 60.--Figure 11.

Fig. 12, Plate 60.--The prostatic lobes are uniformly enlarged, and cause the corresponding part of the urethra to be uniformly contracted, so as closely to embrace the catheter, *d d*, occupying it, and to offer considerable resistance to the passage of the instrument.

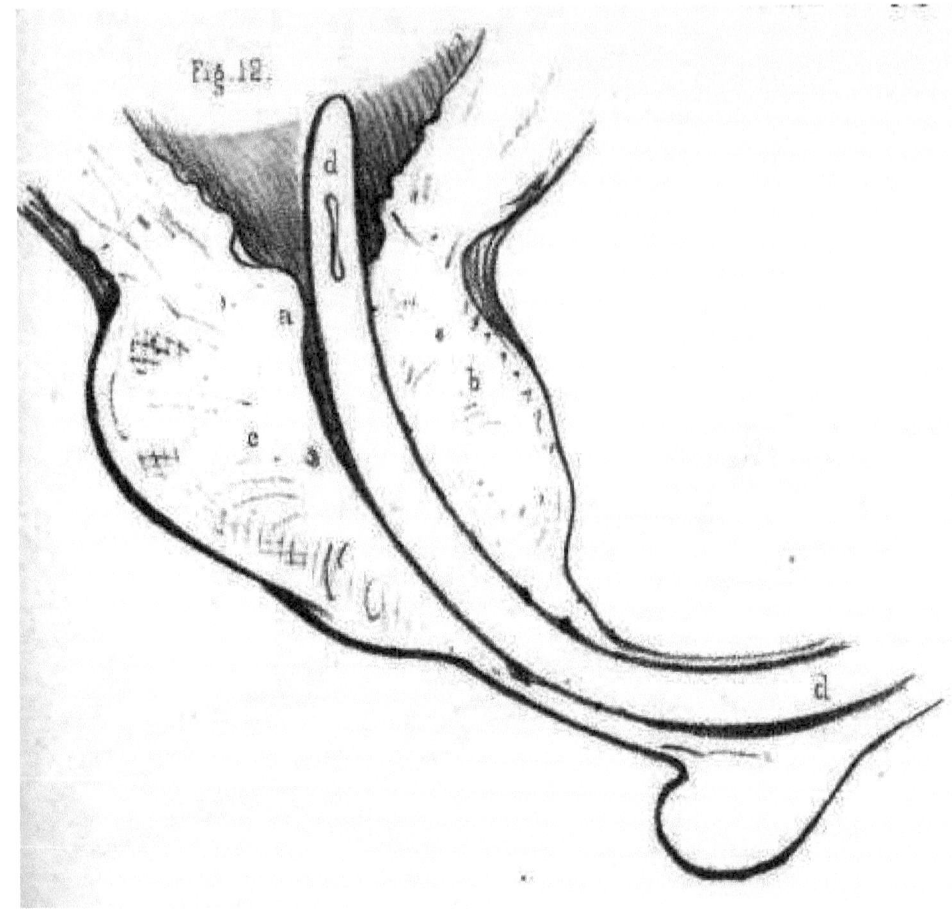

Plate 60.--Figure 12.

Fig. 13, Plate 60.--The prostate, *bc*, is considerably enlarged anteriorly, *b*, in consequence of which the prostatic canal appears more horizontal even than natural. The catheter, *d*, occupying the canal lies nearly straight. The lower wall, *c*, of the prostate is much diminished in thickness. A nipple-shaped process, *a*, is seen to be attached by a pedicle to the back of the upper part, *b*, of the prostate, and to act like a stopper to the neck of the bladder. The body *a* being moveable, it will be perceived how, while the bladder is distended with urine, the pressure from above may block up the neck of the organ with this part, and thus cause complete retention, which, on the introduction of a catheter, becomes readily relieved by the instrument pushing the obstructing body aside.

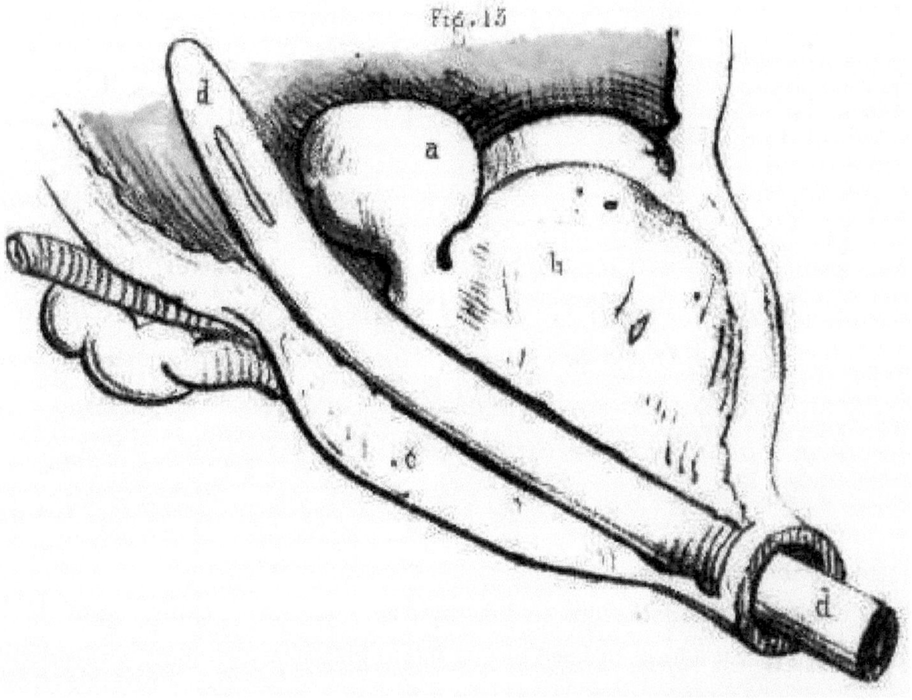

Plate 60.--Figure 13.

DEFORMITIES OF THE PROSTATE; DISTORTIONS AND OBSTRUCTIONS OF THE PROSTATIC URETHRA.

Observations on the nature of the prostate--its signification. Cases of prostate and bulb pouched by catheters. Obstructions of the vesical orifice. Sinuous prostatic canal. Distortions of the vesical orifice. Large prostatic calculus. Sacculated prostate. Triple prostatic urethra. Encrusted prostate. Fasciculated bladder. Prostatic sac distinct from the bladder. Practical remarks. Impaction of a large calculus in the prostate. Practical remarks.

The prostate is liable to such frequent and varied deformities, the consequence of diseased action, whilst, at the same time, its healthy function (if it have any) in the male body is unknown, that it admits at least of one interpretation which may, according to fact, be given of it--namely, that of playing a principal part in effecting some of the most distressing of "the thousand natural ills that flesh is heir to." But heedless of such a singular explanation of a final cause, the practical surgeon will readily confess the fitting application of the interpretation, such as it is, and rest contented with the proximate facts and proofs. As physiologists, however, it behooves us to look further into nature, and search for the *ultimate fact* in her prime moving law. The prostate is peculiar to the male body, the uterus to the female. With the exception of these two organs there is not another which appears in the one sex but has its analogue in the opposite sex; and thus these two organs, the prostate and the uterus, appear by exclusion of the rest to approach the test of comparison, by which their analogy becomes as fully manifested as that between the two quantities, $a-b$, and $a+b$ the only difference which exists depends upon the subtraction or the addition of the quantity, b. The difference between a prostate and a uterus is simply one of quantity, such as we see existing between the male and the female breast. The prostate is to the uterus absolutely what a rudimentary organ is to its fully developed analogue. The one, as being superfluous, is in accordance with nature's law of *nihil supervacaneum nihil frutra*, arrested in its development, and in such a character appears the prostate. This body *is not a gland* any more than is the uterus, but both organs being quantitatively, and hence functionally different, I here once more venture to call down an interpretation of the part from the unfrequented bourne of comparative anatomy, and turning it to lend an interest to the accompanying figures even with a surgical bearing, I remark that the prostatic or rudimentary uterus, like a germ not wholly blighted, is prone to an occasional sprouting or increase beyond its prescribed dimensions--a hypertrophy in barren imitation, as it were, of gestation. [Footnote]

[Footnote: This expression of the fact to which I allude will not, I trust, be extended beyond the limits I assign to it. Though I have every reason to believe, that between the prostate of the male and the uterus of the female, the same amount of analogy exists, as between a coccygeal ossicle and the complete vertebral form elsewhere situated in the spinal series, I am as far from regarding the two former to be in all respects structurally or functionally alike, as I am from entertaining the like idea in respect to the two latter. But still I maintain that between a prostate and a uterus, as between a coccygeal bone and a vertebra, the only difference which exists is one of quantity, and that hence arises the functional difference. A prostate is part of a uterus, just as a coccygeal bone is part (the centrum) of a vertebra. That this is the absolute signification of the prostate I firmly believe, and were this the proper place, I could prove it in detail, by the infallible rule of analogical reasoning. John Hunter has observed that the use of the prostate was not sufficiently known to enable us to form a judgment of the bad consequences of its diseased state. When the part becomes morbidly enlarged, it acts as a mechanical impediment to the passage of urine from the bladder, but from this circumstance we cannot reasonably infer, that while of its normal healthy proportions, its special function is to facilitate the egress of the urine, for the female bladder, though wholly devoid of the prostate, performs its own function perfectly. It appears to me, therefore, that the real question should be, not what is the use of the prostate? but has it any proper function? If the former question puzzled even the philosophy of Hunter, it was because the latter question must be answered in the negative. The prostate has no function proper to itself *per se*. It is a thing distinct from the urinary apparatus, and distinct likewise from the generative organs. It may be hypertrophied or atrophied, or changed in texture, or wholly destroyed by abscess, and yet neither of the functions of these two systems of organs will be impaired, if the part while diseased act not as an obstruction to them. In texture the prostate is similar to an unimpregnated uterus. In form it is, like the uterus, symmetrical. In position it corresponds to the uterus. The prostate has no ducts proper to itself. Those ducts which are said to belong to it (prostatic ducts) are merely mucous cells, similar to those in other parts of the urethral lining membrane. The seminal ducts evidently do not belong to it. The texture of the prostate is not such as appears in glandular bodies generally. In short, the facts which prove what it is not, prove what it actually is--namely, a uterus arrested in its development, and as a sign of that all-encompassing law in nature, which science expresses by the term "unity in variety." This interpretation of the prostate, which I believe to be true to nature, will last perhaps till such time as the microscopists shall discover in its "*secretion*" some species of mannikins, such as may pair with those which they term spermatozoa.]

Fig. 1, Plate 61.--The prostate, *a b*, is here represented thinned in its walls above and below. The lower wall is dilated into a pouch caused by the points of misdirected instruments in catheterism having been rashly forced against it.

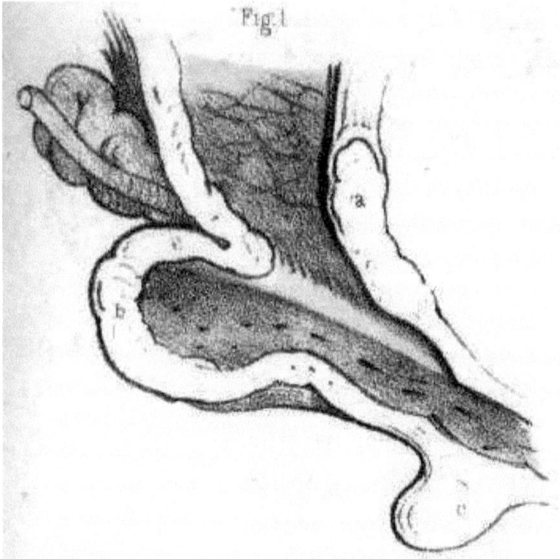

Plate 61.--Figure 1.

Fig. 2, Plate 61.--The prostate, *a b,* is here seen to be somewhat more enlarged than is natural. A tubercle, **b**, surmounts the lower part, *c,* of the prostate, and blocks up the vesical orifice. Catheters introduced by the urethra for retention of urine which existed in this case, have had their points arrested at the bulb, and on being pushed forwards in this direction, have dilated the bulb into the form of a pouch, seen at *d.* The sinus of the bulb, being the lowest part of the urethral canal, is very liable to be distorted or perforated by the points of instruments descending upon it from above and before. [Footnote]

[Footnote: When a stricture exists immediately behind the bulb, this circumstance will, of course, favour the occurrence of the accident. "False passages (observes Mr. Benjamin Phillips) are less frequent here (in the membranous part of the urethra) than in the bulbous portion of the canal. The reason of this must be immediately evident: false passages are ordinarily made in consequence of the difficulty experienced in the endeavour to pass an instrument through the strictured portion of the tube. Stricture is most frequently seated at the point of junction between the bulbous and membranous portions of the canal; consequently, the false passage will be usually anterior to this latter point."--(On the Urethra, its Diseases. &c., p. 15.)]

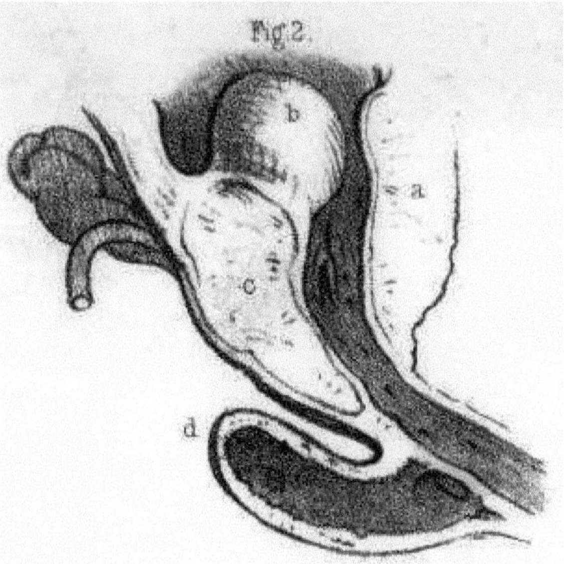

Plate 61.--Figure 2

Fig. 3, Plate 61.--A cyst, *c*, is seen to grow from the left side of the base of the prostate, *a b*, and to form an obstruction at the vesical orifice.

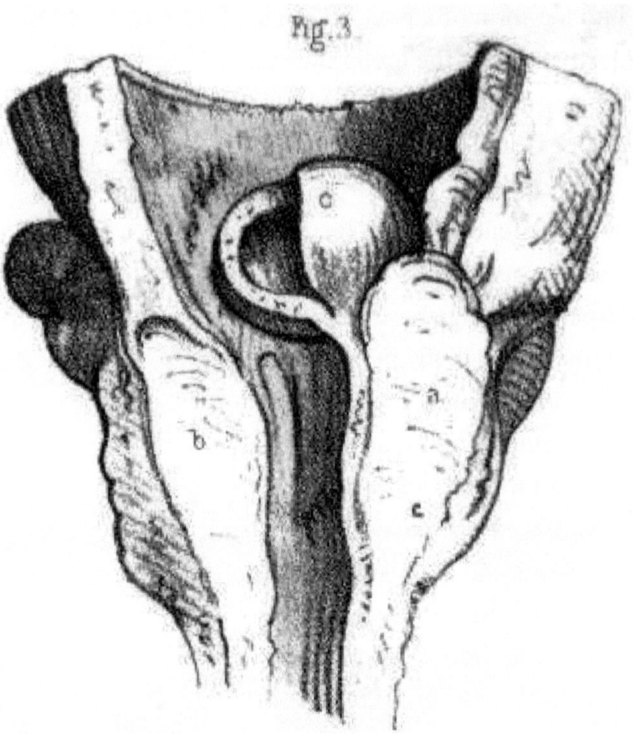

Plate 61.--Figure 3.

Fig. 4, Plate 61.--A globular excrescence, **a**, appears blocking up the vesical orifice, and giving to this the appearance of a crescentic slit, corresponding to the shape of the obstructing body. The prostate, *b b,* is enlarged in both its lateral lobes. A small bougie, *c,* is placed in the prostatic canal and vesical opening.

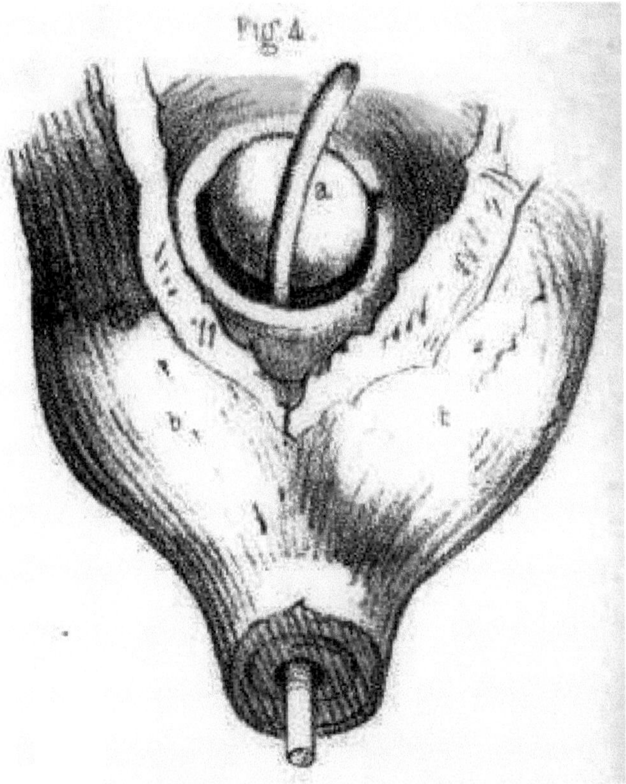

Plate 61.--Figure 4

Fig. 5, Plate 61.--The prostate, *d*, is considerably enlarged, and the vesical orifice is girt by a prominent ring, *b b*, from the right border of which the nipple-shaped body, *a*, projects and occupies the outlet. Owing to the retention of urine caused by this state of the prostate, the ureters, *c c*, have become very much dilated.

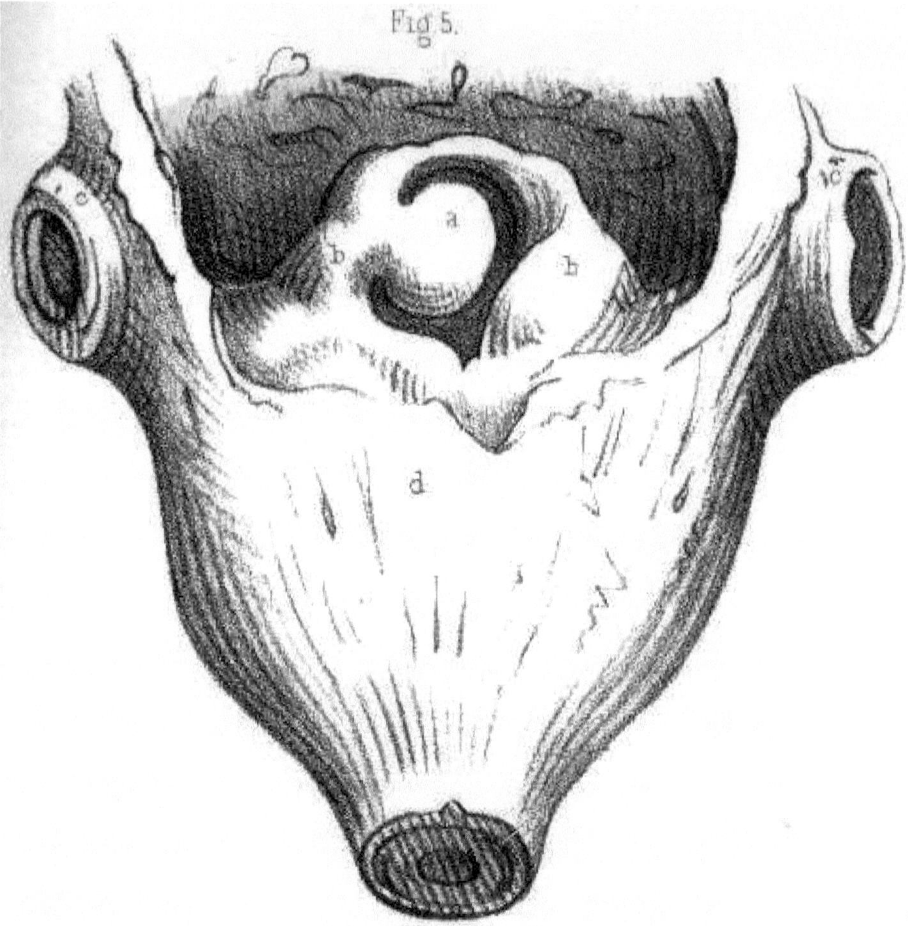

b

Plate 61.--Figure 5.

Fig. 6, Plate 61.--The lateral lobes of the prostate, *c c,* are seen enlarged, and from the inner side and base of each, irregularly shaped masses, *a, b, d,* project, and bend the prostatic urethra first to the right side, then to the left. The part, *a,* resting upon the part, *b,* acts like a valve against the vesical outlet, which would become closed the tighter according to the degree of superincumbent pressure. A flexible catheter would, in such a case as this, be more likely, perhaps, to follow the sinuous course of the prostatic passage than a rigid instrument of metal.

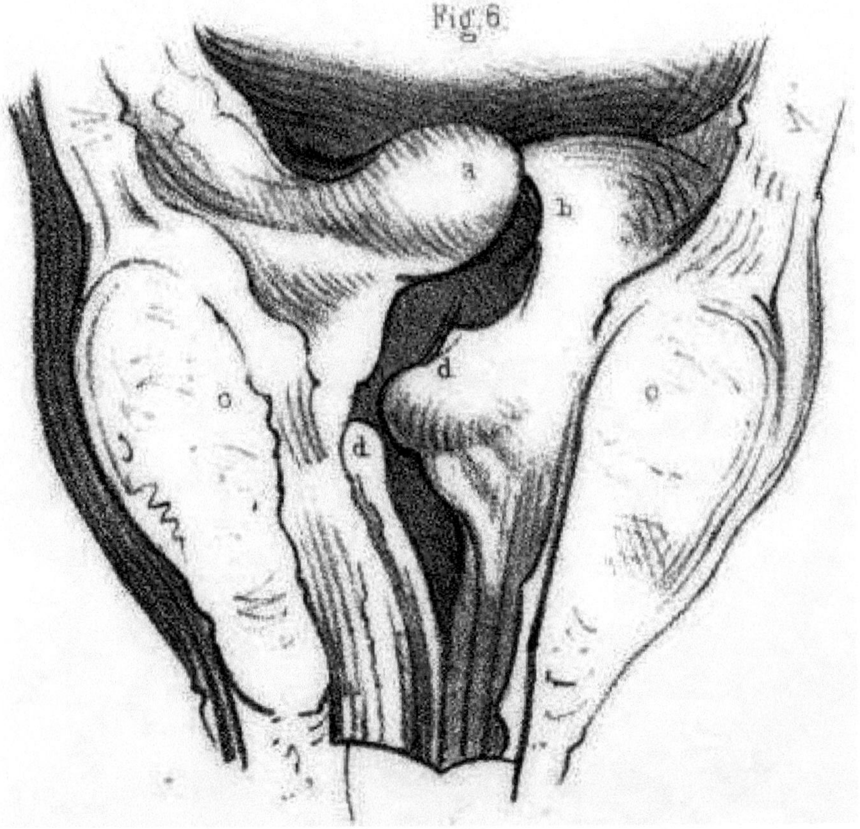

Plate 61.--Figure 6.

Fig. 7, Plate 61.--A globular mass, *a,* of large size, occupies the neck of the bladder, and gives the vesical orifice, *c,* a crescentic shape, convex towards the right side. The two lobes of the prostate, *b,* are much enlarged.

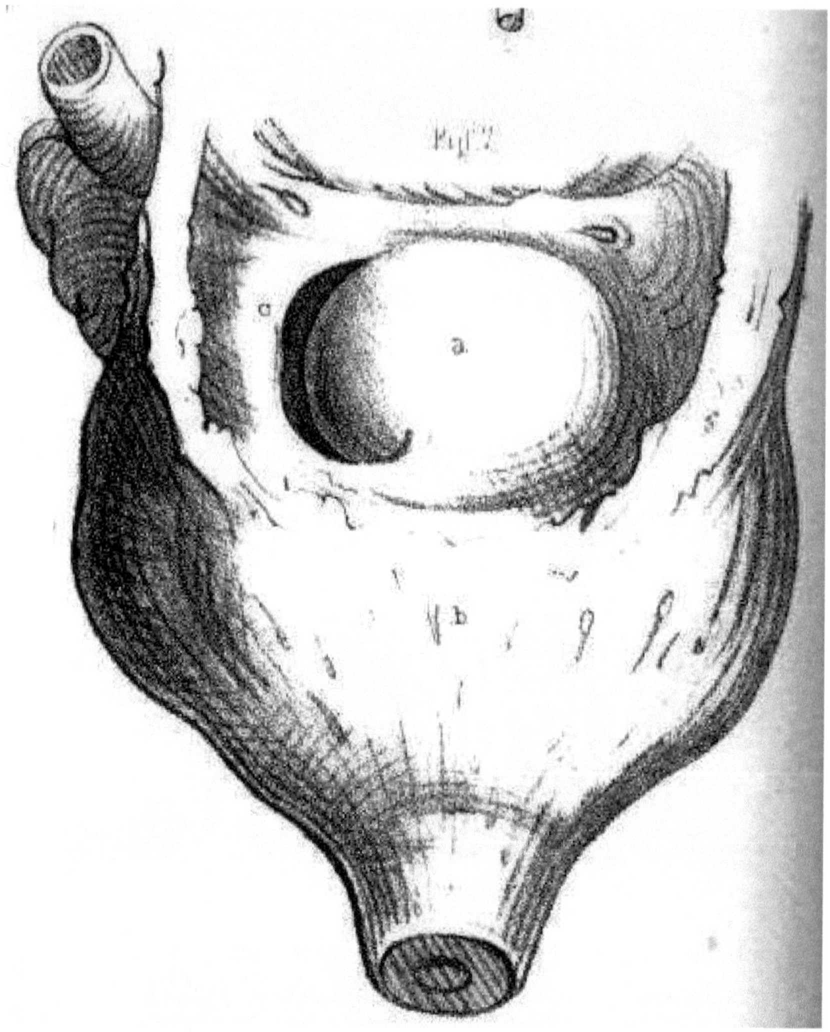

Plate 61.--Figure 7.

Fig. 8, Plate 61.--The lateral lobes, *b b,* of the prostate are irregularly enlarged, and the urinary passage is bent towards the right side, *c*, from the membranous portion, which is central. Surmounting the vesical orifice, *c*, is seen the tuberculated mass, *a*, which being moveable, can be forced against the vesical orifice and thus produce complete retention of urine. In this case, also, a flexible catheter would be more suitable than a metallic one.

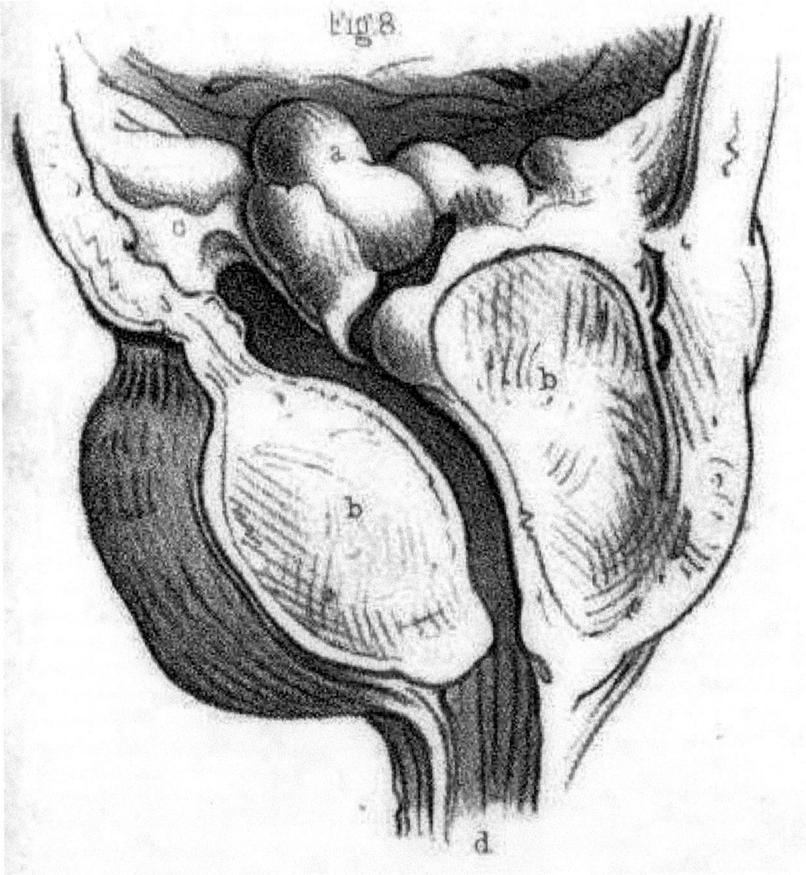

Plate 61.--Figure 8.

Fig. 9, Plate 61.--The lateral lobes, *b b*, of the prostate are enlarged. The third lobe, *a*, projects at the neck of the bladder, distorting the vesical outlet. A small calculus occupies the prostatic urethra, and being closely impacted in this part of the canal, would arrest the progress of a catheter, and probably lead to the supposition that the instrument grated against a stone in the interior of the bladder, in which case it would be inferred that since the urine did not flow through the catheter no retention existed.

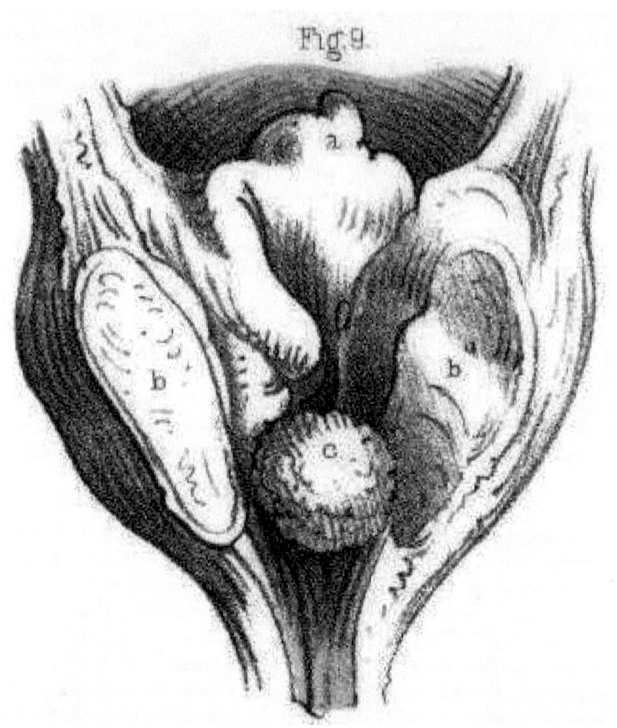

Plate 61.--Figure 9.

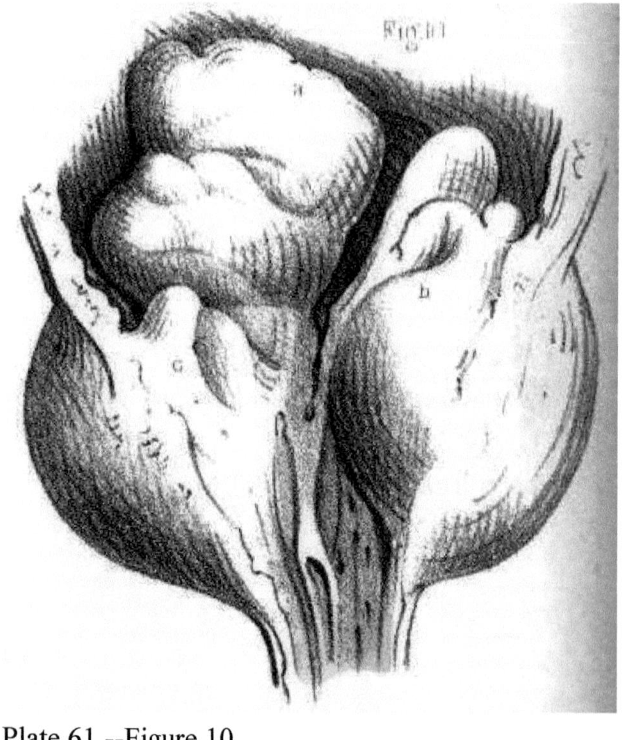

Plate 61.--Figure 10.

Fig. 10, Plate 61.--Both lateral lobes, *b c,* of the prostate appear much increased in size. A large irregular shaped mass, *a,* grows from the base of the right lobe, and distorts the prostatic canal and vesical orifice. When the lobes of the prostate increase in size in this direction, the prostatic canal becomes much more elongated than natural, and hence the instrument which is to be passed for relieving the existing retention of urine should have a wide and long curve to correspond with the form of this part of the urethra. [Footnote]

[Footnote: Both lobes of the prostate are equally liable to chronic enlargement. Home believed the left lobe to be oftener increased in size than the right. Wilson (on the Male Urinary and Genital Organs) mentions several instances of the enlargement of the right lobe. No reason can be assigned why one lobe should be more prone to hypertrophy than the other, even supposing it to be matter of fact, which it is not. But the observations made by Cruveilhier (Anat. Pathol.), that the lobulated projections of the prostate always take place internally at its vesical aspect, is as true as the manner in which he accounts for the fact is plausible. The dense fibrous envelope of the prostate is sufficient to repress its irregular growth externally.]

Fig. 11, Plate 61.--Both lobes of the prostate are enlarged, and from the base of each a mass projects prominently around the vesical orifice, *a b*. The prostatic urethra has been moulded to the shape of the instrument, which was retained in it for a considerable time.

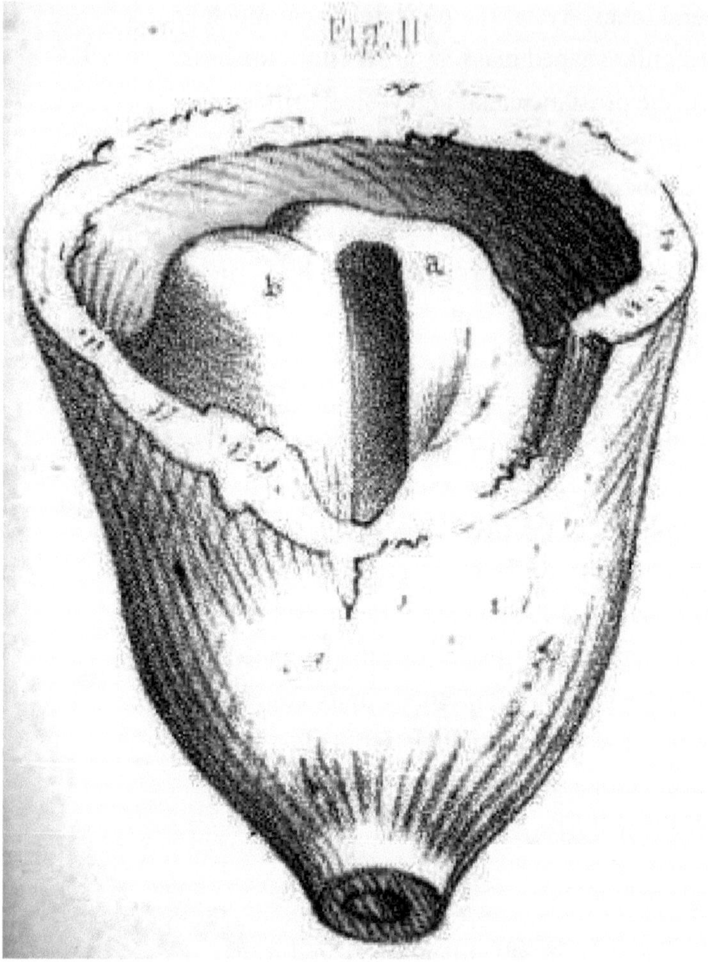

Plate 61.--Figure 11.

Fig. 12, Plate 61.--The prostate, *c b*, is enlarged and dilated, like a sac. Across the neck, *a,* of the bladder the prostate projects in an arched form, and is transfixed by the instrument, *d*. The prostate may assume this appearance, as well from instruments having been forced against it, as from an abscess cavity formed in its substance having received, from time to time, a certain amount of the urine, and retained this fluid under the pressure of strong efforts, made to void the bladder while the vesical orifice was closed above.

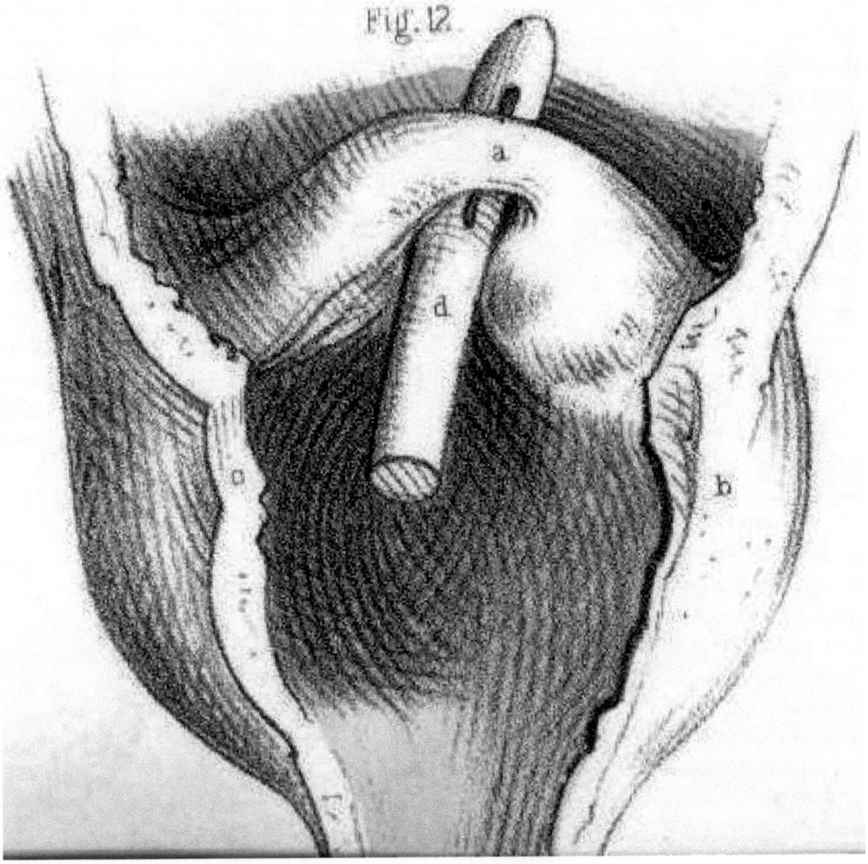

Plate 61.--Figure 12.

Fig. 13, Plate 61.--The lateral lobes, *d e*, of the prostate are enlarged; and, occupying the position of the third lobe, appear as three masses, *a b c,* plicated upon each other, and directed towards the vesical orifice, which they close like valves. The prostatic urethra branches upwards into three canals, formed by the relative position of the parts, *e, c, b, a, d,* at the neck of the bladder. The ureters are dilated, in consequence of the regurgitation of the contents of the bladder during the retention which existed ..

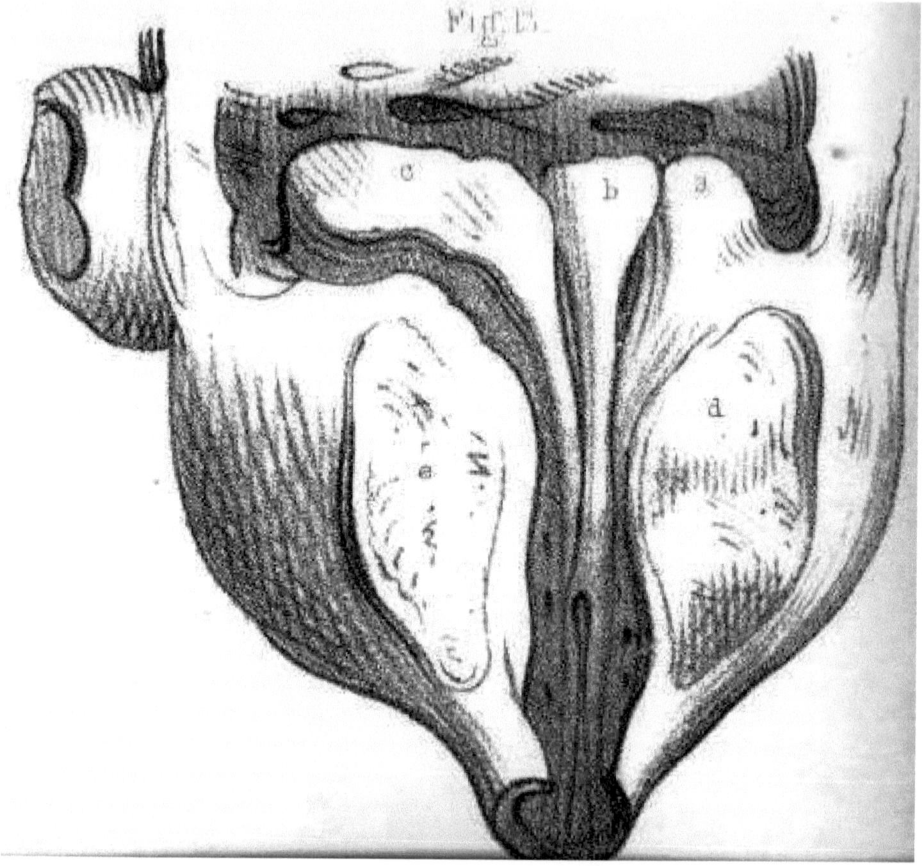

Plate 61.--Figure 13.

Fig. 1, Plate 62, exhibits the lobes of the prostate greatly increased in size. The part, *a b,* girds irregularly, and obstructs the vesical outlet, while the lateral lobes, *c d,* encroach upon the space of the prostatic canal. The walls of the bladder are much thickened.

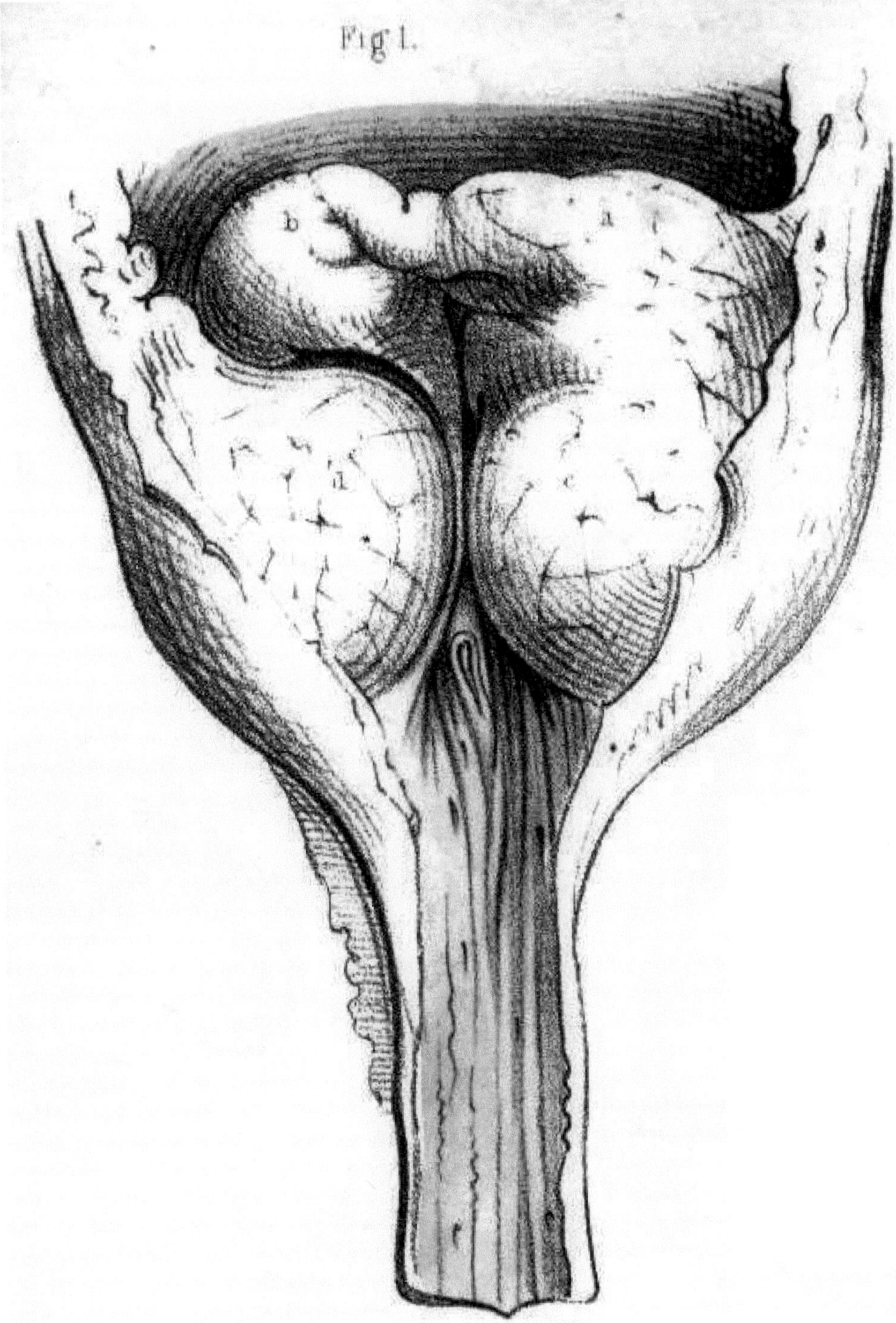

Plate 62.--Figure 1.

Fig. 2, Plate 62.--The three lobes, *a, d, c,* of the prostate are enlarged and of equal size, moulded against each other in such a way that the prostatic canal and vesical orifice appear as mere clefts between them. The three lobes are encrusted on their vesical surfaces with a thick calcareous deposit. The surface of the third lobe, *a,* which has been half denuded of the calcareous crust, *b,* in order to show its real character, appeared at first to be a stone impacted in the neck of the bladder, and of such a nature it certainly would seem to the touch, on striking it with the point of a sound or other instrument.

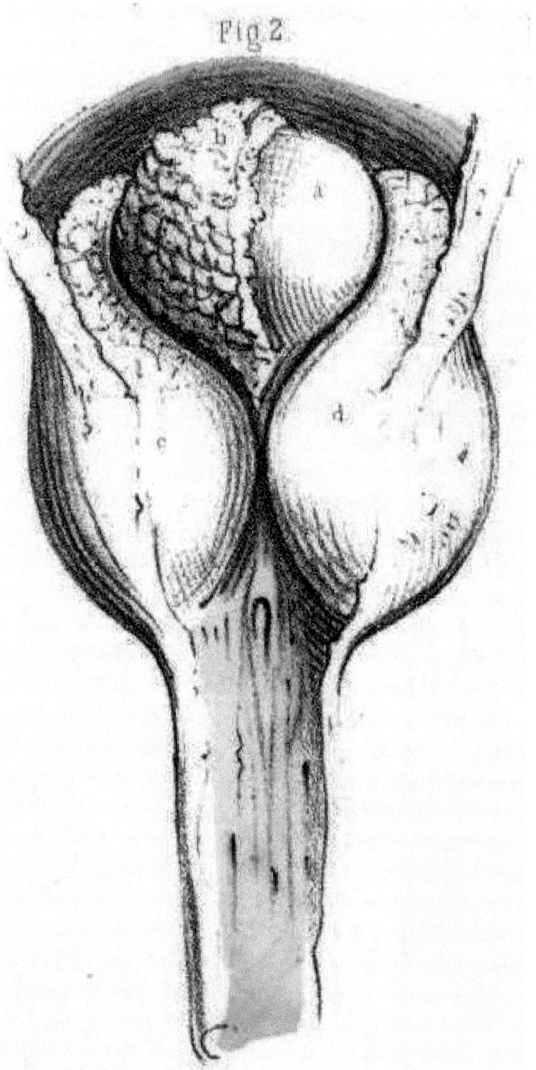

Plate 62.--Figure 2.

Fig. 3, Plate 62, represents the prostate with its three lobes enlarged, and the prostatic canal and vesical orifice narrowed. The walls of the bladder are thickened, fasciculated, and sacculated; the two former appearances being caused by a hypertrophy of the vesical fibres, while the latter is in general owing to a protrusion of the mucous membrane between the fasciculi.

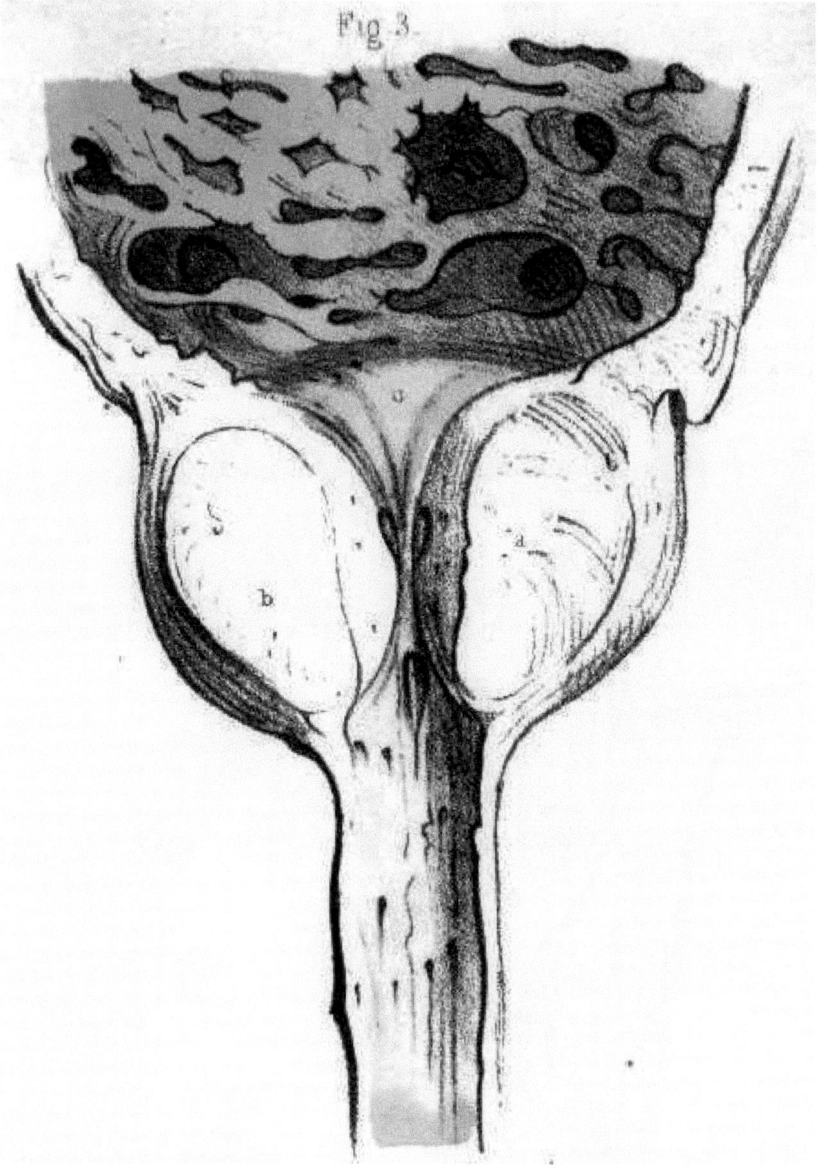

Plate 62.--Figure 3.

Fig. 4, Plate 62.--The prostate presents four lobes, *a, b, c, d,* each being of large size, and projecting far into the interior of the bladder, from around the vesical orifice which they obstruct. The bladder is thickened, and the prostatic canal is elongated. The urethra and the lobes of the prostate have been perforated by instruments, passed for the retention of urine which existed. A stricturing band, *e,* is seen to cross the membranous part of the canal.

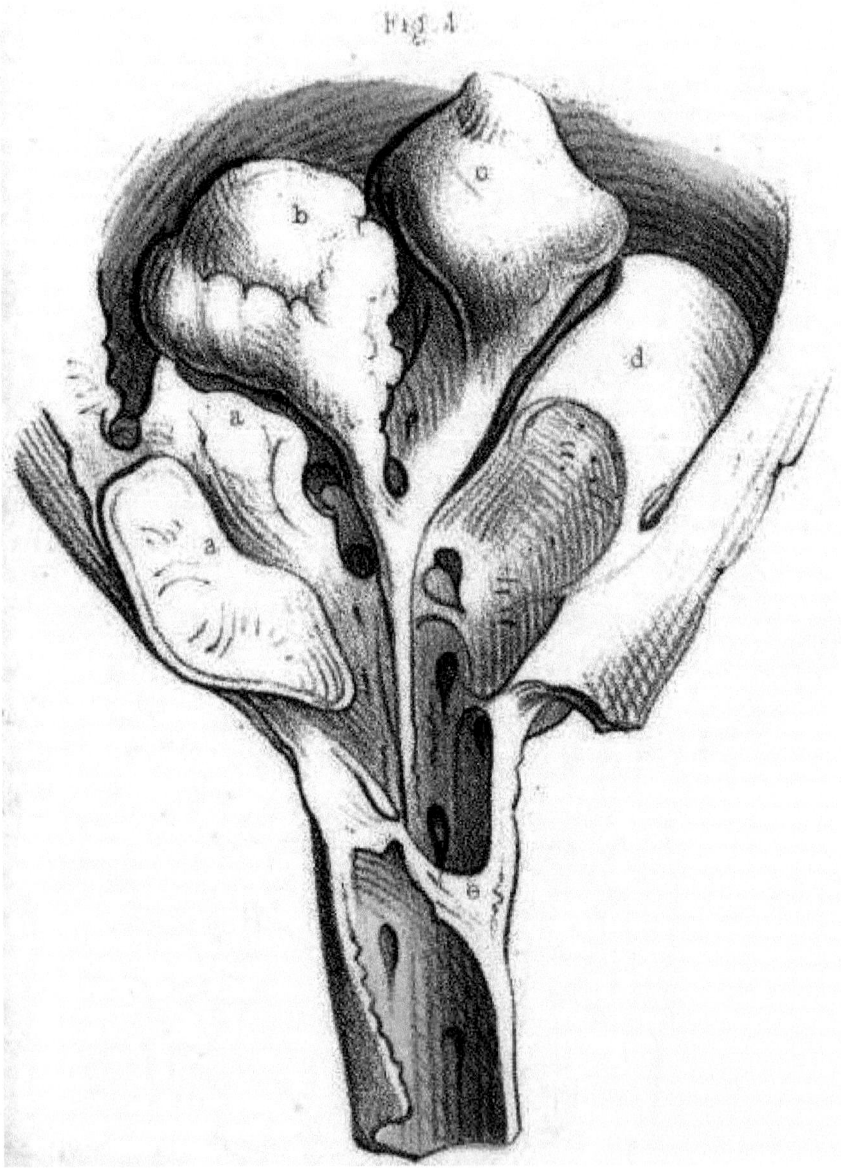

Plate 62.--Figure 4.

Fig. 5, Plate 62.--The prostate, *a a,* is greatly enlarged, and projects high in the bladder, the walls of the latter, *b b,* being very much thickened. The ureters, *c*, are dilated, and perforations made by instruments are seen in the prostate. The prostatic canal being directed almost vertically, and the neck of the bladder being raised nearly as high as the upper border of the pubic symphysis, it must appear that if a stone rest in the bas fond of the bladder, a sound or staff cannot reach the stone, unless by perforating the prostate; and if, while the staff occupies this position, lithotomy be performed, the incisions will not be required to be made of a greater depth than if the prostate were of its ordinary proportions. On the contrary, if the staff happen to have surmounted the prostate, the incision, in order to divide the whole vertical thickness of this body, will require to be made very deeply from the perinaeal surface, and this circumstance occasions what is termed a "deep perinaeum."

Plate 62.--Figure 5.

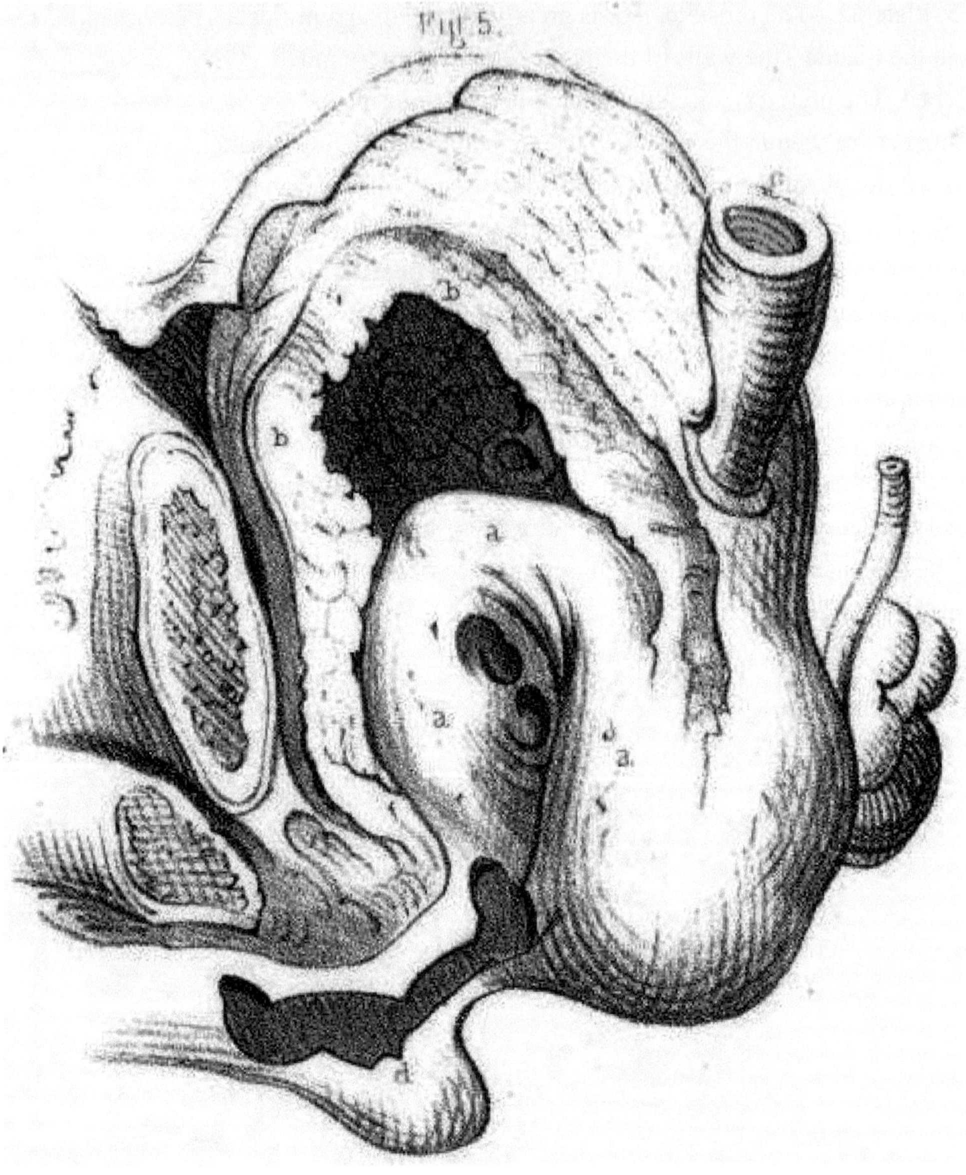

Fig. 6, Plate 62.--The lower half, *c, b, f,* of the prostate, having become the seat of abscess, appears hollowed out in the form of a sac. This sac is separated from the bladder by a horizontal septum, *e e,* the proper base of the bladder, *g g*. The prostatic urethra, between *a e*, has become vertical in respect to the membranous part of the canal, in consequence of the upward pressure of the abscess. The sac opens into the urethra, near the apex of the prostate, at the point *c*; and a catheter passed along the urethra has entered the orifice of the sac, the interior of which the instrument traverses, and the posterior wall of which it perforates. The bladder contains a large calculus, *i.* The bladder and sac do not communicate, but the urethra is a canal common to both. In a case of this sort it becomes evident that, although symptoms may strongly indicate either a retention of urine, or the presence of a stone in the bladder, any instrument taking the position and direction of *d d,* cannot relieve the one or detect the other; and such is the direction in which the instrument must of necessity pass, while the sac presents its orifice more in a line with the membranous part of the urethra than the neck of the bladder is. The sac will intervene between the rectum and the bladder; and on examination of the parts through the bowel, an instrument in the sac will readily be mistaken for being in the bladder, while neither a calculus in the bladder, nor this organ in a state of even extreme distention, can be detected by the touch any more than by the sound or catheter. If, while performing lithotomy in such a state of the parts, the staff occupy the situation of *d d d*, then the knife, following the staff, will open, not the bladder which contains the stone, but the sac, which, moreover, if it happen to be filled with urine regurgigated from the urethra, will render the deception more complete.

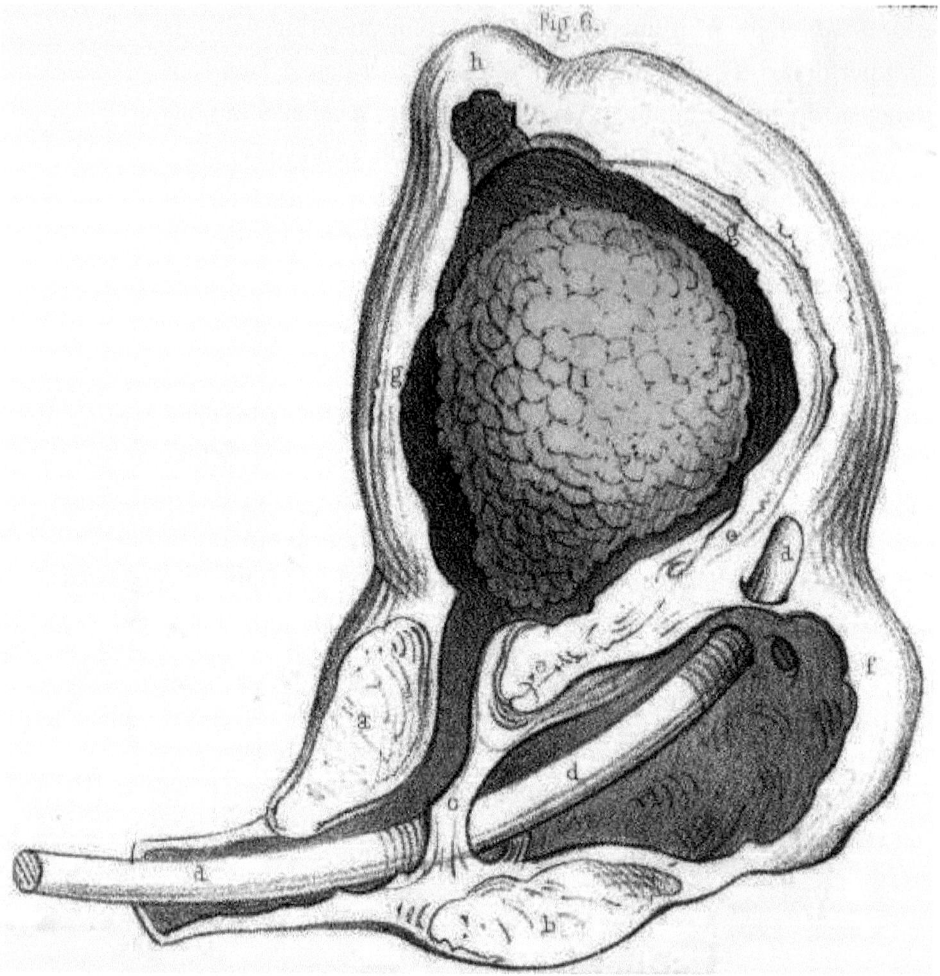

Plate 62.--Figure 6.

Fig. 7, Plate 62.--The walls, *a a*, of the bladder, appear greatly thickened, and the ureters, *b,* dilated. The sides, *c c c,* of the prostate are thinned; and in the prostatic canal are two calculi, *d d,* closely impacted. In such a state of the parts it would be impossible to pass a catheter into the bladder for the relief of a retention of urine, or to introduce a staff as a guide to the knife in lithotomy. If, however, the staff can be passed as far as the situation of the stone, the parts may be held with a sufficient degree of steadiness to enable the operator to incise the prostate upon the stone.

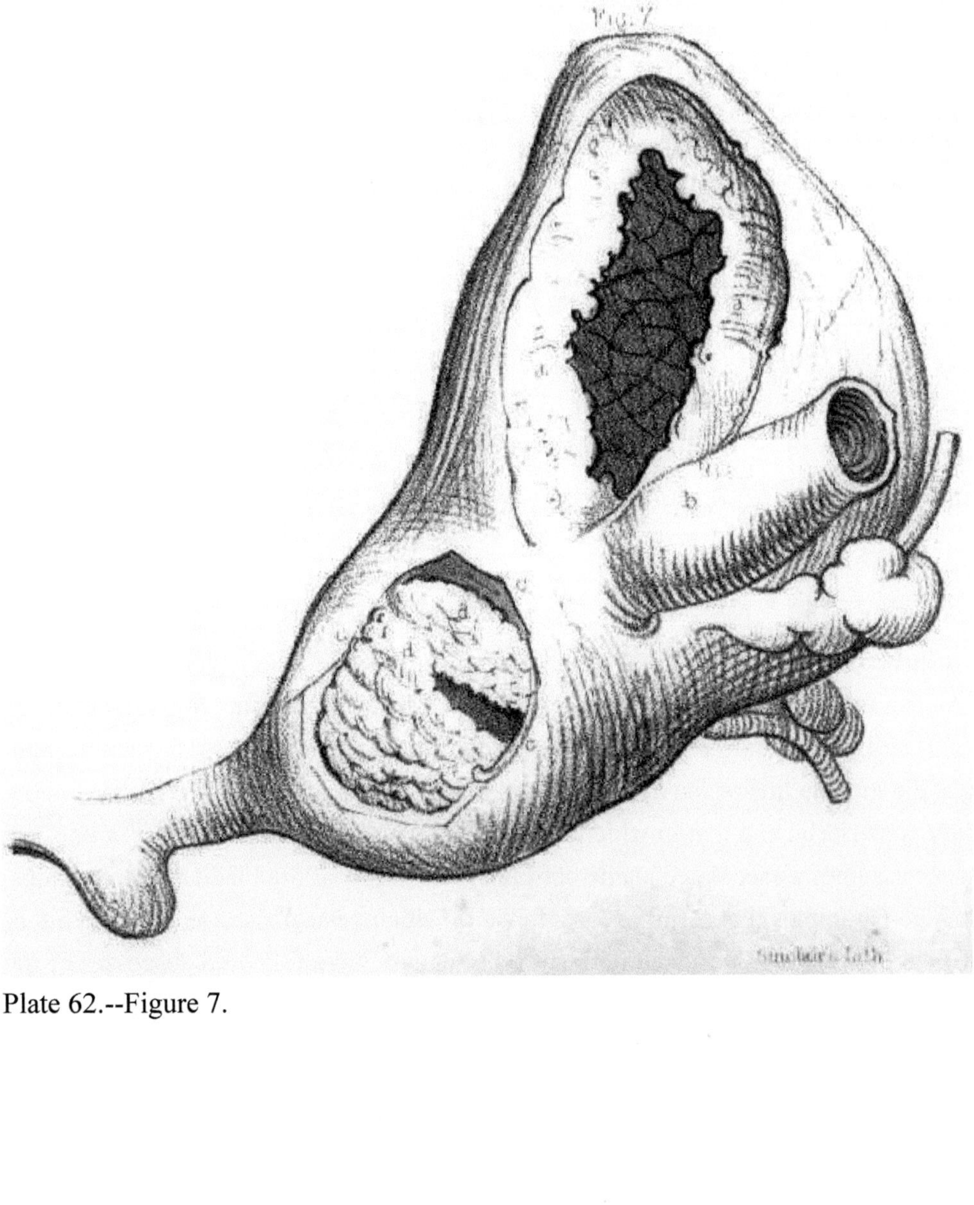

Plate 62.--Figure 7.

DEFORMITIES OF THE URINARY BLADDER; THE OPERATIONS OF SOUNDING FOR STONE; OF CATHETERISM AND OF PUNCTURING THE BLADDER ABOVE THE PUBES.

General remarks on the causes of the various deformities, and of the formation of stone. Lithic diathesis--its signification. The sacculated bladder considered in reference to sounding, to catheterism, to puncturation, and to lithotomy. Polypi in the bladder. Dilated ureters. The operation of catheterism. General rules to be followed. Remarks on the operation of puncturing the bladder above the pubes.

The urinary bladder presents two kinds of deformity--viz., congenital and pathological. As examples of the former may be mentioned that in which the organ is deficient in front, and has become everted and protruded like a fungous mass through an opening at the median line of the hypogastrium; that in which the rectum terminates in the bladder posteriorly; and that in which the foetal urachus remains pervious as a uniform canal, or assumes a sacculated shape between the summit of the bladder and the umbilicus. The pathological deformities are, those in which vesical fistulae, opening either above the pubes, at the perinaeum, or into the rectum, have followed abscesses or the operation of puncturing the bladder in these situations, and those in which the walls of the organ appear thickened and contracted, or thinned and expanded, or sacculated externally, or ridged internally, in consequence of its having been subjected to abdominal pressure while overdistended with its contents, and while incapable of voiding these from some permanent obstruction in the urethral canal.[Footnote] The bladder is liable to become sacculated from two causes--from a hernial protrusion of its mucous membrane through the separated fasciculi of its fibrous coat, or from the cyst of an abscess which has formed a communication with the bladder, and received the contents of this organ. Sacs, when produced in the former way, may be of any number, or size, or in any situation; when caused by an abscess, the sac is single, is generally formed in the prostate, or corresponds to the base of the bladder, and may attain to a size equalling, or even exceeding, that of the bladder itself. The sac, however formed, will be found lined by mucous membrane. The cyst of an abscess, when become a recipient for the urine, assumes after a time a lining membrane similar to that of the bladder. If the sac be situated at the summit or back of the bladder, it will be found invested by peritonaeum; but, whatever be its size, structure, or position, it may be always distinguished from the bladder by being devoid of the fibrous tunic, and by having but an indirect relation to the vesical orifice.

[Footnote: On considering these cases of physical impediments to the passage of urine from the vesical reservoir through the urethral conduit, it seems to me as if these were sufficient to account for the formation of stone in the bladder, or any other part of the urinary apparatus, without the necessity of ascribing it to a constitutional disease, such as that named the *lithic diathesis* by the humoral pathologists.

The urinary apparatus (consisting of the kidneys, ureters, bladder, and urethra) is known to be the principal emunctory for eliminating and voiding the detritus formed by the continual decay of the parts comprising the animal economy. The urine is this detritus in a state of solution. The components of urine are chemically similar to those of calculi, and as the components of the one vary according to the disintegration occurring at the time in the vital alembic, so do those of the other. While, therefore, a calculus is only as urine precipitated and solidified, and this fluid only as calculous matter suspended in a menstruum, it must appear that the lithic diathesis is as natural and universal as structural disintegration is constant and general in operation. As every individual, therefore, may be said to void day by day a dissolved calculus, it must follow that its form of precipitation within some part of the urinary apparatus alone constitutes the disease, since in this form it cannot be passed. On viewing the subject in this light, the question that springs directly is, (while the lithic diathesis is common to individuals of all ages and both sexes,) why the lithic sediment should present in the form of concrement in some and not in others? The principal, if not the sole, cause of this seems to me to be obstruction to the free egress of the urine along the natural passage. Aged individuals of the male sex, in whom the prostate is prone to enlargement, and the urethra to organic stricture, are hence more subject to the formation of stone in the bladder, than youths, in whom these causes of obstruction are less frequent, or than females of any age, in whom the prostate is absent, and the urethra simple, short, readily dilatable, and seldom or never strictured. When an obstruction exists, lithic concretions take place in the urinary apparatus in the same manner as sedimentary particles cohere or crystallize elsewhere. The urine becoming pent up and stagnant while charged with saline matter, either deposits this around a nucleus introduced into it, or as a surplus when the menstruum is insufficient to suspend it. The most depending part of the bladder is that where lithic concretions take place; and if a sacculus exist here, this, becoming a recipient for the matter, will favour the formation of stone.] [End Footnote]

FIG. 1, Plate 63.--The lateral lobes of the prostate, 3, 4, are enlarged, and contract the prostatic canal. Behind them the third lobe of smaller size occupies the vesical orifice, and completes the obstruction. The walls of the bladder have hence become fasciculated and sacculated. One sac, 1, projects from the summit of the bladder; another, 2, containing a stone, projects laterally. When a stone occupies a sac, it does not give rise to the usual symptoms as indicating its presence, nor can it be always detected by the sound.

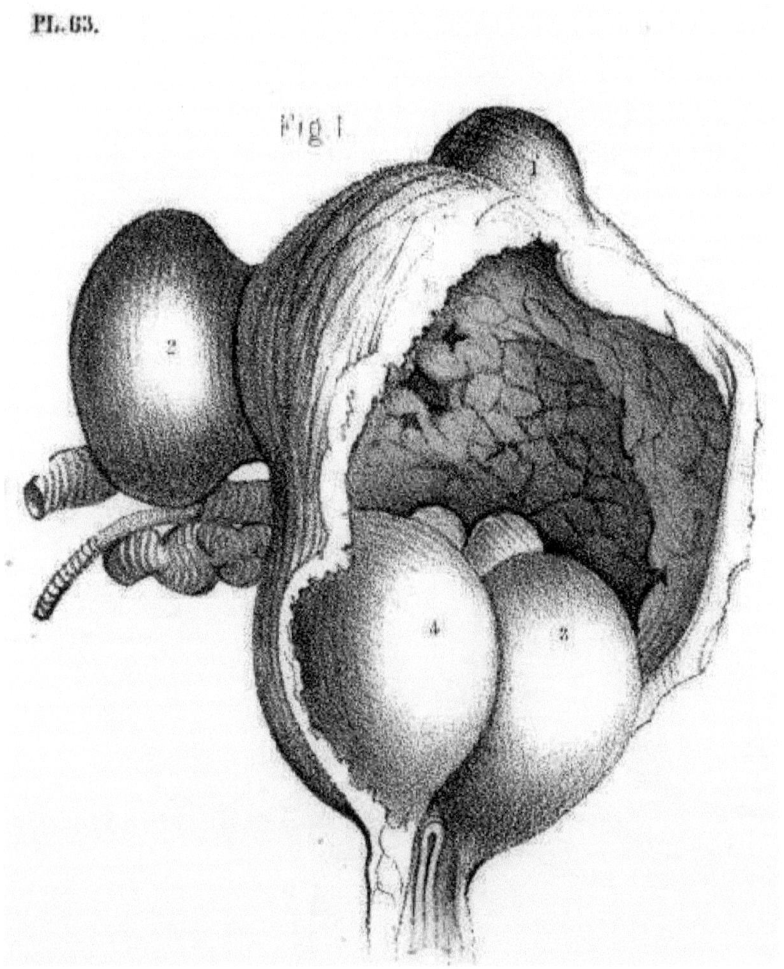

Plate 63,--Figure 1.

FIG. 2, Plate 63.--The prostate, 2, 3, is enlarged, and the middle lobe, 2, appears bending the prostatic canal to an almost vertical position, and obstructing the vesical orifice. The bladder, 1, 1, 1, is thickened; the ureters, 7, are dilated; and a large sac, 6, 6, projects from the base of the bladder backwards, and occupies the recto-vesical fossa. The sac, equal in size to the bladder, communicates with this organ by a small circular opening, 8, situated between the orifices of the ureters. The peritonaeum is reflected from the summit of the bladder to that of the sac. A catheter, 4, appears perforating the third lobe of the prostate, 2, and entering the sac, 5, through the base of the bladder, below the opening, 8. In a case of this kind, a catheter occupying the position 4, 5, would, while voiding the bladder through the sac, make it seem as if it really traversed the vesical orifice. If a stone occupied the bladder, the point of the instrument in the sac could not detect it, whereas, if a stone lay within the sac, the instrument, on striking it here, would give the impression as if it lay within the bladder.

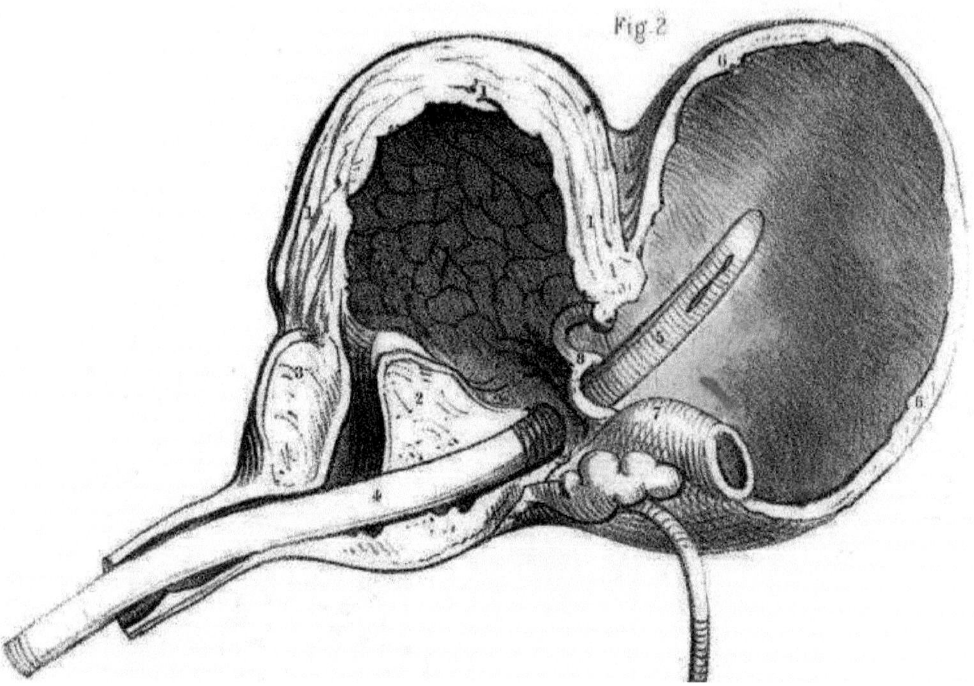

Plate 63,--Figure 2.

FIG. 3, Plate 63.--The urethra being strictured, the bladder has become sacculated. In the bas fond of the bladder appears a circular opening, 2, leading to a sac of large dimensions, which rested against the rectum. In such a case as this, the sac, occupying a lower position than the base of the bladder, must first become the recipient of the urine, and retain this fluid even after the bladder has been evacuated, either voluntarily or by means of instruments. If, in such a state of the parts, retention of urine called for puncturation, it is evident that this operation would be performed with greater effect by opening the depending sac through the bowel, than by entering the summit of the bladder above the pubes.

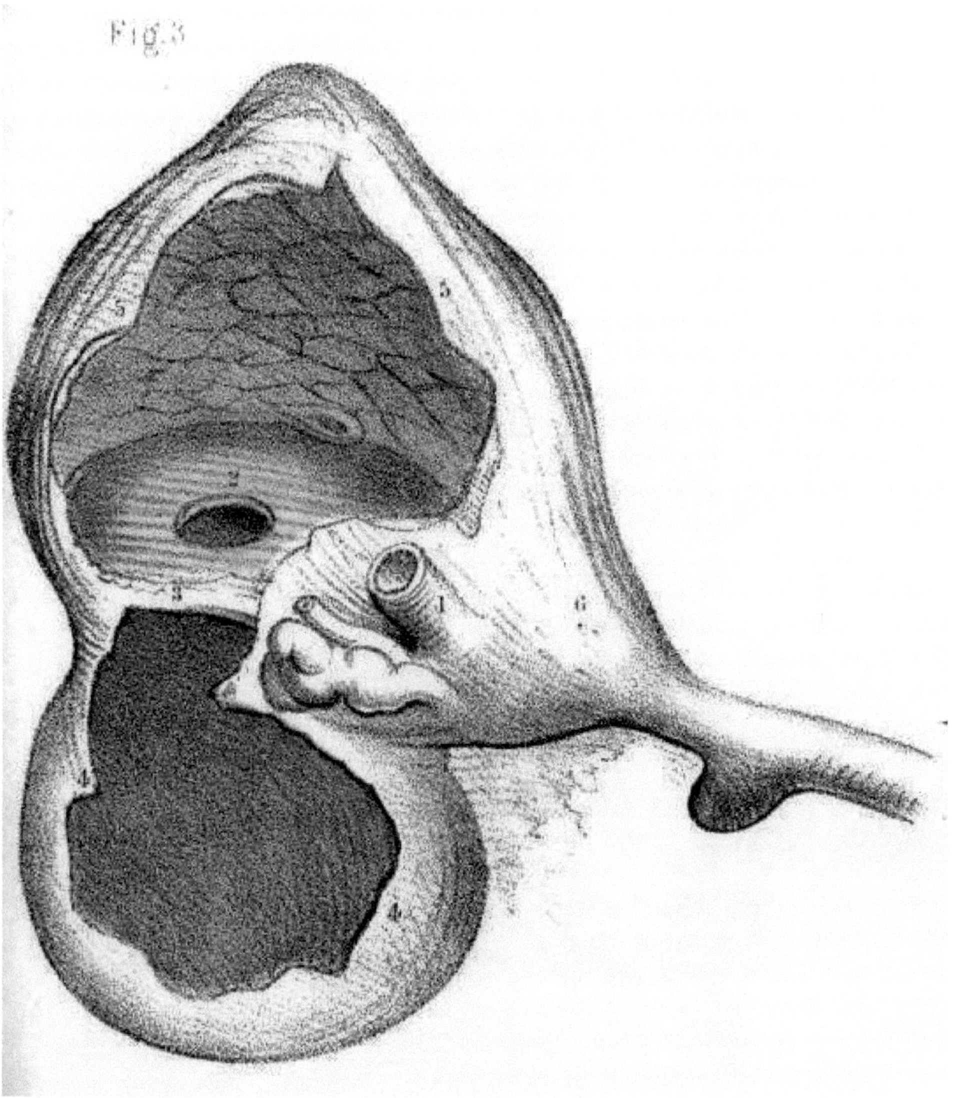

Plate 63,--Figure 3.

FIG. 4, Plate 63.--The vesical orifice is obstructed by two portions, 3, 4, of the prostate, projecting upwards, one from each of its lateral lobes, 6, 6. The bladder is thickened and fasciculated, and from its summit projects a double sac, 1, 2, which is invested by the peritonaeum.

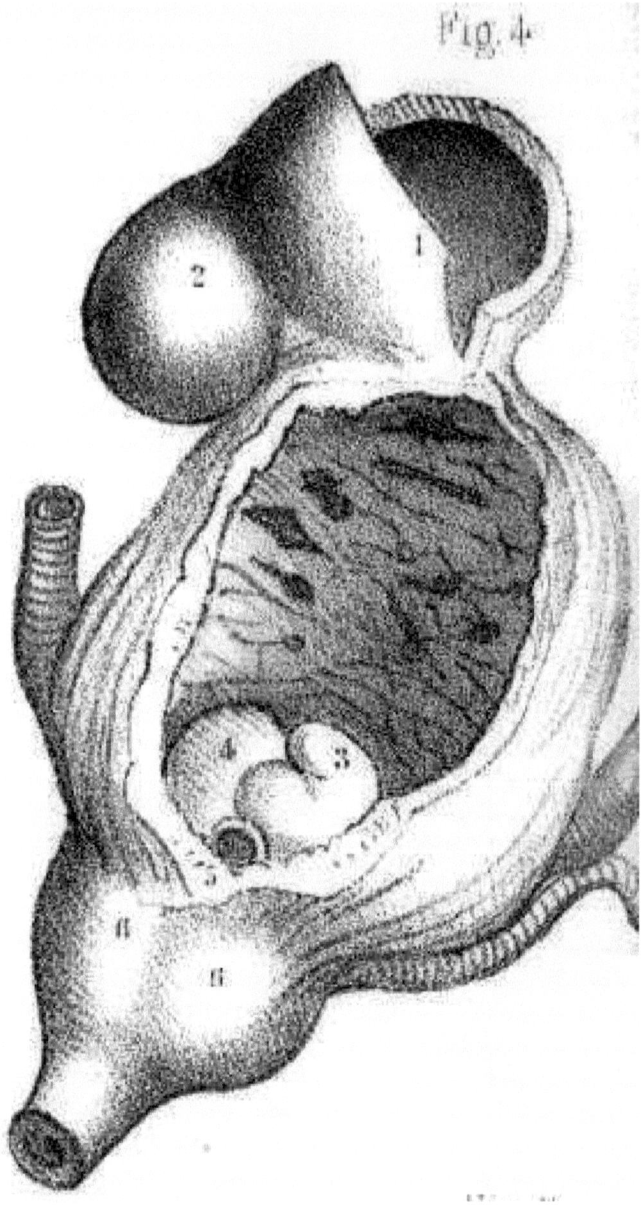

Plate 63,--Figure 4.

FIG. 5, Plate 63.--The prostatic canal is constricted and bent upwards by the third lobe. The bladder is thickened, and its base is dilated in the form of a sac, which is dependent, and upon which rests a calculus. An instrument enters the bladder by perforating the third lobe, but does not come into contact with the calculus, owing to the low position occupied by this body.

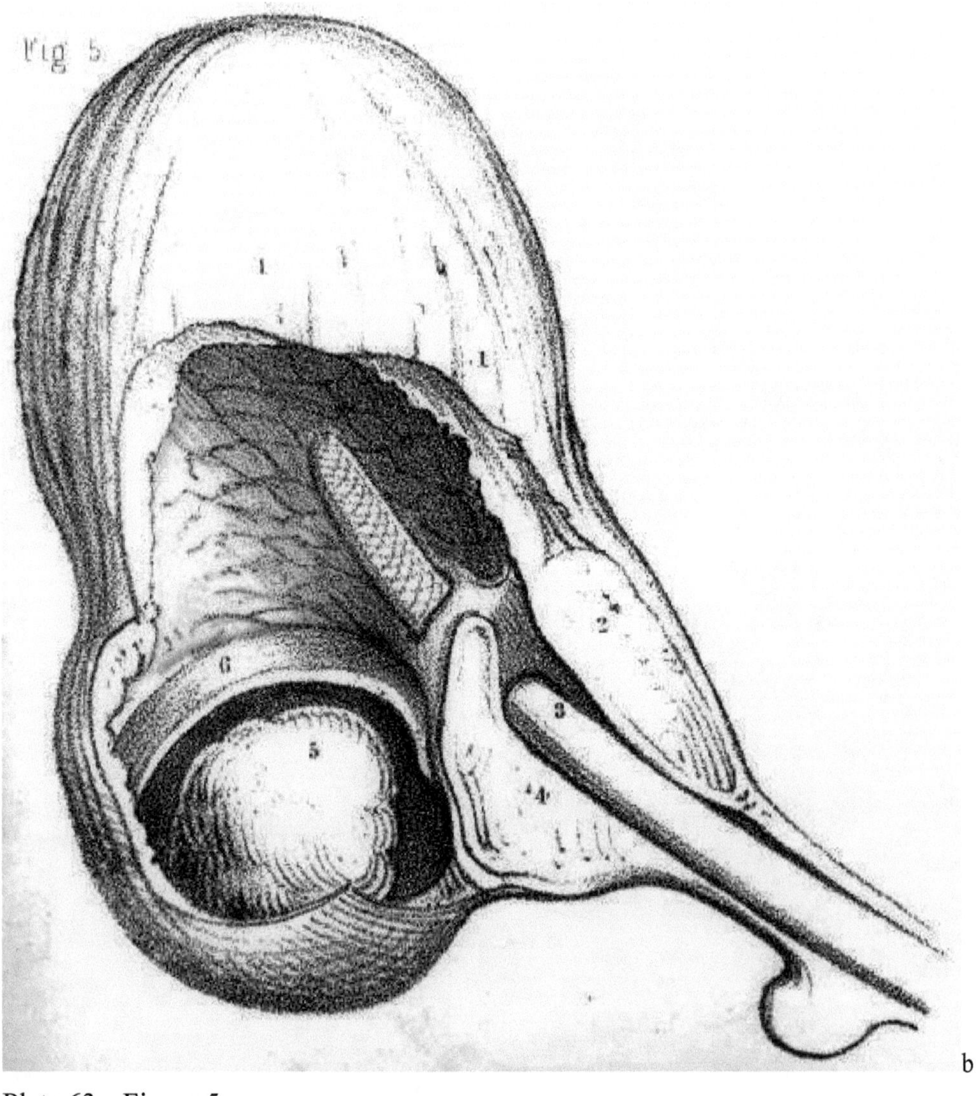

b

Plate 63,--Figure 5.

FIG. 6, Plate 63.--Two sacs appear projecting on either side of the base of the bladder. The right one, 5, contains a calculus, 6; the left one, of larger dimensions, is empty. The rectum lay in contact with the base of the bladder between the two sacs.

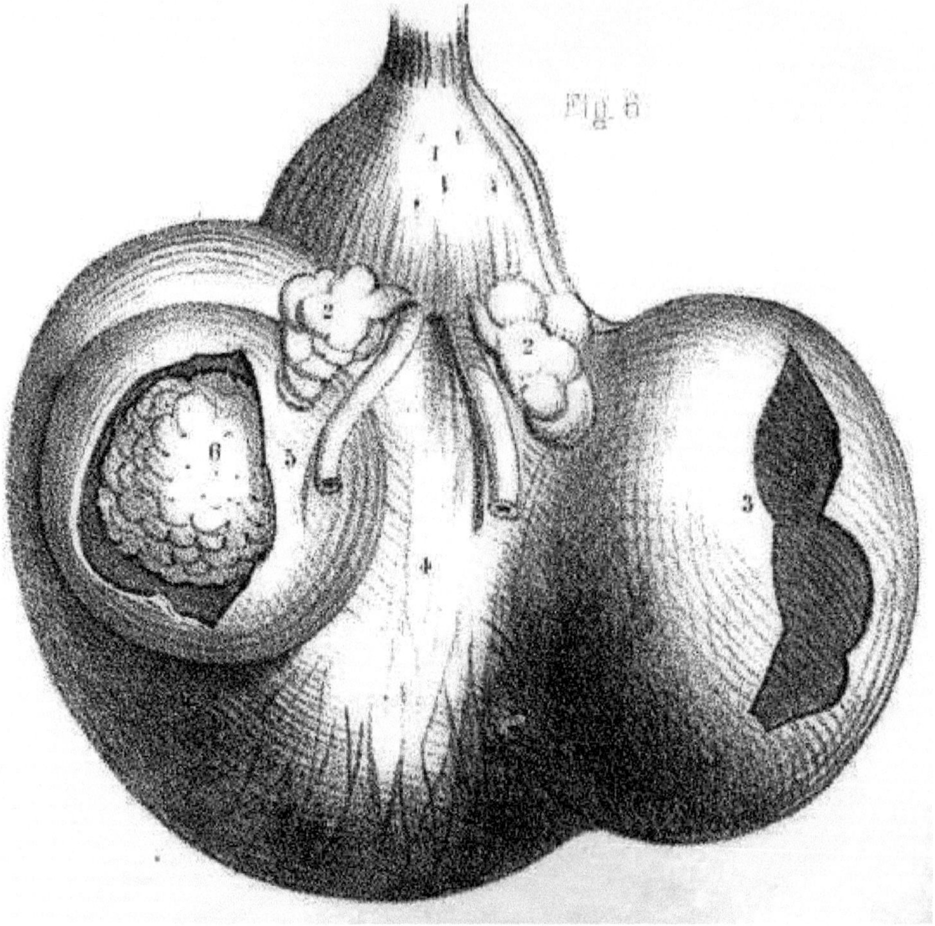

Plate 63,--Figure 6.

FIG. 7, Plate 63.--Four calculi are contained in the bladder. This organ is divided by two septa, 2, 4, into three compartments, each of which, 1, 3, 5, gives lodgment to a calculus; and another, 6, of these bodies lies impacted in the prostatic canal, and becomes a complete bar to the passage of a catheter. Supposing lithotomy to be performed in an instance of this kind, it is probable that, after the extraction of the calculi, 6, 5, the two upper ones, 3, 1, would, owing to their being embedded in the walls of the bladder, escape the forceps.

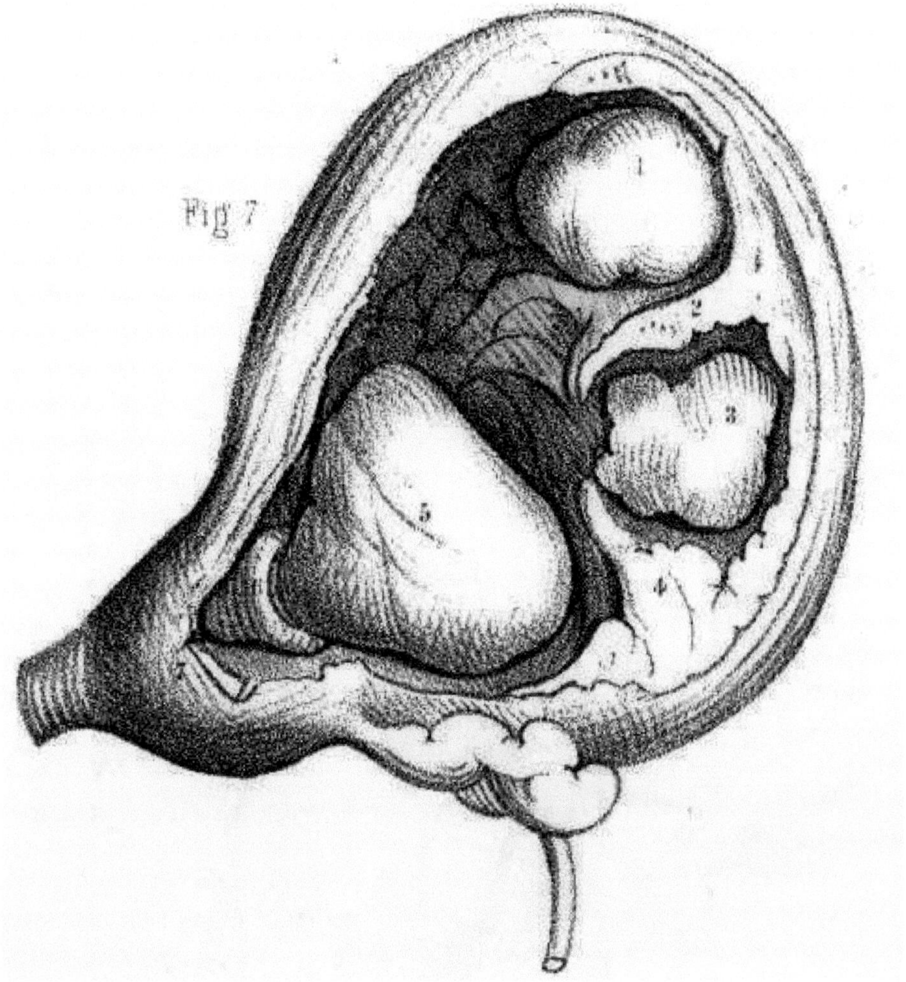

Plate 63,--Figure 7.

FIG. 8, Plate 63.-Two large polypi, and many smaller ones, appear growing from the mucous membrane of the prostatic urethra and vesical orifice, and obstructing these parts. In examining this case during life by the sound, the two larger growths, 1, 2, were mistaken by the surgeon for calculi. Such a mistake might well be excused if they happened to be encrusted with lithic matter.

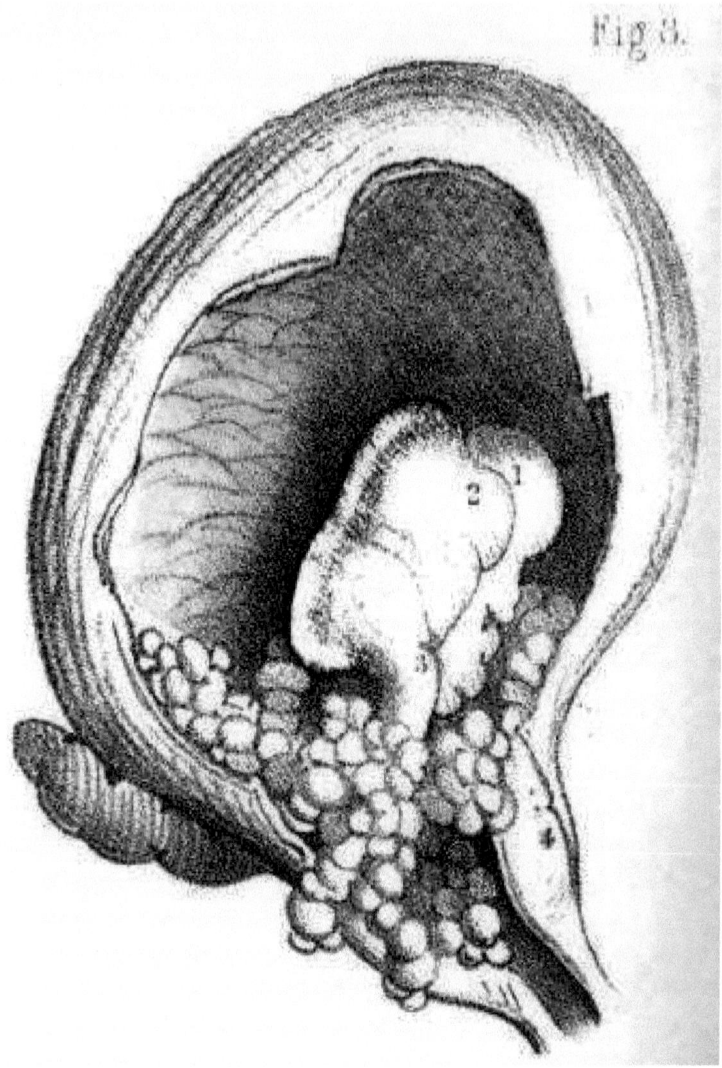

Plate 63,--Figure 8.

FIG. 9, Plate 63.--The base of the bladder, 8, 8, appears dilated into a large uniform sac, and separated from the upper part of the organ by a circular horizontal fold, 2, 2. The ureters are also dilated. The left ureter, 3, 4, opens into the sac below this fold, while the right ureter opens above it into the bladder. In all cases of retention of urine from permanent obstruction of the urethra, the ureters are generally found more or less dilated. Two circumstances combine to this effect--while the renal secretion continues to pass into the ureters from above, the contents of the bladder under abdominal pressure are forced regurgitating into them from below, through their orifices.

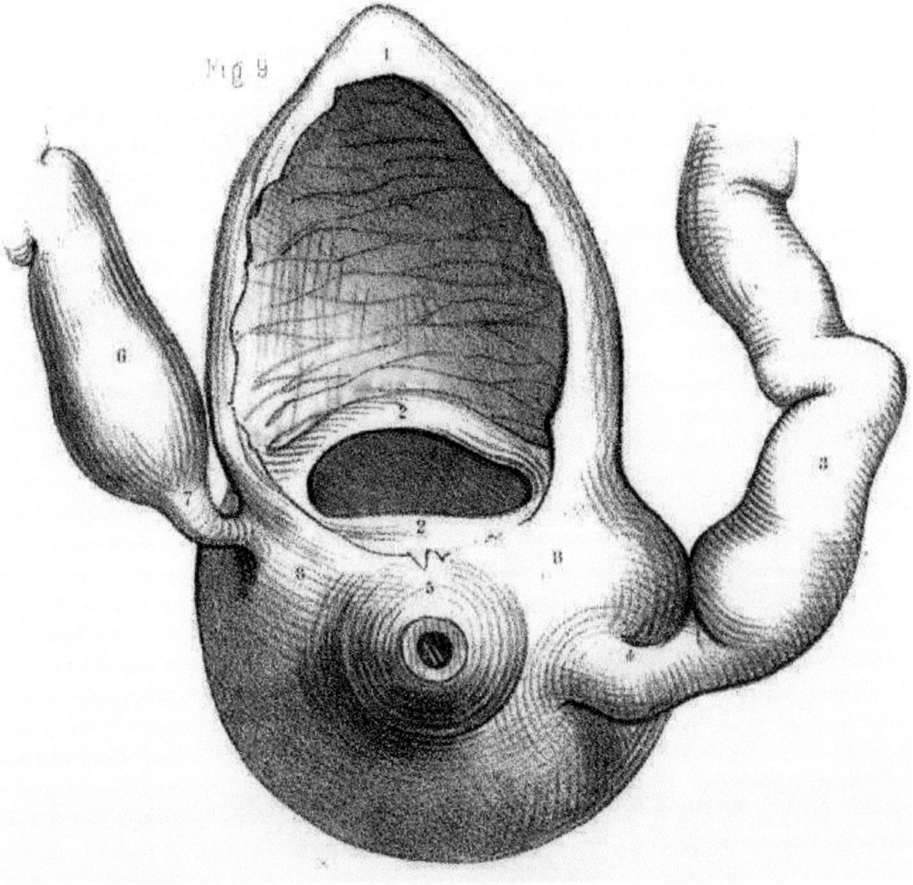

Plate 63,--Figure 9.

FIG. 1, Plate 64.--The bladder, 6, appears symmetrically sacculated. One sac, 1, is formed at its summit, others, 3, 2, project laterally, and two more, 5, 4, from its base. The ureters, 7, 7, are dilated, and enter the bladder between the lateral and inferior sacs.

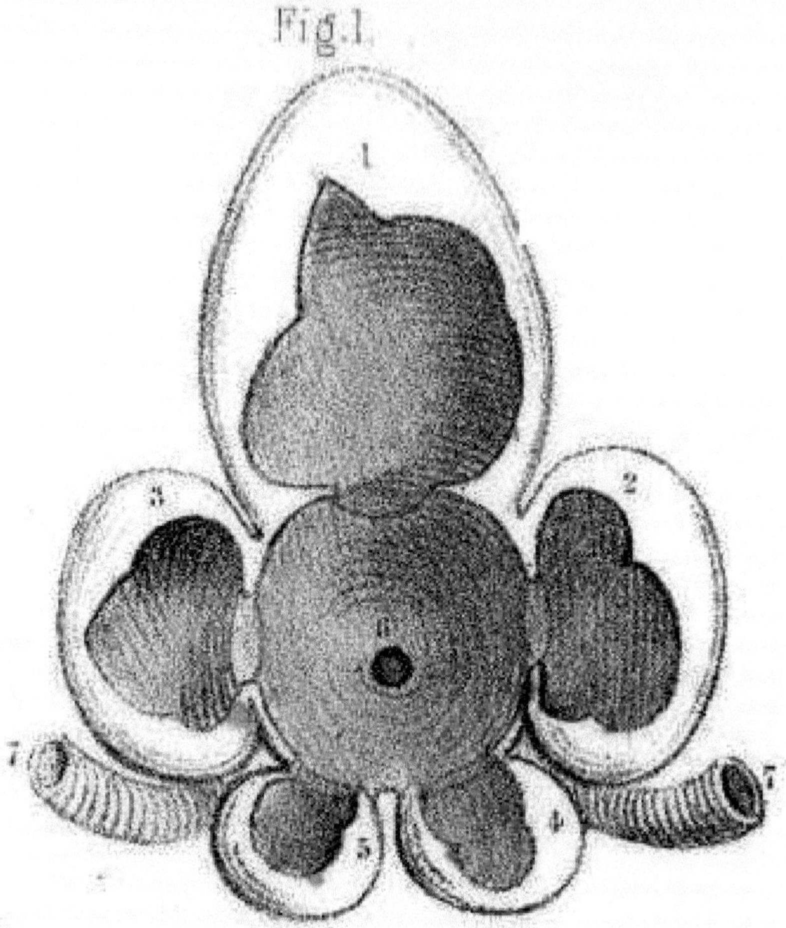

Plate 64,--Figure 1.

Fig. 2, Plate 64.--The prostate is greatly enlarged, and forms a narrow ring around the vesical orifice. Through this an instrument, 12, enters the bladder. The walls of the bladder are thickened and sacculated. On its left side appear numerous sacs, 2, 3, 4, 5, 6, 7, 8, and on the inner surface of its right side appear the orifices of as many more. On its summit another sac is formed. The ureters, 9, are dilated.

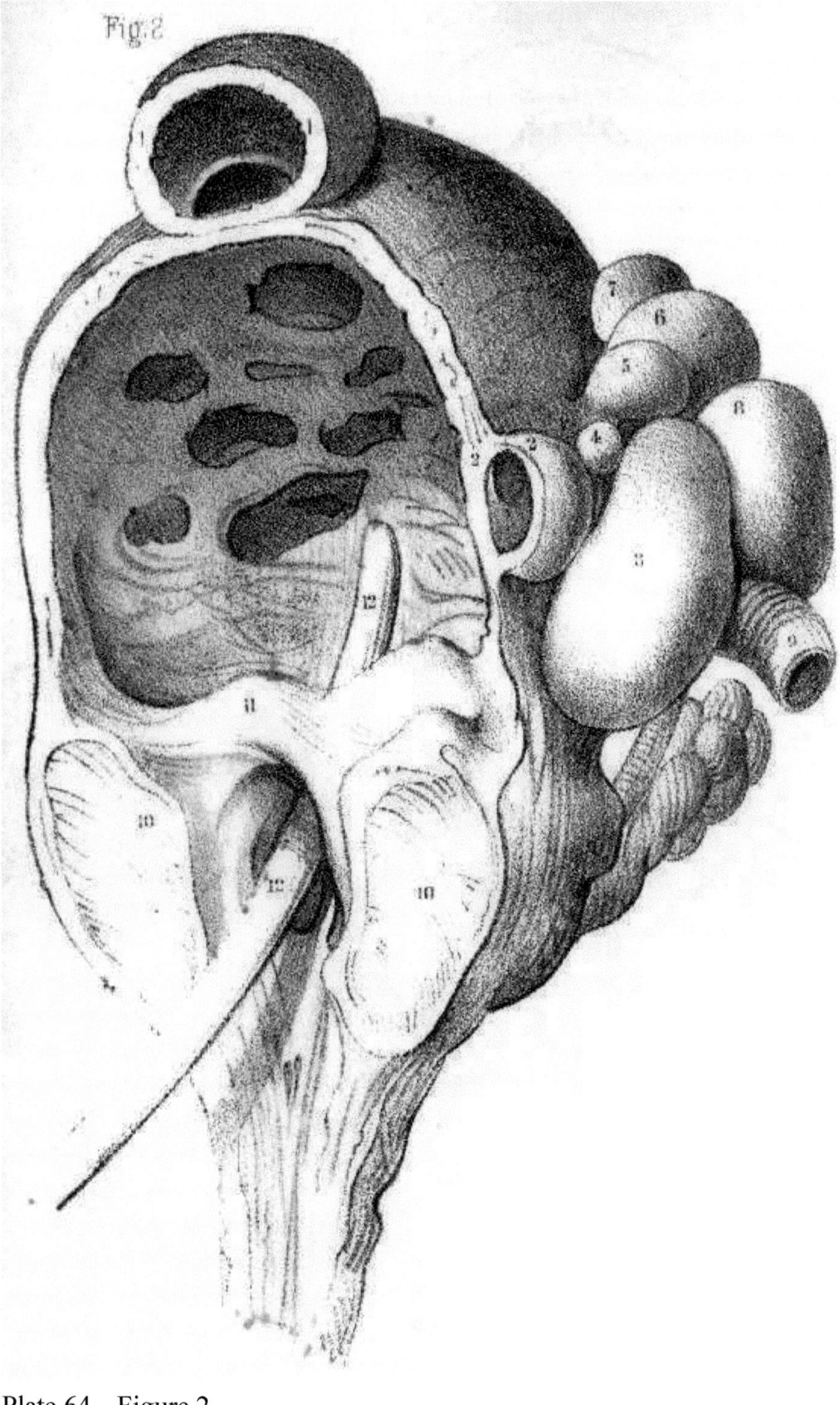

Plate 64,--Figure 2.

FIG. 3, Plate 64.--The prostate is enlarged, its canal is narrowed, and the bladder is thickened and contracted. A calculus, 1, 2, appears occupying nearly the whole vesical interior. The incision in the neck of the bladder in lithotomy must necessarily be extensive, to admit of the extraction of a stone of this size.

PL. 64

Fig. 3

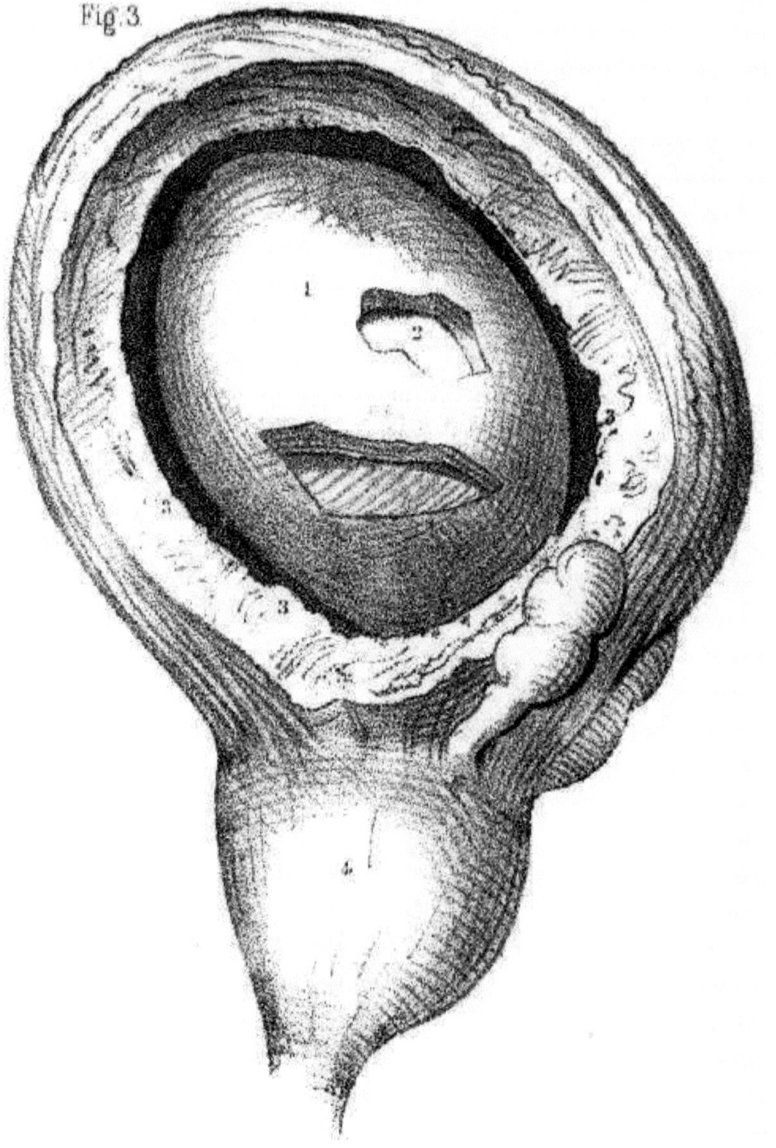

Plate 64,--Figure 3.

FIG. 4, Plate 64.--The prostatic canal is contracted by the lateral lobes, 4, 5; resting upon these, appear three calculi, 1, 2, 3, which nearly fill the bladder. This organ is thickened and fasciculated. In cases of this kind, and that last mentioned, the presence of stone is readily ascertainable by the sound.

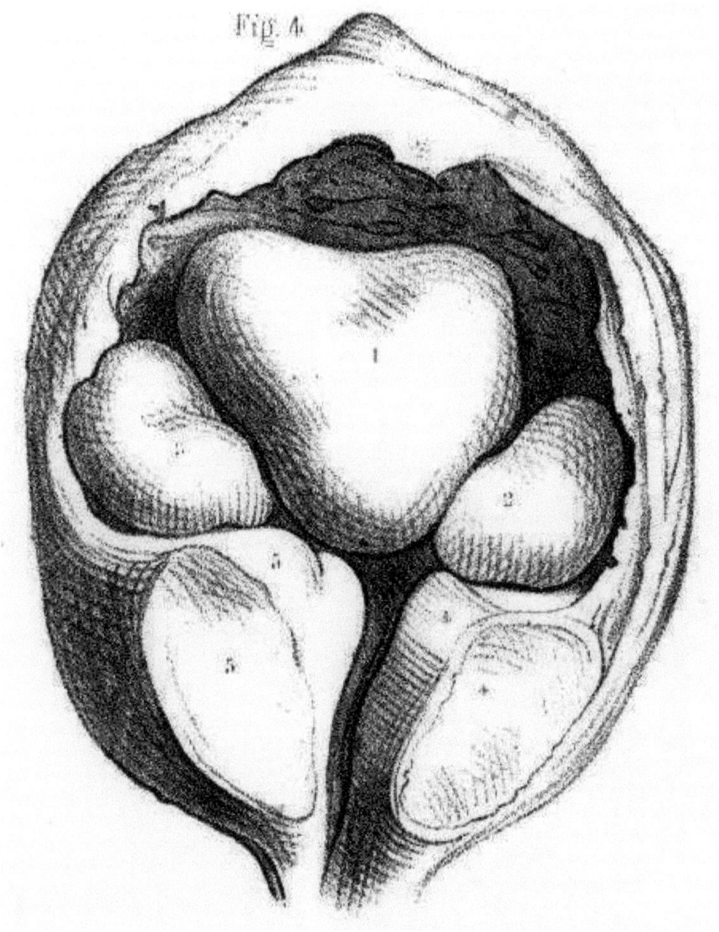

Plate 64,--Figure 4.

FIG. 5, Plate 64.--The three prostatic lobes are enlarged, and appear contracting the vesical orifice. In the walls of the bladder are embedded several small calculi, 2, 2, 2, 2, which, on being struck with the convex side of a sound, might give the impression as though a single stone of large size existed. In performing lithotomy, these calculi would not be within reach of the forceps.

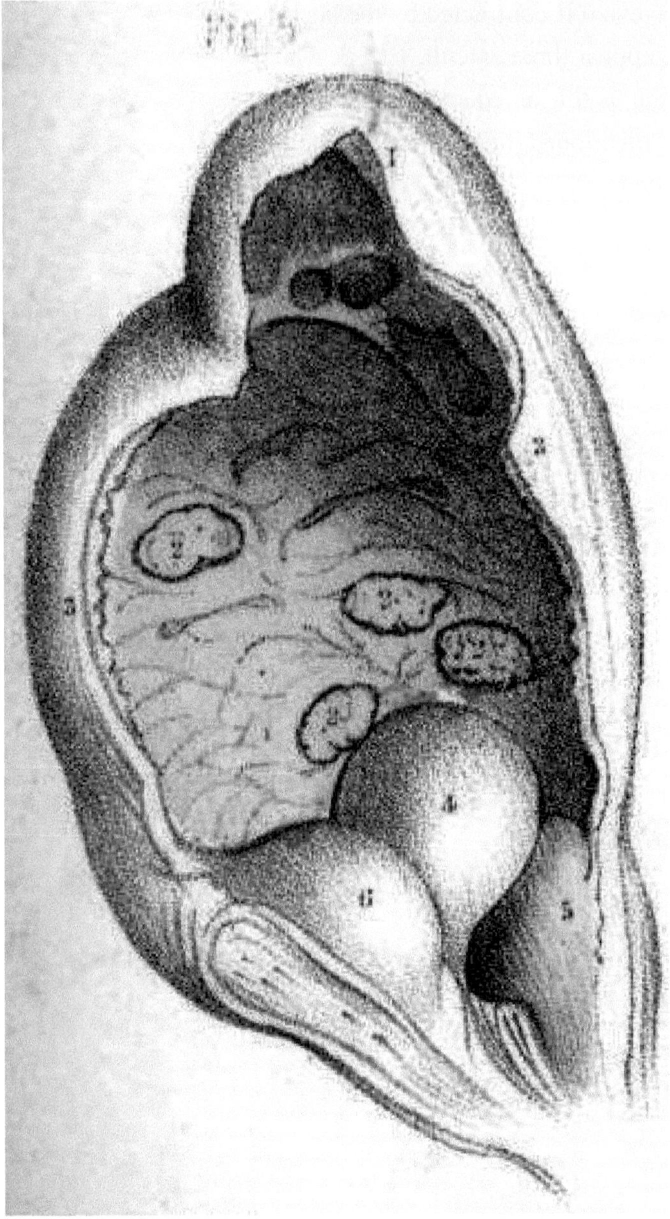

Plate 64,--Figure 5.

FIG. 6, Plate 64.--Two sacculi, 4, 5, appear projecting at the middle line of the base of the bladder, between the vasa deferentia, 7, 7, and behind the prostate, in the situation where the operation of puncturing the bladder per anum is recommended to be performed in retention of urine.

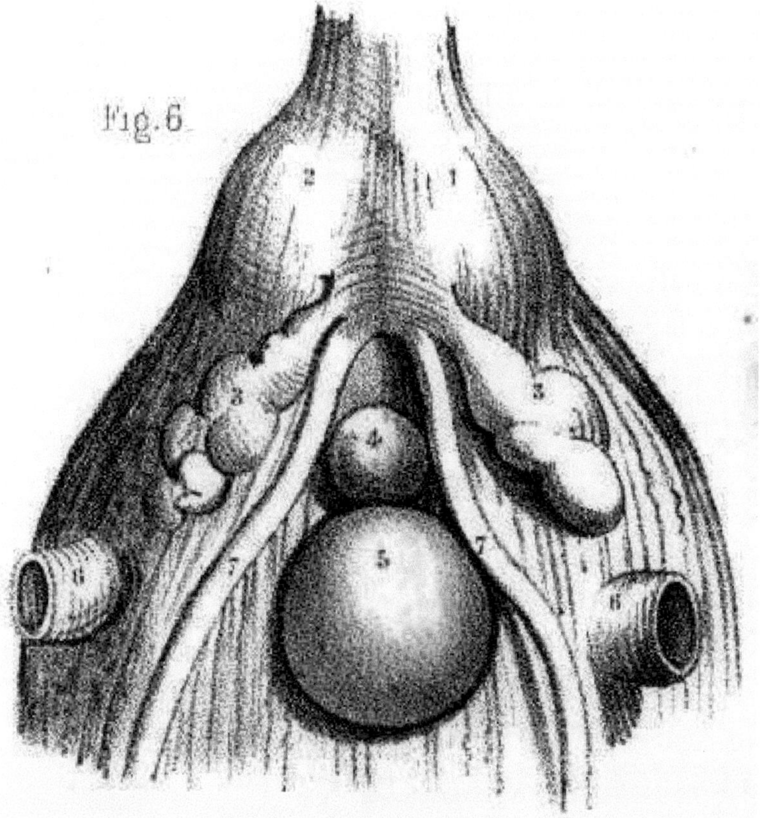

Plate 64,--Figure 6.

FIG. 7, Plate 64.--A sac, 4, is situated on the left side of the bladder, 3, 3, immediately above the orifice of the ureter. In the sac was contained a mass of phosphatic calculus. This substance is said to be secreted by the mucous lining of the bladder, while in a state of chronic inflammation, but there seems nevertheless very good reason for us to believe that it is, like all other calculous matter, a deposit from the urine.

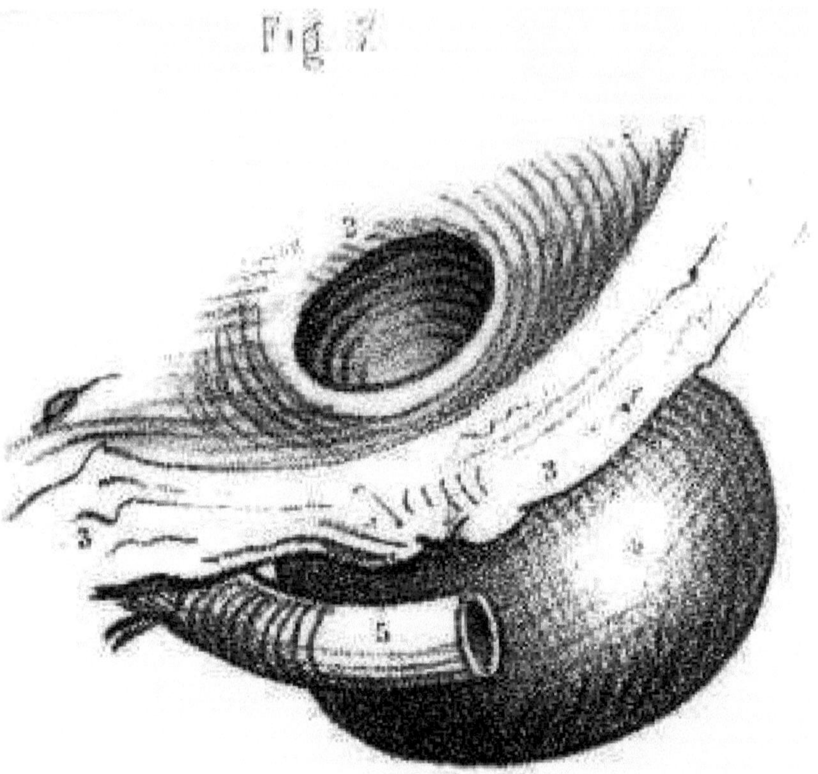

Plate 64,--Figure 7.

FIG. 8, Plate 64, represents, in section, the relative position of the parts concerned in catheterism. [Footnote] In performing this operation, the patient is to be laid supine; his loins are to be supported on a pillow; and his thighs are to be flexed and drawn apart from each other. By this means the perinaeum is brought fully into view, and its structures are made to assume a fixed relative position. The operator, standing on the patient's left side, is now to raise the penis so as to render the urethra, 8, 8, 8, as straight as possible between the meatus, *a*, and the bulb, 7. The instrument (the concavity of its curve being turned to the left groin) is now to be inserted into the meatus, and while being gently impelled through the canal, the urethra is to be drawn forwards, by the left hand, over the instrument. By stretching the urethra, we render its sides sufficiently tense for facilitating the passage of the instrument, and the orifices of the lacunae become closed. While the instrument is being passed along this part of the canal, its point should be directed fairly towards the urethral opening, 6*, of the triangular ligament, which is situated an inch or so below the pubic symphysis, 11. With this object in view, we should avoid depressing its handle as yet, lest its point be prematurely tilted up, and rupture the upper side of the urethra anterior to the ligament. As soon as the instrument has arrived at the bulb, its further progress is liable to be arrested, from these causes:--1st, This portion of the canal is the lowest part of its perinaeal curve, 3, 6, 8, and is closely embraced by the middle fibres of the accelerator urinae muscle. 2nd, It is immediately succeeded by the commencement of the membranous urethra, which, while being naturally narrower than other parts, is also the more usual seat of organic stricture, and is subject to spasmodic constriction by the fibres of the compressor urethrae. 3d, The triangular ligament is behind it, and if the urethral opening of the ligament be not directly entered by the instrument, this will bend the urethra against the front of that dense structure. On ascertaining these to be the causes of resistance, the instrument is to be withdrawn a little in the canal, so as to admit of its being readjusted for engaging precisely the opening in the triangular ligament. As this structure, 6, is attached to the membranous urethra, 6*, which perforates it, both these parts may be rendered tense, by drawing the penis forwards, and thereby the instrument may be guided towards and through the aperture. The instrument having passed the ligament, regard is now to be paid to the direction of the pelvic portion of the canal, which is upwards and backwards to the vesical orifice, 3, *d*, 3. In order that the point of the instrument may freely traverse the urethra in this direction, its handle, *a*, requires to be depressed, *b c*, slowly towards the perinaeum, and at the same time to be impelled steadily back in the line *d, d,* through the pubic arch, 11. If the third lobe of the prostate happen to be enlarged, the vesical orifice will accordingly be more elevated than usual. In this case, it becomes necessary to depress the instrument to a greater extent than is otherwise required, so that its point may surmount the obstacle. But since the suspensory ligament of the penis, 10, and the perinaeal structures prevent the handle being depressed beyond a certain degree, which is insufficient for the object to be attained, the instrument should possess the *prostatic curve, c c,* compared with *c b*.

[Footnote: It may be necessary for me to state that, with the exception of this figure (which is obviously a plan, but sufficiently accurate for the purposes it is intended to serve) all the others representing pathological conditions and congenital deformities of the urethra, the prostate, and the bladder, have been made by myself from *natural specimens* in the museums and hospitals of London and Paris.]

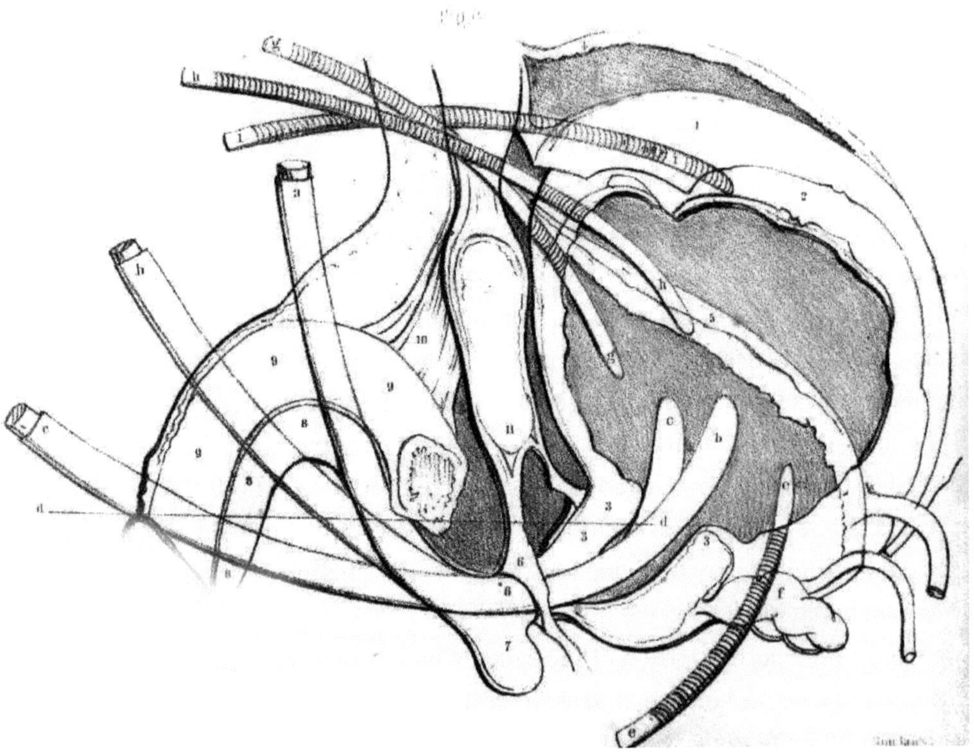

Plate 64,--Figure 8.

In the event of its being impossible to pass a catheter by the urethra, in cases of retention of urine threatening rupture, the base or the summit of the bladder, according as either part may be reached with the greater safety to the peritonaeal sac, will require to be punctured. If the prostate be greatly and irregularly enlarged, it will be safer to puncture the bladder above the pubes, and here the position of the organ in regard to the peritonaeum, 1, becomes the chief consideration. The shape of the bladder varies very considerably from its state of collapse, 3, 3, 5, to those of mediate, 3, 3, 2, 1, and extreme distention, 3, 3, 4. This change of form is chiefly effected by the expansive elevation of its upper half, which is invested by the peritonaeum. As the summit of the bladder falls below, and rises above the level of the upper margin of the pubic symphysis, it carries the peritonaeum with it in either direction. While the bladder is fully expanded, 4, there occurs an interval between the margin of the symphysis pubis and the point of reflexion of the peritonaeum, from the recti muscles, to the summit of the viscus. At this interval, close to the pubes, and in the median line, the trocar may be safely passed through the front wall of the bladder. The instrument should, in all cases, be directed downwards and backwards, *h, h*, in a line pointing to the hollow of the sacrum.

THE SURGICAL DISSECTION OF THE POPLITEAL SPACE, AND THE POSTERIOR CRURAL REGION.

Varieties of the popliteal and posterior crural vessels. Remarks on popliteal aneurism, and the operation for tying the popliteal artery, in wounds of this vessel. Wounds of the posterior crural arteries requiring double ligatures. The operations necessary for reaching these vessels.

On comparing the bend of the knee with the bend of the elbow, as evident a correspondence can be discerned between these two regions, as exists between the groin and the axilla.

Behind the knee-joint, the muscles which connect the leg with the thigh enclose the space named popliteal. When the integuments and subcutaneous substance are removed from this place, the dense fascia lata may be seen binding these muscles so closely together as to leave but a very narrow interval between them at the mesial line. On removing this fascia, B B M M, Plate 65, the muscles part asunder, and the popliteal space as usually described is thereby formed. This region now presents of a lozenge-shaped form, B J D K, of which the widest diameter, D J, is opposite the knee-joint. The flexor muscles, C D J, in diverging from each other as they pass down from the sides of the thigh to those of the upper part of the leg, form the upper angle of this space; whilst its lower angle is described by the two heads of the gastrocnemius muscle, E E, arising inside the flexors, from the condyles of the femur. The popliteal space is filled with adipose substance, in which are embedded several lymphatic bodies and through which pass the principal vessels and nerves to the leg.

In the dissection of the popliteal space, the more important parts first met with are the branches of the great sciatic nerve. In the upper angle of the space, this nerve will be found dividing into the peronaeal, I, and posterior tibial branches, H K. The peronaeal nerve descends close to the inner margin of the tendon, J, of the biceps muscle; and, having reached the outer side of the knee, I*, Plate 66, below the insertion of the tendon into the head of the fibula, winds round the neck of this bone under cover of the peronaeus longus muscle, S, to join the anterior tibial artery. The posterior tibial nerve, H K, Plate 65, descends the popliteal space midway to the cleft between the heads of the gastrocnemius; and, after passing beneath this muscle, to gain the inner side of the vessels, H*, Plate 66, it then accompanies the posterior tibial artery. On the same plane with and close to the posterior tibial nerve in the popliteal space, will be seen the terminal branch of the lesser sciatic nerve, together with a small artery and vein destined for distribution to the skin and other superficial parts on the back of the knee. Opposite the heads of the gastrocnemius, the peronaeal and posterior tibial nerves give off each a branch, both of which descend along the mesial line of the calf, and joining near the upper end of the tendo Achillis, the single nerve here, N, Plate 65, becomes superficial to the fascia, and thence descends behind the outer ankle to gain the external border of the foot, where it divides into cutaneous branches and others to be distributed to the three or four outer toes. In company with this nerve will be seen the posterior saphena vein, L, which, commencing behind the outer ankle, ascends the mesial line of the calf to join the popliteal vein, G, in the cleft between the heads of the gastrocnemius.

On removing next the adipose substance and lymphatic glands, we expose the popliteal vein and artery. The relative position of these vessels and the posterior tibial nerve, may now be seen. Between the heads of the gastrocnemius, the nerve, H, giving off large branches to this muscle, lies upon the popliteal vein, G, where this is joined by the posterior saphena vein. Beneath the veins lies the popliteal artery, F. On tracing the vessels and nerve from this point upwards through the popliteal space, we find the nerve occupying a comparatively superficial position at the mesial line, while the vessels are directed upwards, forwards, and inwards, passing deeply, as they become covered by the inner flexor muscles, C D, to the place where they perforate the tendon of the adductor magnus on the inner side of the lower third of the femur.

The popliteal artery, F, Plate 66, being the continuation of the femoral, extends from the opening in the great adductor tendon at the junction of the middle and lower third of the thigh, to the point where it divides, in the upper, and back part of the leg, at the lower border of the popliteus muscle, L, into the anterior and posterior tibial branches. In order to expose the vessel through this extent, we have to divide and reflect the heads of the gastrocnemius muscle, E E, and to retract the inner flexors. The popliteal artery will now be seen lying obliquely over the middle of the back of the joint. It is deeply placed in its whole course. Its upper and lower thirds are covered by large muscles; whilst the fascia and a quantity of adipose tissue overlies its middle. The upper part of the artery rests upon the femur, its middle part upon the posterior ligament of the joint, and its lower part upon the popliteus muscle. The popliteal vein, G; adheres to the artery in its whole course, being situated on its outer side above, and posterior to it below. The vein is not unfrequently found to be double; one vein lying to either side of the artery, and both having branches of communication with each other, which cross behind the artery. In some instances the posterior saphena vein, instead of joining the popliteal vein, ascends superficially to terminate in some of the large veins of the thigh. Numerous lymphatic vessels accompany the superficial and deep veins into the popliteal space, where they join the lymphatic bodies, which here lie in the course of the artery.

The branches derived from the popliteal artery are the muscular and the articular. The former spring from the vessel opposite those parts of the several muscles which lie in contact with it; the latter are generally five in number--two superior, two inferior, and one median. The two superior articular branches arise from either side of the artery, and pass, the one beneath the outer, the other beneath the inner flexors, above the knee-joint; and the two inferior pass off from it, the one internally, the other externally, beneath the heads of the gastrocnemius below the joint; while the middle articular enters the joint through the posterior ligament. The two superior and inferior articular branches anastomose freely around the knee behind, laterally, and in front, where they are joined by the terminal branches of the anastomotic, from the femoral, and by those of the recurrent, from the anterior tibial. The main vessel, having arrived at the lower border of the popliteus muscle, divides into two branches, of which one passes through the interosseous ligament to become the anterior tibial; while the other, after descending a short way between the bones of the leg, separates into the peronaeal and posterior tibial arteries. In some rare instances the popliteal artery is found to divide above the popliteus muscle into the anterior, or the posterior tibial, or the peronaeal.

The two large muscles, (gastrocnemius and soleus,) forming the calf of the leg, have to be removed together with the deep fascia in order to expose the posterior tibial, and peronaeal vessels and nerves. The fascia forms a sheath for the vessels, and binds them close to the deep layer of muscles in their whole course down the back of the leg. The point at which the main artery, F, Plate 66, gives off the anterior tibial, is at the lower border of the popliteus muscle, on a level with N, the neck of the fibula; that at which the artery again subdivides into the peronaeal, P, and posterior tibial branches, O, is in the mesial line of the leg, and generally on a level with the junction of its upper and middle thirds. From this place the two arteries diverge in their descent; the peronaeal being directed along the inner border of the fibula towards the back of the outer ankle; while the posterior tibial, approaching the inner side of the tibia, courses towards the back of the inner ankle. The gastrocnemius and soleus muscles overlie both arteries in their upper two thirds; but as these muscles taper towards the mesial line where they end in the tendo Achillis, V V, Plate 65, they leave the posterior tibial artery, O, with its accompanying nerve and vein, uncovered in the lower part of the leg, except by the skin and the superficial and deep layers of fasciae. The peronaeal artery is deeply situated in its whole course. Soon after its origin, it passes under cover of the flexor longus pollicis, R, a muscle of large size arising from the lower three fourths of the fibula, N, and will be found overlapped by this muscle on the outer border of the tendo Achillis, as low down as the outer ankle. The two arteries are accompanied by venae comites, which, with the short saphena vein, form the popliteal vein. The posterior tibial artery is closely followed by the posterior tibial nerve. In the popliteal space, this nerve crosses to the inner side of the posterior tibial artery, where both are about to pass under the gastrocnemius muscle, to which they give large branches. Near the middle of the leg, the nerve recrosses the artery to its outer side and in this relative position both descend to a point about midway between the inner ankle and calcaneum, where they appear having the tendons of the tibialis posticus and flexor longus digitorum to their inner side and the tendon of the flexor longus pollicis on their outer side. Numerous branches are given off from the nerve and artery to the neighbouring parts in their course.

The varieties of the posterior crural arteries are these--the tibial vessel, in some instances, is larger than usual, while the peronaeal is small, or absent; and, in others, the peronaeal supplies the place of the posterior tibial, when the latter is diminished in size. The peronaeal has been known to take the position of the posterior tibial in the lower part of the leg, and to supply the plantar arteries. In whatever condition the two vessels may be found, there will always be seen ramifying around the ankle-joint, articular branches, which anastomose freely with each other and with those of the anterior tibial.

The popliteal artery is unfavourably circumstanced for the application of a ligature. It is very deeply situated, and the vein adheres closely to its posterior surface. Numerous branches (articular and muscular) arise from it at short intervals; and these, besides being a source of disturbance to a ligature, are liable to be injured in the operation, in which case the collateral circulation cannot be maintained after the main vessel is tied. There is a danger, too, of injuring the middle branch of the sciatic nerve, in the incisions required to reach the artery; and, lastly, there is a possibility of this vessel dividing higher up than usual. Considering these facts in reference to those cases in which it might be supposed necessary to tie the popliteal artery--such cases, for example, as aneurism of either of the crural arteries, or secondary haemorrhages occurring after amputations of the leg at a time when the healing process was far advanced and the bleeding vessels inaccessible,--it becomes a question whether it would not be preferable to tie the femoral, rather than the popliteal artery. But when the popliteal artery itself becomes affected with aneurism, and when, in addition to the anatomical circumstances which forbid the application of a ligature to this vessel, we consider those which are pathological,--such as the coats of the artery being here diseased, the relative position of the neighbouring parts being disturbed by the tumour, and the large irregular wound which would be required to isolate the disease, at the risk of danger to the health from profuse suppuration, to the limb from destruction of the collateral branches, or to the joint from cicatrization, rendering it permanently bent,--we must acknowledge at once the necessity for tying the femoral part of the main vessel.

When the popliteal artery happens to be divided in a wound, it will be required to expose its bleeding orifices, and tie both these in the wound. For this purpose, the following operation usually recommended for reaching the vessel may be necessary. The skin and fascia lata are to be incised in a direction corresponding to that of the vessel. The extent of the incision must be considerable, (about three inches,) so as the more conveniently to expose the artery in its deep situation. On laying bare the outer margin of the semi-membranosus muscle, while the knee is straight, it now becomes necessary to flex the joint, in order that this muscle may admit of being pressed inwards from over the vessel. The external margin of the wound, including the middle branch of the sciatic nerve, should be retracted outwards, so as to ensure the safety of that nerve, while room is gained for making the deeper incisions. The adipose substance, which is here generally abundant, should now be divided, between the mesial line and the semimembranosus, till the sheath of the vessels be exposed. The sheath should be incised at its inner side, to avoid wounding the popliteal vein. The pulsation of the artery will now indicate its exact position. As the vein adheres firmly to the coats of the artery, some care is required to separate the two vessels, so as to pass the ligature around each end of the artery from without inwards, while excluding the vein. While this operation is being performed in a case of wound of the popliteal artery, the haemorrhage may be arrested by compressing the femoral vessel, either against the femur or the os pubis.

In the operation for tying the posterior tibial artery near its middle, an incision of three or four inches in extent is to be made through the skin and fascia, in a line corresponding with the inner posterior margin of the tibia and the great muscles of the calf. The long saphena vein should be here avoided. The origins of the gastrocnemius and soleus muscles require to be detached from the tibia, and then the knee is to be flexed and the foot extended, so as to allow these muscles to be retracted from the plane of the vessels. This being done, the deep fascia which covers the artery and its accompanying nerve is next to be divided. The artery will now appear pulsating at a situation an inch from the edge of the tibia. While the ligature is being passed around the artery, due care should be taken to exclude the venae comites and the nerve.

DESCRIPTION OF PLATES 65 & 66.

PLATE 65.

A. Tendon of the gracilis muscle.

B B. The fascia lata.

C C. Tendon of the semimembranosus muscle.

D. Tendon of the semitendinosus muscle.

E E. The two heads of the gastrocnemius muscle.

F. The popliteal artery.

G. The popliteal vein joined by the short saphena vein.

H. The middle branch of the sciatic nerve.

I. The outer (peronaeal) branch of the sciatic nerve.

K. The posterior tibial nerve continued from the middle branch of the sciatic, and extending to K, behind the inner ankle.

L. The posterior (short) saphena vein.

M M. The fascia covering the gastrocnemius muscle.

N. The short (posterior) saphena nerve, formed by the union of branches from the peronaeal and posterior tibial nerves.

O. The posterior tibial artery appearing from beneath the soleus muscle in the lower part of the leg.

P. The soleus muscle joining the tendo Achillis.

Q. The tendon of the flexor longus communis digitorum muscle.

R. The tendon of the flexor longus pollicis muscle.

S. The tendon of the peronaeus longus muscle.

T. The peronaeus brevis muscle.

U U. The internal annular ligament binding down the vessels, nerves, and tendons in the hollow behind the inner ankle.

V V. The tendo Achillis.

W. The tendon of the tibialis posticus muscle.

X. The venae comites of the posterior tibial artery.

PLATE 66.

A C D E F G H I indicate the same parts as in Plate 65.

B. The inner condyle of the femur.

K. The plantaris muscle lying upon the popliteal artery.

L. The popliteus muscle.

M M M. The tibia.

N N. The fibula.

O O. The posterior tibial artery.

P. The peronaeal artery.

Q R S T U V W. The parts shown in Plate 65.

X. The astragalus.

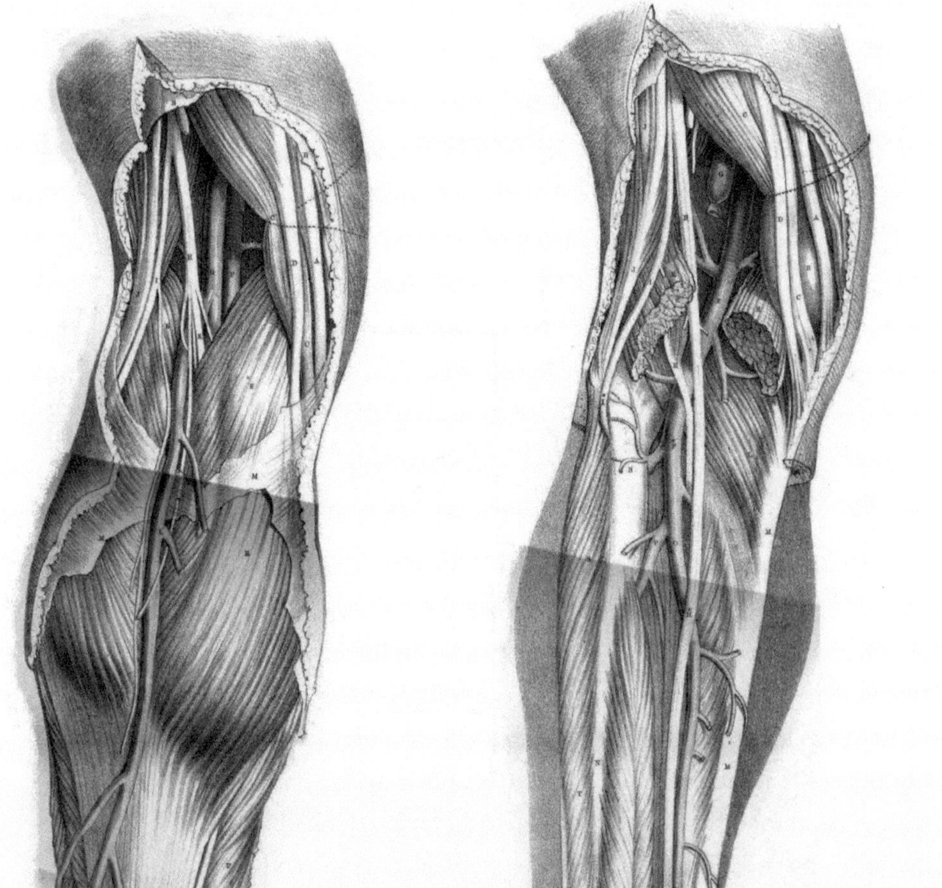

Plates 65, 66

THE SURGICAL DISSECTION OF THE ANTERIOR CRURAL REGION; THE ANKLES AND THE FOOT.

Varieties of the anterior and posterior tibial and the peronaeal arteries. The operations for tying these vessels in several situations. Practical observations on wounds of the arteries of the leg and foot.

Beneath the integuments and subcutaneous adipose tissue on the fore part of the leg and foot, the fascia H H, Plate 67, Figure 2, is to be seen stretched over the muscles and sending processes between them, thus encasing each of these in a special sheath.

The fascia is here of considerable density. It is attached on the inner side of the leg to the spine of the tibia, D, Plate 67, Figure 2, and on the outer side it passes over the peronaeal muscles to those forming the calf. Between the extensor communis digitorum, B *b*, and the peronaeus longus, F, it sends in a strong process to be attached to the fibula, E. In front of the ankle joint, the fascia is increased in density, constituting a band (anterior annular ligament) which extends between the malleoli, forms sheaths for the several extensor tendons, and binds these down in front of the joint. From the lower border of the annular ligament, the fascia is continued over the dorsum of the foot, forming sheaths for the tendons and muscles of this part. Behind the inner malleolus, *d*, Plate 67, Figure 1, the fascia attached to this process and to the inner side of the os calcis appears as the internal annular ligament, which being broad and strong, forms a kind of arch, beneath which in special sheaths the flexor tendons, and the posterior tibial vessel and nerve, pass to the sole of the foot. On tracing the fascia from the front to the back of the leg, it will be seen to divide into two layers--superficial and deep; the former passes over the muscles of the calf and their common tendon (tendo Achillis) to which it adheres, while the latter passes between these muscles and the deep flexors. The deep layer is that which immediately overlies the posterior tibial and peronaeal vessels and nerves. While exposing the fascia on the forepart of the leg and dorsum of the foot, we meet with the musculo-cutaneous branch of the peronaeal nerve, which pierces the fascia at about the middle of the limb, and descends superficially in a direction between the fibula, and the extensor longus digitorum muscle, and after dividing into branches a little above the outer ankle, these traverse in two groups the dorsum of the foot, to be distributed to the integuments of the five toes. On the inner side of the tibia, D, Plate 67, Figure 1, will be seen the internal or long saphena vein, B B, which commencing by numerous branches on the dorsal surface of the foot ascends in front of the inner ankle, *d*, to gain the inner side of the leg, after which it ascends behind the inner side of the knee and thigh, till it terminates at the saphenous opening, where it joins the femoral vein. In its course along the lower part of the thigh, the leg and the foot, this vein is closely accompanied by the long saphenous nerve, derived from the anterior crural, and also by a group of lymphatics.

By removing the fascia from the front of the leg and foot, we expose the several muscles and tendons which are situated in these parts. In the upper part of the leg the tibialis anticus, A, Plate 67, Figure 2, and extensor-communis muscle, B, are adherent to the fascia which covers them, and to the intermuscular septum which divides them. In the lower part of the leg where these muscles and the extensor pollicis, C, terminate in tendons, *a b c,* they are readily separable from one another. The tibialis anticus lies along the outer side of the tibia, from which, and from the head of the fibula and interosseous ligament, it arises tendinous and fleshy. This muscle is superficial in its whole length; its tendon commencing about the middle of the leg, passes in a separate loose sheath of the annular ligament in front of the inner ankle, to be inserted into the inner side of the cuneiform bone and base of the metatarsal bone of the great toe. The extensor communis digitorum lies close to the outer side of the anterior tibial muscle, and arises from the upper three-fourths of the fibula, from the interosseous ligament and intermuscular septum. At the lower part of the leg, this muscle ends in three or four flat tendons, which pass through a ring of the annular ligament, and extending forwards, *b b b b*, over the dorsum of the foot, become inserted into the four outer toes. The peronaeus tertius or anterior, is that part of the common extensor muscle which is inserted into the base of the fifth metatarsal bone. On separating the anterior tibial and common extensor muscles, we find the extensor pollicis, C *c*, which, concealed between the two, arises from the middle of the fibula, and the interosseous ligament; its tendon passes beneath the annular ligament in front of the ankle joint, and after traversing the inner part of the dorsum of the foot, becomes inserted into the three phalanges of the great toe. Beneath the tendons of the extensor communis on the instep, will be seen the extensor digitorum brevis, K K, lying in an oblique direction, between the upper and outer part of the os calcis, from which it arises, and the four inner toes, into each of which it is inserted by a small flat tendon, which joins the corresponding tendon of the long common extensor.

The anterior tibial artery, L, Plate 67, Figure 2, extends from the upper part of the interosseous ligament which it perforates, to the bend of the ankle, whence it is continued over the dorsum of the foot. In the upper third of the leg, the anterior tibial artery lies deeply situated between the tibialis anticus, and flexor communis muscles. Here it will be found, close in front of the interosseous ligament, at about an inch and-a-half in depth from the anterior surface, and removed from the spine of the tibia at an interval equal to the width of the tibialis anticus muscle. In its course down the leg, the vessel passes obliquely from a point close to the inner side of the neck of the fibula, to midway between the ankles. In its descent, it becomes gradually more superficial. In the middle of the leg, the vessel passes between the extensor longus pollicis, and the tibialis anticus muscles. Above, beneath, and below the annular ligament, this artery will be found to pass midway between the extensor pollicis tendon, and those of the extensor communis, and to hold the same relation to these parts in traversing the dorsum of the foot, till it gains the interval between the two inner metatarsal bones, where it divides into two branches, one of which passes forwards in the first interdigital space, while the other sinks between the bones, to inosculate with the plantar arteries. The innermost tendon of the short common extensor crosses in front of the dorsal artery of the foot near its termination. Between the ankle and the first interosseous space the artery lies comparatively superficial, being here covered only by the skin and fascia and cellular membrane. Two veins accompany the anterior tibial artery and its continuation on the dorsum of the foot. The anterior tibial nerve, a branch of the peronaeal, joins the outer side of the artery, about the middle of the leg, and accompanies it closely in this position, till both have passed beneath the annular ligament. On the dorsum of the foot the nerve will be found to the inner side of the artery.

The branches of the anterior tibial artery are articular and muscular. From its upper end arises the recurrent branch which anastomoses in front of the knee with the articular branches of the popliteal artery. Near the ankle, arise on either side of the vessel two malleolar branches, internal and external, the former communicating with branches of the posterior tibial, the latter with those of the peronaeal. Numerous muscular branches arise, at short intervals, from the vessel in its passage down the leg. Tarsal, metatarsal, and small digital branches spring from the dorsal artery of the foot. The anterior tibial artery is rarely found to deviate from its usual course; in some cases it appears of less or of greater size than usual. When this vessel appears deficient, its place is usually supplied by some branch of the peronaeal or posterior tibial, which pierces the interosseous ligament from behind.

The anterior tibial artery when requiring a ligature to be applied to it in any part of its course, may be exposed by an incision, extending for three or four inches, (more or less, according to the depth of the vessel) along the outer border of the tibialis anticus muscle. The fibrous septum between this muscle and the extensor communis, will serve as a guide to the vessel in the upper third of the leg, where it lies deeply on the interosseous ligament. In the middle of the leg, the vessel is to be sought for between the anterior tibial and extensor longus pollicis muscles. In the lower part of the leg, and on the dorsum of the foot, it will be found between the extensor longus pollicis, and extensor communis tendons, the former being taken as a guide for the incision. In passing the ligature around this vessel at either of these situations, care is required to avoid including the venae comites and the accompanying nerve.

The sole of the foot is covered by a hard and thick integument, beneath which will be seen a large quantity of granulated adipose tissue so intersected by bands of fibrous structure as to form a firm, but elastic cushion, in the situations particularly of the heel and joints of the toes. On removing this structure, we expose the plantar fascia, B, Plate 68, Figure 1, extending from the os calcis, A, to the toes. This fascia is remarkably strong, especially its middle and outer parts, which serve to retain the arched form of the foot, and thereby to protect the plantar structures from superincumbent pressure during the erect posture. The superficial plantar muscles become exposed on removing the plantar fascia, to which they adhere. In the centre will be seen the thick fleshy flexor digitorum brevis muscle, B, arising from the inferior part of the os calcis, and passing forwards to divide into four small tendons, *b b b b,* for the four outer toes. On the inner side of the foot appears the abductor pollicis, D, arising from the inner side of the os calcis and internal annular ligament, and passing to be inserted with the flexor pollicis brevis, H, into the sesamoid bones and base of the first phalanx of the great toe. On the external border of the foot is situated the abductor minimi digiti, C, arising from the outer side of the os calcis, and passing to be inserted with the flexor brevis minimi digiti into the base of the first phalanx of the little toe. When the flexor brevis digitorum muscle is removed, the plantar arteries, L M, and nerves, are brought partially into view; and by further dividing the abductor pollicis, D, their continuity with the posterior tibial artery and nerves, K L, Plate 67, Figure 1, behind the inner ankle may be seen.

The plantar branches of the posterior tibial artery are the internal and external, both of which are deeply placed between the superficial and deep plantar muscles. The internal plantar artery is much the smaller of the two. The external plantar artery, L, Plate 68, Figure 1, is large, and seems to be the proper continuation of the posterior tibial. It corresponds, in the foot, to the deep palmar arch in the hand. Placed at first between the origin of the abductor pollicis and the calcaneum, the external plantar artery passes outwards between the short common flexor, B, and the flexor accessorius, E, to gain the inner borders of the muscles of the little toe; from this place it curves deeply inwards between the tendons of the long common flexor of the toes, F *f f,* and the tarso-metatarsal joints, to gain the outer side of the first metatarsal bone, H, Plate 68, Figure 2. In this course it is covered in its posterior half by the flexor brevis digitorum, and in its anterior half by this muscle, together with the tendons of the long flexor, F, Plate 68, Figure 1, of the toes and the lumbricales muscles, *i i i i.* From the external plantar artery are derived the principal branches for supplying the structures in the sole of the foot. The internal plantar nerve divides into four branches, for the supply of the four inner toes, to which they pass between the superficial and deep flexors. The external plantar nerve, passing along the inner side of the corresponding artery, sends branches to supply the outer toe and adjacent side of the next, and then passes, with the artery, between the deep common flexor tendon and the metatarsus, to be distributed to the deep plantar muscles.

The posterior tibial artery may be tied behind the inner ankle, on being laid bare in the following way:--A curved incision (the concavity forwards) of two inches in length, is to be made midway between the tendo Achillis and the ankle. The skin and superficial fascia having been divided, we expose the inner annular ligament, which will be found enclosing the vessels and nerve in a canal distinct from that of the tendons. Their fibrous sheath having been slit open, the artery will be seen between the venae comites, and with the nerve, in general, behind it.

When any of the arteries of the leg or the foot are wounded, and the haemorrhage cannot be commanded by compression, it will be necessary to search for the divided ends of the vessel in the wound, and to apply a ligature to both. The expediency of this measure must become fully apparent when we consider the frequent anastomoses existing between the collateral branches of the crural arteries, and that a ligature applied to *any one* of these above the seat of injury will not arrest the recurrent circulation through the vessels of the foot.

DESCRIPTION OF PLATES 67 & 68.

PLATE 67.

FIGURE 1.

A. The tendon of the tibialis anticus muscle.

B B. The long saphena vein.

C C. The tendon of the tibialis posticus muscle.

D. The tibia; *d*, the inner malleolus.

E E. The tendon of the flexor longus digitorum muscle.

F. The gastrocnemius muscle; *f*, the tendo Achillis.

G. The soleus muscle.

H. The tendon of the plantaris muscle.

I I. The venae comites.

K K. The posterior tibial artery.

L L. The posterior tibial nerve.

FIGURE 2.

A. The tibialis anticus muscle; *a*, its tendon.

B. The extensor longus digitorum muscle; *b b b b*, its four tendons.

C C. The extensor longus pollicis muscle.

D D. The tibia.

E. The fibula; *e*, the outer malleolus.

F F. The tendon of the peronaeus longus muscle.

G G. The peronaeus brevis muscle; i, the peronaeus tertius.

H H. The fascia.

K. The extensor brevis digitorum muscle; *k k*, its tendons.

L L. The anterior tibial artery and nerve descending to the dorsum of the foot.

PL. 67.

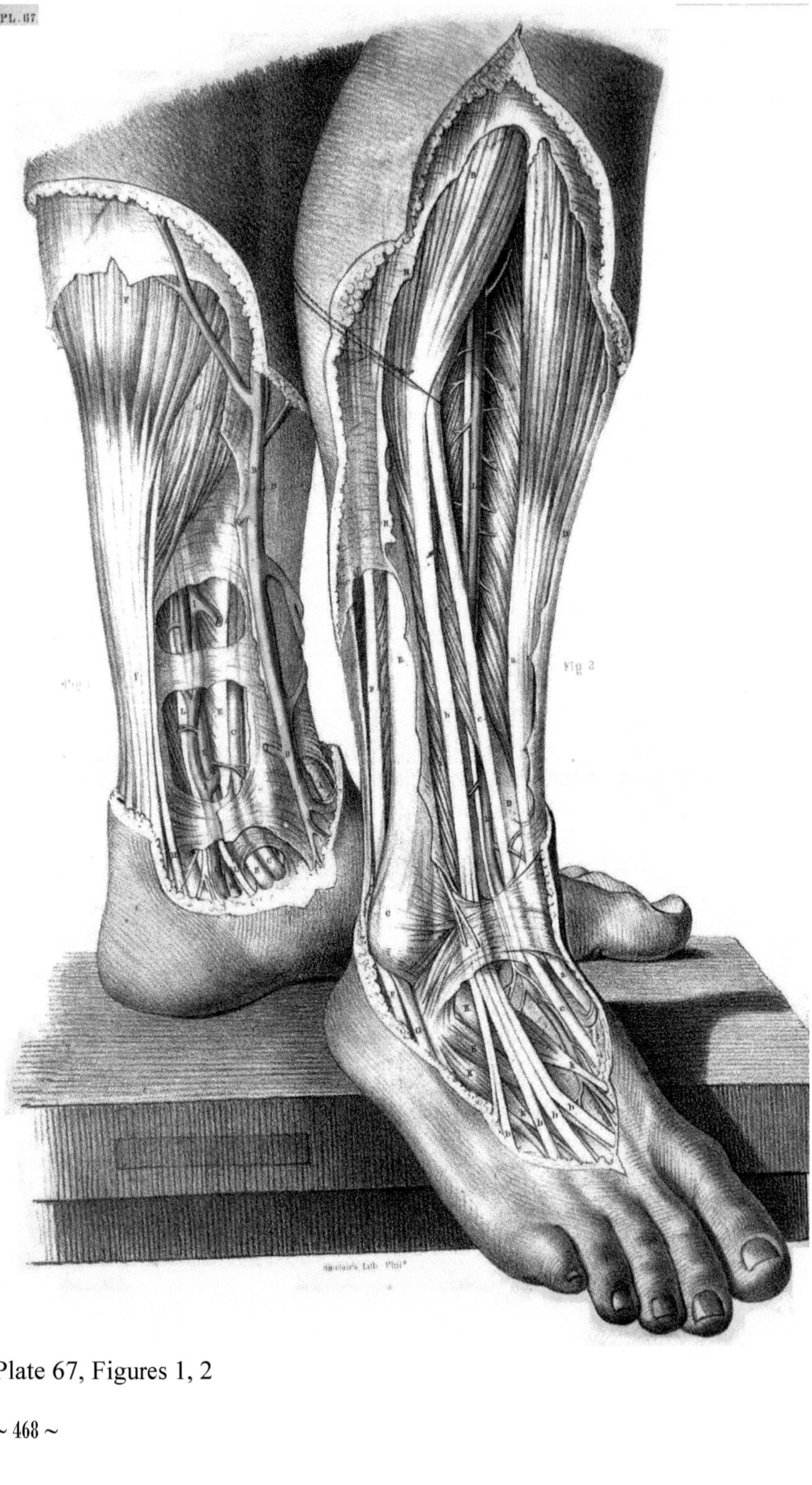

Plate 67, Figures 1, 2

PLATE 68.

FIGURE 1.

A. The calcaneum.

B. The plantar fascia and flexor brevis digitorum muscle cut; *b b b*, its tendons.

C. The abductor minimi digiti muscle.

D. The abductor pollicis muscle.

E. The flexor accessorius muscle.

F. The tendon of the flexor longus digitorum muscle, subdividing into *f f f f*, tendons for the four outer toes.

G. The tendon of the flexor pollicis longus muscle.

H. The flexor pollicis brevis muscle.

i i i i. The four lumbricales muscles.

K. The external plantar nerve.

L. The external plantar artery.

M. The internal plantar nerve and artery.

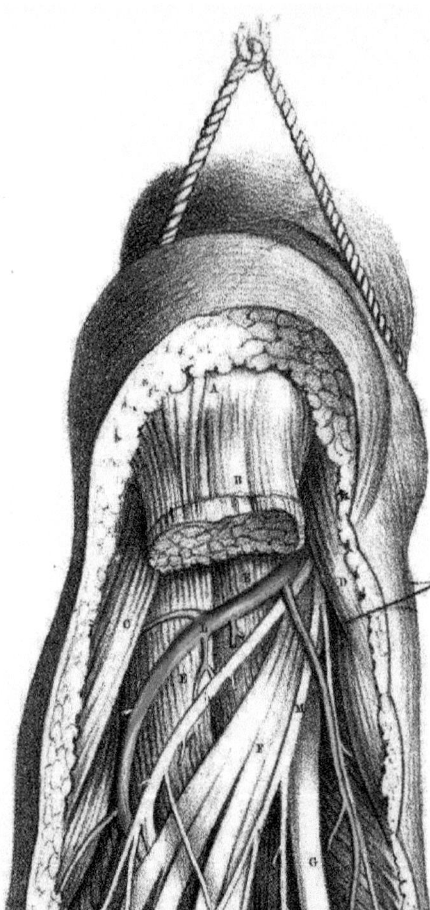

Fig. 1

Plate 68, Figure 1

FIGURE 2.

A. The heel covered by the integument.

B. The plantar fascia and flexor brevis digitorum muscle cut; *b b b*, the tendons of the muscle.

C. The abductor minimi digiti.

D. The abductor pollicis.

E. The flexor accessorius cut.

F. The tendon of the flexor digitorum longus cut; *f f f*, its digital ends.

G. The tendon of the flexor pollicis.

H. The head of the first metatarsal bone.

I. The tendon of the tibialis posticus.

K. The external plantar nerve.

L L. The arch of the external plantar artery.

M M M M. The four interosseous muscles.

N. The external plantar nerve and artery cut.

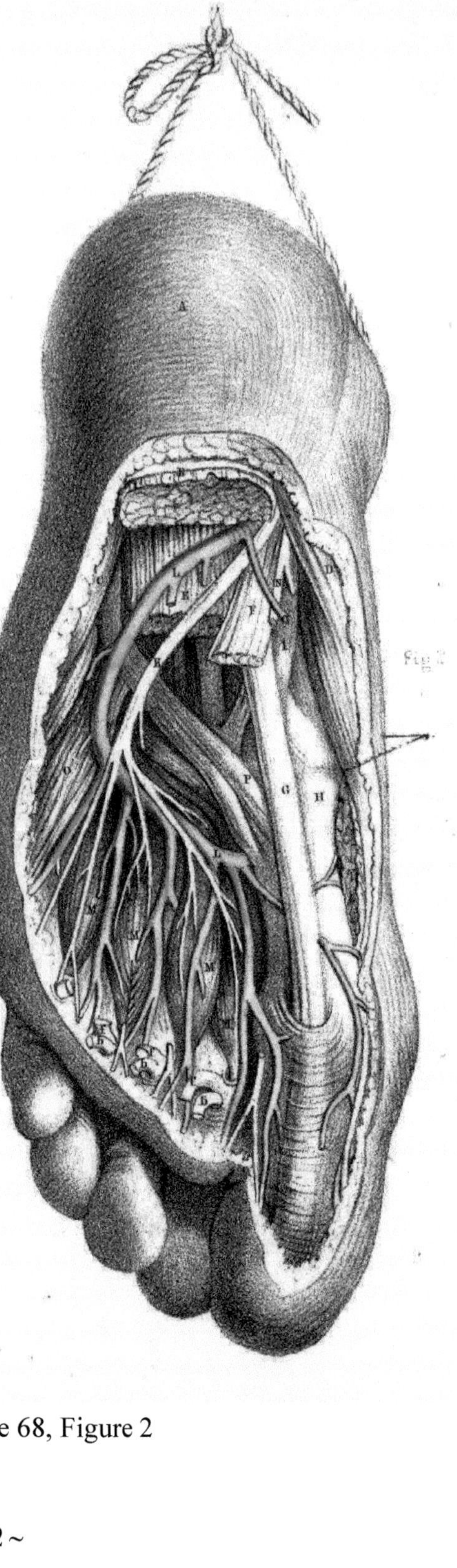

Plate 68, Figure 2

CONCLUDING COMMENTARY:

ON THE FORM AND DISTRIBUTION OF THE VASCULAR SYSTEM AS A WHOLE; ANOMALIES; RAMIFICATION; ANASTOMOSIS.

The double heart. Universal systemic capillary anastomosis. Its division, by the median line, into two great lateral fields—those subdivided into two systems or provinces--viz., pulmonary and systemic. Relation of pulmonary and systemic circulating vessels. Motions of the heart. Circulation of the blood through the lungs and system. Symmetry of the hearts and their vessels. Development of the heart and primary vessels. Their stages of metamorphosis simulating the permanent conditions of the parts in lower animals. The primitive branchial arches undergoing metamorphosis. Completion of these changes. Interpretation of the varieties of form in the heart and primary vessels. Signification of their normal condition. The portal system no exception to the law of vascular symmetry. Signification of the portal system. The liver and spleen as homologous organs,--as parts of the same whole quantity. Cardiac anastomosing vessels. Vasa vasorum. Anastomosing branches of the systemic aorta considered in reference to the operations of arresting by ligature the direct circulation through the arteries of the head, neck, upper limbs, pelvis, and lower limbs. The collateral circulation. Practical observations on the most eligible situations for tying each of the principal vessels, as determined by the greatest number of their anastomosing branches on either side of the ligature, and the largest amount of the collateral circulation that may be thereby carried on for the support of distal parts.

I.--The heart, in all stages of its development, is to the vascular system what the point of a circle is to the circumference--namely, at once *the beginning and the end.* The heart, occupying, it may be said, the centre of the thorax, circulates the blood in the same way, by similar channels, to an equal extent, in equal pace, and at the same period of time, through both sides of the body. In its adult normal condition, the heart presents itself as a double or symmetrical organ. The two hearts, though united and appearing single, are nevertheless, as to their respective cavities, absolutely distinct. Each heart consists again of two compartments--an auricle and a ventricle. The two auricles are similar in structure and form. The two ventricles are similar in the same respects. A septum divides the two auricles, and another--the two ventricles. Between the right auricle and ventricle, forming the right heart, there exists a valvular apparatus (tricuspid), by which these two compartments communicate; and a similar valve (bicuspid) admits of communication between the left auricle and ventricle. The two hearts being distinct, and the main vessels arising from each respectively being distinct likewise, it follows that the capillary peripheries of these vessels form the only channels through which the blood issuing from one heart can enter the other.

II.--As the aorta of the left heart ramifies throughout all parts of the body, and as the countless ramifications of this vessel terminate in an equal number of ramifications of the principal veins of the right heart, it will appear that between the systemic vessels of the two hearts respectively, the capillary anastomotic circulation reigns *universal.*

III.--The body generally is marked by the median line, from the vertex to the perinaeum, into corresponding halves. All parts excepting the main bloodvessels in the neighbourhood of the heart are naturally divisible by this line into equals. The vessels of each heart, in being distributed to both sides of the body alike, cross each other at the median line, and hence they are inseparable according to this line, unless by section. If the vessels proper to each heart, right and left, ramified alone within the limits of their respective sides of the body, then their capillary anastomosis could only take place along the median line, and here in such case they might be separated by median section into two distinct systems. But as each system is itself double in branching into both sides of the body, the two would be at the same time equally divided by vertical section. From this it will appear that the vessels belonging to *each* heart form a symmetrical system, corresponding to the sides of the body, and that the capillary anastomosis of these systemic veins and arteries is divisible into *two great fields*, one situated on either side of the median line, and touching at this line.
IV.--The vessels of the right heart do not communicate at their capillary peripheries, for its veins are systemic, and its arteries are

pulmonary. The vessels of the left heart do not anastomose, for its veins are pulmonary, and its arteries are systemic. The arteries of the right and left hearts cannot anastomose, for the former are pulmonary, and the latter are systemic; and neither can the veins of the right and left hearts, for a similar reason. Hence, therefore, there can be, between the vessels of both hearts, but *two provinces of anastomosis*--viz., that of the lungs, and that of the system. In the lungs, the arteries of the right heart and the veins of the left anastomose. In the body generally (not excepting the lungs), the arteries of the left heart, and the veins of the right, anastomose; and thus in the pulmonary and the systemic circulation, each heart plays an equal part through the medium of its proper vessels. The pulmonary bear to the systemic vessels the same relation as a lesser circle contained within a greater; and the vessels of each heart form the half of each circle, the arteries of the one being opposite the veins of the other.

V.--The two hearts being, by the union of their similar forms, as one organ in regard to place, act, by an agreement of their corresponding functions, as one organ in respect to time. The action of the auricles is synchronous; that of the ventricles is the same; that of the auricles and ventricles is consentaneous; and that of the whole heart is rhythmical, or harmonious--the diastole of the auricles occurring in harmonical time with the systole of the ventricles, and *vice versa*. By this correlative action of both hearts, the pulmonary and systemic circulations take place synchronously; and the phenomena resulting in both reciprocate and balance each other. In the pulmonary circulation, the blood is aerated, decarbonized, and otherwise depurated; whilst in the systemic circulation, it is carbonized and otherwise deteriorated.

VI.--The circulation through the lungs and the system is carried on through vessels having the following form and relative position, which, as being most usual, is accounted normal. The two brachio-cephalic veins joining at the root of the neck, and the two common iliac veins joining in front of the lumbar vertebrae, form the superior and inferior venae cavae, by which the blood is returned from the upper and lower parts of the body to the right auricle, and thence it enters the right ventricle, by which it is impelled through the pulmonary artery into the two lungs;

and from these it is returned (aerated) by the pulmonary veins to the left auricle, which passes it into the left ventricle, and by this it is impelled through the systemic aorta, which branches throughout the body in a similar way to the systemic veins, with which the aortic branches anastomose generally. On viewing together the system of vessels proper to each heart, they will be seen to exhibit in respect to the body a figure in doubly symmetrical arrangement, of which the united hearts form a duplex centre. At this centre, which is the theatre of metamorphosis, the principal abnormal conditions of the bloodvessels appear; and in order to find the signification of these, we must retrace the stages of development.

VII.--From the first appearance of an individualized centre in the vascular area of the human embryo, that centre (punctum saliens) and the vessels immediately connected with it, undergo a phaseal metamorphosis, till such time after birth as they assume their permanent character. In each stage of metamorphosis, the embryo heart and vessels typify the normal condition of the organ in one of the lower classes of animals. The several species of the organ in these classes are parallel to the various stages of change in the human organ. In its earliest condition, the human heart presents the form of a simple canal, similar to that of the lower Invertebrata, the veins being connected with its posterior end, while from its anterior end a single artery emanates. The canal next assumes a bent shape, and the vessels of both its ends become thereby approximated. The canal now being folded upon itself in heart-shape, next becomes constricted in situations, marking out the future auricle and ventricle and arterial bulb, which still communicate with each other. From the artery are given off on either side symmetrically five branches (branchial arches), which arch laterally from before, outwards and backwards, and unite in front of the vertebrae, forming the future descending aorta. In this condition, the human heart and vessels resemble the Piscean pipe. The next changes which take place consist in the gradual subdivision, by means of septa, of the auricle and ventricle respectively into two cavities. On the separation of the single auricle into two, while the ventricle as yet remains single, the heart presents that condition which is proper to the Reptilian class. The interauricular and interventricular septa, by gradual development from without inwards, at length meet and coalesce, thereby dividing the two cavities into four--two auricles and two ventricles--a condition proper to the Avian and Mammalian classes generally. In the centre of the interauricular septum of the human heart, an aperture (*foramen ovale*) is left as being necessary to the foetal circulation. While the septa are being completed, the arterial bulb also becomes divided by a partition formed in its interior in such a manner as to adjust the two resulting arteries, the one in connexion with the right, the other with the left ventricle. The right ventricular artery (pulmonary aorta) so formed, has assigned to it the fifth (posterior) opposite pair of arches, and of these the right one remaining pervious to the point where it gives off the right pulmonary branch, becomes obliterated beyond this point to that where it joins the descending aorta, while the left arch remains pervious during foetal life, as the *ductus arteriosus* still communicating with the descending aorta, and giving off at its middle the left pulmonary branch. The left ventricular artery (systemic aorta) is formed of the fourth arch of the left side, while the opposite arch (fourth right) is altogether obliterated. The third and second arches remain pervious on both sides, afterwards to become the right and left brachio-cephalic arteries. The first pair of arches, if not converted into the vertebral arteries, or the thyroid axes, are altogether metamorphosed. By these changes the heart and primary arteries assume the character in which they usually present themselves at birth, and in all probability the primary veins corresponded in form, number, and distribution with the arterial vessels, and underwent, at the same time, a similar mode of metamorphosis. One point in respect to the original symmetrical character of the primary veins is demonstrable--namely, that in front of the aortic branches the right and left brachio-cephalic veins, after joining by a cross branch, descend separately on either side of the heart, and enter (as two superior venae cavae) the right auricle by distinct orifices. In some of the lower animals, this double condition of the superior veins is constant, but in the human species the left vein below the cross branch (left brachio-cephalic) becomes obliterated, whilst the right vein (vena cava superior) receives the two brachio-cephalic veins, and in this condition remains throughout life. After birth, on the commencement of respiration, the *foramen ovale* of the interauricular septum closes, and the *ductus arteriosus* becomes impervious. This completes the stages of metamorphosis, and changes the course of the simple foetal circulation to one of a more complex order--viz., the systemic-pulmonary characteristic of the normal state in the adult body.

VIII.--Such being the phases of metamorphosis of the primary (branchial) arches which yield the vessels in their normal adult condition, we obtain in this history an explanation of the signification not only of such of their anomalies as are on record, but of such also as are *potential* in the law of development; a few of them will suffice to illustrate the meaning of the whole number:--l*st*, The interventricular as well as the interauricular septum may be arrested in growth, leaving an aperture in the centre of each; the former condition is *natural to the human foetus*, the latter *to the reptilian class,* while both would be *abnormal in the human adult. 2nd*. The heart may be *cleft at its apex* in the situation of the interventricular septum--a condition *natural to the Dugong*, A similar cleavage may divide the *base of the heart* in the situation of the interauricular septum. 3*rd*. The *partitioning of the bulbus arteriosus* may occur in such a manner as to assign to the two aortae a relative position, the *reverse* of that which they *normally* occupy--the *pulmonary aorta* springing from the *left ventricle* and the *systemic aorta* arising from the *right,* and giving off from its arch the *primary branches* in the usual order. [Footnote 1] 4*th*. As the *two aortae* result from a *division* of the *common primary vessel (bulbus arteriosus)*, an *arrest* in the growth of the partition would leave them still as *one vessel*, which (supposing the ventricular septum remained also incomplete) would then arise from a *single ventricle*. 5*th.* The *ductus arteriosus* may remain *pervious*, and while co-existing with the proper *aortic arch, two arches* would then appear on the *left side*. 6*th*. The *systemic normal aortic arch* may be obliterated as far up as the *innominate branch*, and while the *ductus arteriosus* remains *pervious,* and leading from the pulmonary artery to the descending part of the aortic arch, this vessel would then present the appearance of *a branch* ascending from the left side and giving off the brachio-cephalic arteries. The *right ventricular artery* would then, through the medium of the *ductus arteriosus*, supply both the lungs and the system. Such a state of the vessels would require (in order that the circulation of a mixed blood might be carried on) that the two ventricles freely communicate. 7*th*. If the *fourth arch* of the *right side* remained pervious opposite the *proper aortic arch*, there would exist *two aortic arches* placed *symmetrically*, one on either side of the vertebral column, and, joining below, would *include in their circle* the trachea and oesophagus. 8th. If the *fifth arch* of the *right side* remained *pervious* opposite the *open ductus arteriosus*, both vessels would present a similar arrangement, as *two symmetrical ducti arteriosi* co-existing with symmetrical aortic arches. 9*th*. If the vessels appeared *co-existing* in the *two conditions* last mentioned, they would represent *four aortic arches, two on either side of the vertebral column.* 10*th*. If the *fourth right arch, instead of the fourth left* (aorta), remained *pervious*, the *systemic aortic arch* would then be turned to the *right side* of the vertebral column, and have the trachea and oesophagus on its *left*. 11*th*. When the *bulbus arteriosus* divides itself into *three parts*, the *two lateral parts*, in becoming connected with the *left ventricle*, will represent a *double ascending systemic aorta*, and having the *pulmonary artery* passing *between them* to the lungs. 12*th*. When of the *two original superior venae cavae* the *right one* instead of the left suffers metamorphosis, the *vena cava superior* will then appear on the *left side* of the *normal aortic arch.* [Footnote 2] Of these malformations, some are rather frequently met with, others very seldom, and others cannot exist compatible with life after birth. Those which involve a more or less imperfect discharge of the blood-aerating functions of the lungs, are in those degrees more or less fatal, and thus nature aborting as to the fitness of her creation, cancels it.

[Footnote 1: This physiological truth has, I find, been applied by Dr. R. Quain to the explanation of a numerous class of malformations connected with the origins of the great vessels from the heart, and of their primary branches. See *The Lancet*, vol. I. 1842.]

[Footnote 2: For an analysis of the occasional peculiarities of these primary veins in the human subject, see an able and original monograph in the *Philosophical Transactions,* Part 1., 1850, entitled, "On the Development of the Great Anterior Veins in Man and Mammalia." By John Marshall, F.R.C.S., &c.]

IX.--The *portal system of veins* passing *to* the liver, and the hepatic veins passing *from* this organ to join the inferior vena cava, exhibit in respect to the median line of the body an example of a-symmetry, since appearing on the right side, they have no counterparts on the left. As the law of symmetry seems to prevail universally in the development of organized beings, forasmuch as every lateral organ or part has its counterpart, while every central organ is double or complete, in having two similar sides, then the portal system, as being an exception to this law, is as a natural note of interrogation questioning the signification of that fact, and in the following observations, it appears to me, the answer may be found. Every artery in the body has its companion vein or veins. The inferior vena cava passes sidelong with the aorta in the abdomen. Every branch of the aorta which ramifies upon the abdominal parietes has its accompanying vein returning either to the vena cava or the vena azygos, and entering either of these vessels at a point on the same level as that at which itself arises. The renal vessels also have this arrangement. But all the other veins of the abdominal viscera, instead of entering the vena cava opposite their corresponding arteries, unite into a single trunk (vena portae), which enters the liver. The special purpose of this destination of the portal system is obvious, but the function of a part gives no explanation of its form or relative position, whether singular or otherwise. On viewing the vessels in presence of the general law of symmetrical development, it occurs to me that the *portal* and *hepatic veins form one continuous system*, which taken in *the totality*, represents the *companion veins of the arteries of the abdominal viscera.* The liver under this interpretation appears as a gland *developed midway* upon these veins, and *dismembering them* into a mesh of countless capillary vessels, (a condition necessary for all processes of secretion,) for the special purpose of decarbonizing the blood. In this great function the liver is an organ correlative or compensative to the lungs, whose office is similar. The secretion of the liver (bile) is fluidform; that of the lungs is aeriform. The bile being necessary to the digestive process, the liver has a duct to convey that product of its secretion to the intestines. The trachea is as it were the duct of the lungs. In the liver, then, *the portal* and *hepatic veins* being continuous *as veins,* the two systems, notwithstanding their apparent distinctness, caused by the intervention of the hepatic lobules, may be regarded as the *veins corresponding with the arteries of the coeliac axis, and the two mesenteric.* The hepatic artery and the hepatic veins evidently do *not* pair in the sense of *afferent* and *efferent,* with respect to the liver, both these vessels having destinations as different as those of the bronchial artery and the pulmonary veins in the lungs. The bronchial artery is attended by its vein proper, while the vein which corresponds to the hepatic artery joins either the hepatic or portal veins traversing the liver, and in this position escapes notice.[Footnote]

[Footnote: In instancing these facts, as serving under comparison to explain how the hepatic vessels constitute no *radical* exception to the law of symmetry which presides over the development and distribution of the vascular system as a whole, I am led to inquire in what respect (if in any) the liver as an organ forms an exception to this general law either in shape, in function, or in relative position. While seeing that every central organ is single and symmetrical by the union of two absolutely similar sides, and that each lateral pair of organs is double by the disunion of sides so similar to each other in all respects that the description of either side serves for the other opposite, it has long since seemed to me a reasonable inference that, since the liver on the right has no counterpart *as a liver* on the left, and that, since the spleen on the left has no counterpart *as a spleen* on the right, so these two organs (the liver and spleen) must themselves correspond to each other, and as such, express their respective significations. Under the belief that every exception (even though it be normal) to a general law or rule, is, like the anomaly itself, alone explicable according to such law, and expressing a fact not more singular or isolated from other parallel facts than is one form from another, or from all others constituting the graduated scale of being, I would, according to the light of this evidence alone, have no hesitation in stating that the liver and spleen, as opposites, represent corresponding organs, even though they appeared at first view more dissimilar than they really are. In support of this analogy of both organs, which is here, so far as I am aware, originally enunciated for anatomical science, I record the following observations:--1*st.* Between the opposite parts of the *same* organic entity (between the opposite leaves of the same plant, for example), nature manifests no such absolute difference in any case as exists between the leaf of a plant and of a book. 2*ndly.* When between two opposite parts of the *same* organic form there appears any differential character, this is simply the result of a modification or metamorphosis of one of the two perfectly similar originals or archetypes, but never carried out to such an extreme degree as to annihilate all trace of their analogy. 3*rdly.* The liver and the spleen are opposite parts; and as such, they are associated by arteries which arise by a single trunk (coeliac axis) from the aorta, and branch right and left, like indices pointing to the relationship between both these organs, in the same manner as the two emulgent arteries point to the opposite renal organs. 4*thly*. The liver is divided into two lobes, right and left; the left is less than the right; that quantity which is wanting to the left lobe is equal to the quantity of a spleen; and if in idea we add the spleen to the left lobe of the liver, both lobes of this organ become quantitatively equal, and the whole liver symmetrical; hence, as the liver *plus* the spleen represents the whole structural quantity, so the liver *minus* the spleen signifies that the two organs now dissevered still relate to each other as parts of the same whole. 5*thly*. The liver, as being *three-fourths* of the whole, possesses the *duct* which emanates at the centre of all glandular bodies. The spleen, as being *one-fourth* of the whole, is *devoid* of the duct. The liver having the duct, is functional as a gland, while the spleen having no duct, cannot serve any such function. If, in thus indicating the function which the spleen does *not* possess, there appears no proof positive of the function which it *does,* perhaps the truth is, that as being the ductless portion of the whole original hepatic quantity, it exists as a thing degenerate and functionless, for it seems that the animal economy suffers no loss of function when deprived of it. 6*thly.* In early foetal life, the left lobe of the liver *touches* the spleen on the left side; but in the process of abdominal development, the two organs become separated from each other right and left. 7*thly.* In animals devoid of the spleen, the liver appears of a *symmetrical* shape, both its lobes being *equal*; for that quantity which in other animals has become splenic, is in the former still hepatic. 8*thly*. In cases of transposition of both organs, it is the *right* lobe of the liver--that *nearest the spleen,* now on the right side--which is the *smaller* of the two lobes, proving that whichever lobe be in this condition, the *spleen,* as being opposite to it, *represents the minus hepatic quantity*. From these, among other facts, I infer that the spleen is the representative of the liver on the left side, and that as such, its signification being manifest, there exists no exception to the law of animal symmetry. "Tam miram uniformitatem in planetarum systemate, necessario fatendum est intelligentia et concilio fuisse effectam. Idemque dici possit de uniformitate illa quae est in corporibus animalium. Habent videlicet animalia pleraque omnia, bina latera, dextrum et sinistrum, forma consimili: et in lateribus illis, a posteriore quidem corporis sui parte, pedes binos; ab anteriori autem parte, binos armos, vel pedes, vel alas, humeris affixos: interque humeros collum, in spinam excurrens, cui affixum est caput; in eoque capite binas aures, binos oculos, nasum, os et linguam; similiter posita omnia, in omnibus fere animalibus." --*Newton, Optices, sive de reflex, &c*. p. 411.]

X.--*The heart*, though being itself the recipient, the prime mover, and the dispenser of the blood, does not depend either for its growth, vitality, or stimulus to action, upon the blood under these uses, but upon the blood circulating through vessels which are derived from its main systemic artery, and disposed in capillary ramifications through its substance, in the manner of the nutrient vessels of all other organs. The two *coronary arteries* of the heart arise from the systemic aorta immediately outside the semilunar valves, situated in the root of this vessel, and in passing right and left along the auriculo-ventricular furrows, they send off some branches for the supply of the organ itself, and others by which both vessels anastomose freely around its base and apex. The *vasa cordis* form an anastomotic circulation altogether isolated from the vessels of the other thoracic organs, and also from those distributed to the thoracic parietes. The coronary arteries are accompanied by veins which open by distinct orifices (*foramina Thebesii*) into the right auricle. Like the heart itself, its main vessels do not depend for their support upon the blood conveyed by them, but upon that circulated by the small arteries (*vasa vasorum*) derived either from the vessel upon which they are distributed, or from some others in the neighbourhood. These little arteries are attended by veins of a corresponding size (*venules*) which enter the venae comites, thus carrying out the general order of vascular distribution to the minutest particular. Besides the larger nerves which accompany the main vessels, there are delicate filaments of the cerebro-spinal and sympathetic system distributed to their coats, for the purpose, as it is supposed, of governing their "contractile movements." The *vasa vasorum* form an anastomosis as well upon the inner surface of the sheath as upon the artery contained in this part; and hence in the operation for tying the vessel, the rule should be to disturb its connexions as little as possible, otherwise its vitality, which depends upon these minute branches, will, by their rupture, *be destroyed in the situation of the ligature, where it is most needed.*

XI.--*The branches of the systemic aorta* form frequent anastomoses with each other in all parts of the body. *This anastomosis occurs chiefly amongst the branches of the main arteries proper to either side*. Those branches of the opposite vessels which join at the median line are generally of very small size. There are but few instances in which a large blood vessel *crosses the central line from its own side to the other*. Anastomosis at the median line between opposite vessels happens either by a *fusion of their sides* lying parallel, as for example (and the only one) that of the two vertebral arteries on the basilar process of the occipital bone; or else by a direct *end-to-end union*, of which the lateral pair of cerebral arteries, forming the *circle of Willis*, and the two labial arteries, forming the coronary, are examples. The branches of the main arteries of one side form numerous anastomoses in the muscles and in the cellular and adipose tissue generally. Other special branches derived from the parent vessel above and below the several joints ramify and anastomose so very freely over the surfaces of these parts, and seem to pass in reference to them out of their direct course, that to effect this mode of distribution appears to be no less immediate a design than to support the structures of which the joints are composed.

XII.--*The innominate artery.* When this vessel is tied, the free direct circulation through the principal arteries of the right arm, and the right side of the neck, head, and brain, becomes arrested; and the degree of strength of the recurrent circulation depends solely upon the amount of anastomosing points between the following arteries of the opposite sides. The small terminal branches of the two occipital, the two auricular, the two superficial temporal, and the two frontal, inosculate with each other upon the sides, and over the vertex of the head; the two vertebral, and the branches of the internal carotid, at the base and over the surface of the brain; the two facial with each other, and with the frontal above and mental below, at the median line of the face; the two internal maxillary by their palatine, pharyngeal, meningeal, and various other branches upon the surface of the parts to which they are distributed; and lastly, the two superior thyroid arteries inosculate around the larynx and in the thyroid body. By these anastomoses, it will be seen that the circulation is restored to the branches of the common carotid almost solely. In regard to the subclavian artery, the circulation would be carried on through the anastomosing branches of the two inferior thyroid in the thyroid body; of the two vertebral, in the cranium and upon the cervical vertebrae; of the two internal mammary, with each other behind the sternum, and with the thoracic branches of the axillary and the superior intercostal laterally; lastly, through the anastomosis of the ascending cervical with the descending branch of the occipital, and with the small lateral offsets of the vertebral.

XIII.--*The common carotid arteries,* Of these two vessels, the left one arising, in general, from the arch of the aorta, is longer than the right one by the measure of the innominate artery from which the right arises. When either of the common carotids is tied, the circulation will be maintained through the anastomosing branches of the opposite vessels as above specified. When the vertebral or the inferior thyroid branch arises from the middle of the common carotid, this vessel will have an additional source of supply if the ligature be applied to it below the origin of such branch. In the absence of the innominate artery, the right as well as the left carotid will be found to spring directly from the aortic arch.

XIV.--*The subclavian arteries.* When a ligature is applied to the inner third of this vessel within its primary branches, the collateral circulation is carried on by the anastomoses of the arteries above mentioned; but if the vertebral or the inferior thyroid arises either from the aorta or the common carotid, the sources of arterial supply in respect to the arm will, of course, be less numerous. When the outer portion of the subclavian is tied between the scalenus and the clavicle, while the branches arise from its inner part in their usual position and number, the collateral circulation in reference to the arm is maintained by the following anastomosing branches:--viz., those of the superficialis colli, and the supra and posterior scapular, with those of the acromial thoracic; the subscapular, and the anterior and posterior circumflex around the shoulder-joint, and over the dorsal surface of the scapula; and those of the internal mammary and superior intercostal, with those of the thoracic arteries arising from the axillary. Whatever be the variety as to their mode or place of origin, the branches emanating from the subclavian artery are constant as to their destination. The length of the inner portion of the right subclavian will vary according to the place at which it arises, whether from the innominate artery, from the ascending, or from the descending part of the aortic arch.

XV.--*The axillary artery.* As this vessel gives off throughout its whole length, numerous branches which inosculate principally with the scapular, mammary, and superior intercostal branches of the subclavian, it will be evident that, in tying it above its own branches, the anastomotic circulation will with much greater freedom be maintained in respect to the arm, than if the ligature be applied below those branches. Hence, therefore, when the axillary artery is affected with aneurism, thereby rendering it unsafe to apply a ligature to this vessel, it becomes not only pathologically, but anatomically, the more prudent measure to tie the subclavian immediately above the clavicle.

XVI.--*The brachial artery,* When this artery is tied immediately below the axilla, the collateral circulation will be weakly maintained, in consequence of the small number of anastomosing branches arising from it above and below the seat of the ligature. The two circumflex humeri alone send down branches to inosculate with the small muscular offsets from the middle of the brachial artery. When tied in the middle of the arm between the origins of the superior and inferior profunda arteries, the collateral circulation will depend chiefly upon the anastomosis of the former vessel with the recurrent branch of the radial, and of muscular branches with each other. When the ligature is applied to the lower third of the vessel, the collateral circulation will be comparatively free through the anastomoses of the two profundi and anastomotic branches with the radial, interosseous, and ulnar recurrent branches. If the artery happen to divide in the upper part of the arm into either of the branches of the forearm, or into all three, a ligature applied to any one of them will, of course, be insufficient to arrest the direct circulation through the forearm, if this be the object in view.

XVII.--The *radial artery.* If this vessel be tied in any part of its course, the collateral circulation will depend principally upon the free communications between it and the ulnar, through the medium of the superficial and deep palmar arches and those of the branches derived from both vessels, and from the two interossei distributed to the fingers and back of the hand.

XVIII.--The *ulnar artery.* When this vessel is tied, the collateral circulation will depend upon the anastomosis of the palmar arches, as in the case last mentioned. While the radial, ulnar, and interosseous arteries spring from the same main vessel, and are continuous with each other in the hand, they represent the condition of a circle of which, when either side is tied, the blood will pass in a current of almost equal strength towards the seat of the ligature from above and below--a circumstance which renders it necessary to tie both ends of the vessel in cases of wounds.

XIX.--The *common iliac artery*. When a ligature is applied to the middle of this artery, the direct circulation becomes arrested in the lower limb and side of the pelvis corresponding to the vessel operated on. The collateral circulation will then be carried on by the anastomosis of the following branches--viz., those of the lumbar, the internal mammary, and the epigastric arteries of that side with each other, and with their fellows in the anterior abdominal parietes; those of the middle and lateral sacral; those of the superior with the middle and inferior haemorrhoidal; those of the aortic and internal iliac uterine branches in the female; and of the aortic and external iliac spermatic branches in the male. The anastomoses of these arteries with their opposite fellows along the median line, are much less frequent than those of the arteries of the neck and head.

XX.--*The external iliac artery.* This vessel, when tied at its middle, will have its collateral circulation carried on by the anastomoses of the internal mammary with the epigastric; by those of the ilio-lumbar with the circumflex ilii; those of the internal circumflex femoris, and superior perforating arteries of the profunda femoris, with the obturator, when this branch arises from the internal iliac; those of the gluteal with the external circumflex; those of the latter with the sciatic; and those of both obturators, with each other, when arising--the one from the internal, the other from the external iliac. Not unfrequently either the epigastric, obturator, ilio-lumbar, or circumflex ilii, arises from the middle of the external iliac, in which case the ligature should be placed above such branch.

XXI.--The *common femoral artery*. On considering the circles of inosculation formed around the innominate bone between the branches derived from the iliac arteries near the sacro-iliac junction, and those emanating from the common femoral, above and below Poupart's ligament, it will at once appear that, in respect to the lower limb, the collateral circulation will occur more freely if the ligature be applied to the main vessel (external iliac) than if to the common femoral below its branches.

XXII.--The *superficial femoral artery.* When a ligature is applied to this vessel at the situation where it is overlapped by the sartorius muscle, the collateral circulation will be maintained by the following arteries:--the long descending branches of the external circumflex beneath the rectus muscle, inosculate with the muscular branches of the anastomotica magna springing from the lower third of the main vessel; the three perforating branches of the profunda inosculate with the latter vessel, with the sciatic, and with the articular and muscular branches around the knee-joint.

XXIII.--The *popliteal artery.* When any circumstance renders it necessary to tie this vessel in preference to the femoral, the ligature should be placed above its upper pair of articular branches; for by so doing a freer collateral circulation will take place in reference to the leg. The ligature in this situation will lie between the anastomotic and articular arteries, which freely communicate with each other.

XXIV.--The *anterior and posterior tibial and peronoeal arteries.* As these vessels correspond to the arteries of the forearm, the observations which apply to the one set apply also to the other. [Footnote]

[Footnote: For a complete history of the general vascular system, see *The Anatomy of the Arteries of the Human Body*, by Richard Quain, F.R.S., &c., in which work, besides the results of the author's own great experience and original observations, will be found those of Haller's, Scarpa's, Tiedemann's, &c., systematically arranged with a view to operative surgery.]

* END *

This book has been produced in line with the EU GPSR guidelines about the safety of products.

The General Product Safety Regulation is the European Union's updated framework for ensuring that all consumer products, including books, are safe for consumers.

This book has been printed by Libri Plureos GmbH. The printer has issued safety certificates for the materials - like ink, paper and glue - being used.

The product identifier is: 9786256310117

The author is responsible for the content of the book, is the publisher of the works and bears full responsibility for it.

The book has been produced via Bookmundo. Bookmundo enables any author to share their stories with the rest of the world via printed books and ebooks and a broad distribution network.

Bookmundo will act as an intermediary in regard to questions about safety and will address them to the printer / author. Should there be any question in regard to the safety of the product, please contact us.

Bookmundo
Delftsestraat 33
3013AE Rotterdam
The Netherlands
info@bookmundo.com